New Phenomena in Lepton-Hadron Physics

NATO ADVANCED STUDY INSTITUTES SERIES

A series of edited volumes comprising multifaceted studies of contemporary scientific issues by some of the best scientific minds in the world, assembled in cooperation with NATO Scientific Affairs Division.

Series B: Physics

RECENT VOLUMES IN THIS SERIES

Volume 39 – Hadron Structure and Lepton–Hadron Interactions – *Cargèse* 1977
edited by Maurice Lévy, Jean-Louis Basdevant, David Speiser, Jacques Weyers, Raymond Gastmans, and Jean Zinn-Justin

Volume 40 – Kinetics of Ion–Molecule Reactions
edited by Pierre Ausloos

Volume 41 – Fiber and Integrated Optics
edited by D. B. Ostrowsky

Volume 42 – Electrons in Disordered Metals and at Metallic Surfaces
edited by P. Phariseau, B. L. Györffy, and L. Scheire

Volume 43 – Recent Advances in Group Theory and Their Application to Spectroscopy
edited by John C. Donini

Volume 44 – Recent Developments in Gravitation – *Cargèse* 1978
edited by Maurice Lévy and S. Deser

Volume 45 – Common Problems in Low- and Medium-Energy Nuclear Physics
edited by B. Castel, B. Goulard, and F. C. Khanna

Volume 46 – Nondestructive Evaluation of Semiconductor Materials and Devices
edited by Jay N. Zemel

Volume 47 – Site Characterization and Aggregation of Implanted Atoms in Materials
edited by A. Perez and R. Coussement

Volume 48 – Electron and Magnetization Densities in Molecules and Crystals
edited by P. Becker

Volume 49 – New Phenomena in Lepton-Hadron Physics
edited by Dietrich E. C. Fries and Julius Wess

This series is published by an international board of publishers in conjunction with NATO Scientific Affairs Division

A Life Sciences B Physics	Plenum Publishing Corporation London and New York
C Mathematical and Physical Sciences	D. Reidel Publishing Company Dordrecht and Boston
D Behavioral and Social Sciences	Sijthoff International Publishing Company Leiden
E Applied Sciences	Noordhoff International Publishing Leiden

New Phenomena in Lepton-Hadron Physics

Edited by
Dietrich E.C. Fries and Julius Wess
Kernforschungszentrum und Universität
Karlsruhe, Federal Republic of Germany

PLENUM PRESS • NEW YORK AND LONDON
Published in cooperation with NATO Scientific Affairs Division

Library of Congress Cataloging in Publication Data

Nato Advanced Study Institute on New Phenomena in Lepton-Hadron Physics, Karlsruhe, 1978.
New Phenomena in lepton-hadron physics.

(Nato advanced study institutes series: Series B, Physics; v. 49)
"Lectures presented at the NATO Advanced Study Institute on New Phenomena in Lepton-Hadron Physics, held in Karlsruhe, West Germany, September 4–16, 1978."
Includes index.
1. Leptons–Congresses. 2. Hadrons–Congresses. I. Fries, Dietrich, E. C. II. Title. III. Series.
QC793.5.L422N37 1978 539.7'211 79-19005

Lectures presented at the NATO Advanced Study Institute on New Phenomena in Lepton-Hadron Physics, held in Karlsruhe, West Germany, September 4–16, 1978

ISBN 978-1-4684-7667-5 ISBN 978-1-4684-7665-1 (eBook)
DOI 10.1007/978-1-4684-7665-1

Softcover reprint of the hardcover 1st edition 1979
A Division of Plenum Publishing Corporation
227 West 17th Street, New York, N.Y. 10011

Preface

The NATO Advanced Summer Institute 1978 was held at Karlsruhe from Sept. 4 to Sept. 16. The title of the school "New Phenomena in Lepton and Hadron Physics" relates to the present very exciting phase in particle physics. An impressive amount of experimental data has been collected in support of a fundamental new picture of the subnuclear world, - a picture which has found its theoretical formulation in Quantum Chromodynamics and Gauge theories.

It is a general philosophy of the ASI to address the courses mainly to young and learning scientists, hence our major objective was to offer systematic reviews of both, the experimental situation and the basic theoretical concepts of the field. This volume contains the written versions of the major lectures delivered during the course.

In addition several lectures and seminars had been scheduled in which also more original and specialized subjects were discussed by invited speakers and participants of the school. Not all of these contributions are contained in this book.

The organizers felt that the school would be particularly successful for students and teachers if there was opportunity to continue the scientific discussion outside of the lecture hall and beyond lecture hours. This opportunity existed through many informal or social events and it was widely used by students and teachers; the location of the lecture hall in the beautiful Karlsruhe Schloßpark had perhaps some share in this. If there has existed throughout the school a lively atmosphere of scientific and social participation then we owe this to the dedicated staff of lecturers and an always interested and responsive audience.

The organisation of the Summer Institute would not have been possible without the devoted help of the organizers R. Fischer, H. Genz, W. Schmidt and H.M. Staudenmaier, and the colleagues and students of the Institute of Theoretical Physics of the University of Karlsruhe. We would like to express our warm gratitude to all of them. We like to thank our Conference secretaries Vera Lallemand

and Dr. Rakshanda Faizi for their charming assistance.
Major parts of the book were typed by Mrs V. Lallemand and Evelyne Xanke, whose diligent work we gratefully acknowledge.
Finally we wish to acknowledge the help and support of the Scientific Affairs division of the North Atlantic Treaty Organisation, represented by Dr. Kesler, as well as the support given to us by the Bundesministerium für Forschung und Technologie, the Land Baden-Württemberg, the University and the Kernforschungszentrum Karlsruhe.

Dietrich Fries
Julius Wess

Contents

Gauge Theories 1
C. Jarlskog

Drell-Yan Versus Experiment 55
F. Vannucci

Parton Models 83
O. Nachtmann

The Last Heavy Lepton and the Next 99
T. F. Walsh

Heavy Lepton Phenomenology 115
M. L. Perl

Quarkonium 161
M. Krammer and H. Krasemann

Chromodynamics and Deep Inelastic Processes 215
G. Altarelli

Deep Inelastic Lepton Scattering 263
W. S. C. Williams

Topless Model for Grand Unification 303
Y. Achiman and B. Stech

Asymptotic Freedom: Infrared Instability 315
P. Minkowski

Exotic Quark States 337
P. H. Dondi

Weak Neutral Currents Unveiled 371
H. Faissner

Charm Spectroscopy 433
G. Buschhorn

Index 435

GAUGE THEORIES

Cecilia Jarlskog

Department of Physics
University of Bergen
Bergen, Norway

1. INTRODUCTORY REMARKS

In these lectures, an introduction to the unified gauge theories, of weak and electromagnetic interactions, is given. Strong interactions are treated only briefly; they constitute the theme of Professor Nachtmann's lectures.

The idea of unifying seemingly different forces of nature is not new. More than a century ago, James Clerk Maxwell (1831-1879) succeeded in unifying electricity and magnetism. Within the last decade most of the effort has gone into unifying weak forces with electromagnetism. Such a unification was suggested, by Schwinger, to be realized through a triplet of vector fields, whose universal coupling would generate both the weak and electromagnetic forces (J. Schwinger, Ann. of Phys. 2, 407 (1957)). Salam and Ward (Nuovo Cimento 11, 568 (1959)) suggested that the vector bosons should be introduced à la Yang and Mills. It is interesting to note that at that time the muon-neutrino had not yet revealed itself; the known leptons e, μ, ν_e, were supposed to form a triplet.

The Schwinger model was extended by Glashow (Nucl. Phys. 22, 579 (1961)) who, in addition to the triplet of vector bosons, introduced a new neutral vector boson. In the Glashow model, the two neutral vector bosons, denoted by Z^3 and Z^s, mix. The linear combinations $Z^3\cos\theta + Z^s\sin\theta$ and $-Z^3\sin\theta + Z^s\cos\theta$ were assumed to correspond to the physical particles (photon and Z^o, in present terminology). Nowadays, θ is usually referred to as the Weinberg angle. The model predicted neutral currents. Glashow prophesized: "For no choice of θ is the interaction of neutral current small compared with weak interactions involving charged currents".

The model was constructed for leptons; we know that it is not easy to detect the leptonic neutral current, even when it is not small.

In the Glashow model of 1961 the vector meson masses were put in by hand and therefore the model is not renormalizable. The next major step was taken by Weinberg (Phys. Rev. Lett. 19, 1264 (1967)) who employed a spontaneous symmetry breaking mechanism to generate masses. He introduced four Higgs particles, whereof three disappear by giving masses to the intermediate bosons and one (denoted by φ^0) remains. The model relates the masses of the neutral and charged intermediate bosons, $M_W = \cos\theta \; M_W \rightarrow M_W = \cos\theta \; M_Z$. Weinberg guessed that the model may be renormalizable.

Salam and Ward had also been working on introducing masses via spontaneous symmetry breaking (see the talk by A. Salam in the proceedings of the 1968 Noble Symposium, Ed. N. Svartholm).

The extension to hadrons, for four quarks, was given by Weinberg (Phys. Rev. D5, 1412 (1972)) who utilized the so-called GIM-mechanism (Glashow, Iliopoulos and Maiani, Phys. Rev. D2, 185 (1970)).

In 1960's the unified models and Yang-Mills type theories were not particularly popular (however, I remember Veltman giving a series of lectures in a Copenhagen summer school in the late 60's).

The unified theories became fashionable only after t'Hooft (Amsterdam International Conference 1971) demonstrated that the spontaneously broken gauge theories are renormalizable. Two years after, in 1973, neutral currents were discovered at CERN. Since then the unified theories have become more and more the standard framework for describing nature.

Nowadays, it is believed that, the weak and electromagnetic interaction are described by a gauge theory based on the (non-abelian Lie) group SU(2) x U(1) and strong interactions follow a SU(3) in colour. One goes further and advocates a unification of all nongravitational interactions, for example as in the SU(5) model of Georgi and Glashow (Phys. Rev. Lett. 32, 438 (1974)).

Of course, we have not yet seen any intermediate vector bosons (excepting the photon) nor any Higgs particles, which are the backbone of the modern unified theories; we have not had enough energy to produce any W or Z^0, etc. The fact that the experiments are in an amazingly good agreement with the simplest gauge models makes us believe that we are on the right track; with enough energy we shall be able to admire the glorious intermediate bosons.

2. INGREDIENTS OF GAUGE THEORIES

Gauge theories assume that the Lagrangian formalism provides a correct language in describing nature which is supposed to consist of three classes of particles:

a) The fundamental constituents of matter (leptons and quarks).
b) Intermediate particles.
c) Higgs particles.

a) The fundamental constituents of matter are fermions with spin $\frac{1}{2}$. There are two types of constituents: leptons and quarks (hadrons).

The lepton family contains the observed ones ν_e, e^-, ν_μ, μ^-, ν_τ and τ^- (and of course their antiparticles). The evidence for ν_τ, although indirect, is substantial. Leptons are integrally charged.

The quark (hadron) family contains the "observed" quarks up (u), down (d), strange (s), charm (c) and beauty or bottom (b). The need for an additional quark, called truth or top (t) is urgent. We hope that Petra will discover the truth.

In the conventional approach, each quark exists in three varieties (colours). For example, there are three up quarks, etc.

b) In gauge theories, the existence of the constituents of matter, together with the Holy Principle of the local gauge invariance (discussed below), leads to the existence of intermediate vector (spin 1) bosons. Their job is to mediate interactions. The family of mediators contains Maxwell's photon (the only member directly seen so far) which mediates electromagnetism. We believe that there are further members, such as $W^\pm$, Z^0 which mediate weak interactions at low energies; gluons G_i, i=1, ..., 8 who mediate strong interactions, etc. Mediators are integrally or fractionally charged depending on the theory.

c) The Higgs particles are theoretically needed bêtes noires. Some day, perhaps we will learn to live without them, but so far their existence is indispensable. They are believed to be the origin of all masses except their own. We shall discuss their properties in section 10.

3. SYMMETRIES AND CONSERVATION LAWS

The Holy Principle in gauge theories is the Principle of Local Gauge Invariance, as we shall discuss shortly. Before we do so, let us remind ourselves, that the concept of invariance (or covariance), which is very essential in physics, is neatly formulated in Lagrangian language. For example, the principle that the laws of physics are independent of the particular Lorentz frame chosen requires the Lagrangian density to be a scalar quantity; the equations of motion must be covariant, etc.

[Exercise: State the physical principles leading to the laws of energy-momentum conservation, angular momentum conservation and parity conservation (if it had been true). How are these conservation laws guaranteed in the Lagrangian formalism?]

Not all conservation laws have to do with space and time. There are a rather large number of physically established conservation laws, which pertain to "internal" properties. Let us give a few examples:

a) Conservation of charge.
b) Conservation of baryon number B, and lepton numbers L_e, L_μ and L_τ.
c) It is commonly believed that colour is an exact symmetry, and only the colour singlet states are observable. There are however theories in which the colour leaks out.

Experimentally, there is, up to now, no evidence against the conservation laws a - c. Approximate conservation laws have also been observed in nature, for example

d) conservation of strange-, charm-, beauty-(ness) by strong and electromagnetic interactions. These are violated by weak interactions.
e) Isospin symmetry in strong interactions.
f) SU(3)-symmetry, again in strong interactions.

We leave it to our readers to ponder on the physical principles responsible for the symmetries above. We may ask whether the conservation laws a) and b) are exact? Are all physical states, which go through our bubble chambers, calorimeters, etc, colour singlets?

We shall return to some of these questions later on in these lectures. Let us conclude this section by simply noting that the experimentally established conservation laws are easily incorporated as symmetries into our Lagrangians, even if we do not understand their origin.

4. LOCAL GAUGE INVARIANCE AND QUANTUM ELECTRODYNAMICS (QED)

We shall now construct the simplest example of a gauge theory employing the principle of local gauge invariance. Suppose that we had only a single massless fermion f. Then the free Lagrangian, describing f is given by

$$\mathcal{L}_o = i\,\bar{\psi}_f(x)\gamma_\mu \frac{\partial}{\partial x_\mu} \psi_f(x), \tag{4-1}$$

Note that $i\frac{\partial}{\partial x_\mu}$ is the four momentum operator in quantum mechanics. We often refer to this term as the kinetic energy term, in analogy with L = T - V, where the mass term goes into V. $\mathcal{L}_o$ is invariant under the <u>global</u> gauge transformation

$$\psi_f(x) \rightarrow e^{i\Lambda}\psi_f(x), \tag{4-2}$$

where Λ is a constant, viz. $\mathcal{L}_o \rightarrow e^{-i\Lambda}\mathcal{L}_o e^{i\Lambda} = \mathcal{L}_o$. This transformation is called global, because Λ is a universal constant, independent of space and time.

The Holy Principle in gauge theories <u>is</u> that the above invariance should hold <u>locally</u>, i.e., for $\Lambda = \Lambda(\bar{x},t)$, where Λ is a real function of the space and time. In other words the phase of $\psi_f(x)$ is unobservable and may be chosen at will. If we accept this law, we must modify our Lagrangian, because

$$\psi_f(x) \rightarrow e^{i\Lambda(x)}\psi_f(x) \tag{4-3}$$

implies

$$\mathcal{L}_o \rightarrow \mathcal{L}_o - \bar{\psi}_f(x)\,\gamma_\mu\,\psi_f(x)\,\frac{\partial\Lambda(x)}{\partial x_\mu}$$

whereby the Lagrangian is not invariant and the phase matters. Since in the additional term $\frac{\partial\Lambda}{\partial x_\mu}$ is a vector, we introduce a vector field $A_\mu(x)$, and replace $\mathcal{L}_o$ with $\mathcal{L}_1$ given by

$$\mathcal{L}_1 = i\,\bar{\psi}_f(x)\,\gamma_\mu\left[\frac{\partial}{\partial x_\mu} - i\,c\,A^\mu(x)\right]\psi_f(x) \tag{4-4}$$

where c is an arbitrary constant. The local gauge transformation is now

$$\psi_f(x) \rightarrow e^{i\Lambda(x)}\psi_f(x),\ \bar{\psi}_f(x) \rightarrow e^{-i\Lambda(x)}\bar{\psi}_f(x),\ A^\mu(x) \rightarrow A^\mu(x) + \delta A^\mu. \tag{4-5}$$

Requiring $\mathcal{L}_1$ to be invariant gives

$$\delta A^\mu = \frac{1}{c}\frac{\partial\Lambda(x)}{\partial x_\mu}\ . \tag{4-5a}$$

This simple example illustrates the major feature of gauge theories: the requirement of local gauge invariance (in our case under phase transformation U(1) implies the existence of the intermediate (gauge) bosons (here the A^μ field). A^μ may be interpreted as the photon field (whereby $c = e\,Q_f$; $e \equiv$ unit of charge and $Q_f \equiv$ the charge of the fermion f) if we are allowed to add a kinetic energy term, for the photon, in accordance with the gauge principle. The term

$$\mathcal{L}_o^{(\gamma)} = \frac{-1}{4} F_{\mu\nu}F^{\mu\nu} \, , \quad F_{\mu\nu} = \partial_\mu A_\nu - \partial_\nu A_\mu \tag{4-6}$$

has the desired properties. Notice that the photon mass must vanish identically, because a mass term $\frac{1}{2}\, m_\gamma^2\, A_\mu A^\mu$ violates the rule (4-5) and is therefore forbidden. A fermion mass term, on the other hand, may be added $-\, m_f\, \bar{\psi}_f(x)\, \psi_f(x)$, since it is invariant under (4-5). Collecting all term we have the Lagrangian of QED

$$\mathcal{L} = i\, \bar{\psi}_f(x)\, \gamma_\mu \left[\frac{\partial}{\partial x_\mu} - i\, e\, Q_f A^\mu\right] \psi_f(x) - \frac{1}{4} F_{\mu\nu}\, F^{\mu\nu} -$$

$$-\, m_f \bar{\psi}_f(x) \psi_f(x) \; . \tag{4-7}$$

Putting f = electron, $Q_f = -1$, gives QED for electrons which has been the most successful theory we have ever had. The fine structure constant $\alpha = \frac{e^2}{4\pi} = \frac{1}{137}$ is fortunately small; furthermore, the perturbative approach makes sence (the theory is renormalizable). The basic interaction in QED is represented by the diagram

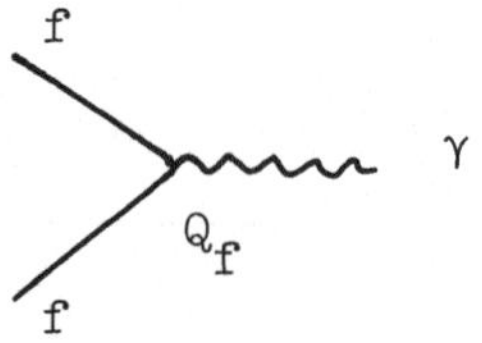

The constants $|e|Q_f$ and m_f are not predicted by the theory but determined from experiment. QED is an example of an abelian (named after the Norwegian mathematician Niels Henrik Abel (1802-1849) gauge theory, i.e., the Λ's (if we had several of them) are just c-numbers and commute with each other. Note also that the gauge boson (γ) carries no charge and the coupling constant Q_f is not universal. The latter is seen by taking several basic fermions such as u, d, e, μ, ... and going through the arguments above, viz.,

$$\mathcal{L}_o \to \mathcal{L}_1 = i \sum \bar{\psi}_j(x)\gamma_\mu \left[\frac{\partial}{\partial x_\mu} - ieQ_j\, A^\mu(x)\right] \psi_j(x) \tag{4-8}$$

$$\psi_j(x) \to e^{i\Lambda_j(x)}\psi_j(x)\ , \quad A^\mu(x) \to A^\mu(x) + \delta A^\mu(x) \tag{4-9}$$

Substituting (4-9) in (4-8) together with the requirement of invariance yields

$$\Lambda_j(x) = Q_j\ \Lambda(x)\ , \quad \delta A^\mu(x) = \frac{1}{e}\frac{\partial\Lambda(x)}{\partial x_\mu}\ , \tag{4-10}$$

where $\Lambda(x)$ is arbitrary. The point is that the phases $\Lambda_j(x)$ of the fermion fields are no longer unobservable and arbitrary; we must choose them proportional to their empirical charges! Of course, there is nothing in the theory which tell us why $Q_u = 2/3$, $Q_d = -1/3$, $Q_e = -1$, etc. The gauge group U(1) has this ugly arbitrariness, i.e., it allows an arbitrary coupling constant for each fermion (or group of fermions, see below).

5. WEAK INTERACTIONS, QUANTUM FLAVOUR DYNAMICS (QFD)

In physics slang, each kind of particle is said to distinguish itself from other kinds by having a particular flavour. The observed leptonic flavours are ν_e, e, ν_μ, ..., and the hadronic flavours are supposed to be u, d, s, ...

Electromagnetism is flavour blind (viz. 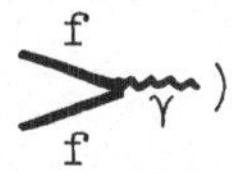

).

"Flavour dynamics" shows up in weak interactions, where the charged currents are flavour-changing. For example an electron-neutrino is turned into an electron, etc.

In this section we discuss the conventional theory of weak interaction, for four quark flavours and four leptons, i.e., the theory that we believed in until 1975, before the τ was discovered.

The charged-current weak interactions were described by the Cabibbo-theory supplemented by the Glashow-Iliopoulos-Maiani (GIM) mechanism

$\mathcal{L}^{c.c.} = \hat{g}\sum_{i,j}$ 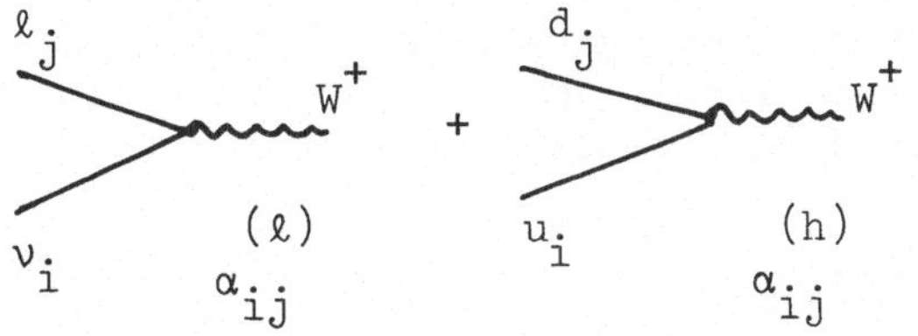

+ h.c. =

$$\hat{g}\sum_{i,j=1,2}\left\{\alpha^{(\ell)}_{ij}\,\bar{\ell}_j\,\gamma^\lambda(1-\gamma_5)\nu_i + \alpha^{(h)}_{ij}\,\bar{d}_j\,\gamma^\lambda(1-\gamma_5)u_i\right\}W^+_\lambda + \text{h.c.} \tag{5-1}$$

Here W^+_λ, ν_i, etc. refer to the field operators for W^+, neutrino, etc.; $\hat{g}$ is a constant $\hat{g}^2/M_W^2 = \frac{G}{\sqrt{2}}$, M_W = mass of $W^\pm$, G = Fermi constant. Furthermore

$$\nu_1 = \nu_e \quad \nu_2 = \nu_\mu \quad u_1 = u \quad u_2 = c$$

$$\ell_1 = e \quad \ell_2 = \mu \quad d_1 = d \quad d_2 = s \ .$$

The α's are given by

$$\alpha_{ij}^{(\ell)} = \delta_{ij} \ , \ \alpha_{ij}^{(h)} = \begin{pmatrix} \cos\theta_c & \sin\theta_c \\ -\sin\theta_c & \cos\theta_c \end{pmatrix}_{ij} \quad , \ \theta_c = \text{Cabibbo angle.}$$

The elegant form (5-1) had gradually emerged from several decades of research on weak interactions. Each piece has a raison d'être and $\mathcal{L}^{cc}$, whenever applicable, gives a good description of data.

After the discovery of the τ-lepton and the quark flavour b, it is believed that we should modify (5-1) by introducing $\nu_3 = \nu_\tau$, $\ell_3 = \tau$, $u_3 = t$, $d_3 = b$ (where t is a to be discovered quark flavour) and let the sum over i and j run from 1 to 3. Furthermore $\alpha_{ij}^{(\ell)} = \delta_{ij}$ and $\alpha^{(h)}$ is a general three by three unitary matrix. We shall discuss the four and six quark models in detail later on.

After the discovery of neutral currents in 1973, we know for sure that the $\mathcal{L}^{cc}$ above cannot explain all weak interactions. As we shall see in the next section, the gauge theories provide a natural scene for neutral currents. What is much more, they allow the unification of weak and electromagnetic interactions which is, of course, marvelous. Moreover, it has been known for quite some time that the theory described by (5-1) is plagued by awful diseases, viz., some physical cross sections and transition probabilities are infinitely large and uncalculable. Gauge theories provide a framework in which the diseases are cured, one gets sensible expressions for cross sections, etc. Has Nature solved her problems by utilizing some local gauge theory? We hope so, and are encouraged by data which are in amazing agreement with the simple gauge model described in the next few sections.

6. CONSTRUCTION OF THE STANDARD SU(2) x U(1) MODEL

In this section we describe the construction of the standard model of weak and electromagnetic interactions, the so-called SU(2) x U(1) model. The model is "standard" in the sence that it seems to explain all data on neutral currents (a task which no other model can fulfill) provided we forget the confused situation in atomic physics experiments.

Let us, for simplicity, assume that there were only two elementary fermions in nature*, denoted by f and f' and that $Q_f = Q_{f'} + 1$, where Q is the electric charge. These assumptions are based on the observation that nature exhibits several such couples. For example (f,f') could be (ν_e,e), (ν_μ,μ), (u,d), etc. We now construct a gauge theory, proceeding along the main road laid out in section 4 but have to make appropriate detours and modifications. As in section 4 we introduce the kinetic term for the (massless) f and f'

$$\mathcal{L}_o(f,f') = i\left[\bar{f}\,\gamma\,\frac{\partial}{\partial x}\,f + \bar{f}'\gamma\,\frac{\partial}{\partial x}\,f'\right]\,,\quad f = \psi_f(x)\,,\quad f' = \psi_{f'}(x) \tag{6-1}$$

We remember (see eq. (5-1)) that the charged currents encountered in nature are of V-A form. In order to account for this fact, we decompose the fields into L = left and R = right components

$$f = \frac{1}{2}(1-\gamma_5)f + \frac{1}{2}(1+\gamma_5)f = f_L + f_R\,, \tag{6-2}$$

$$f_{L,R} = \frac{1}{2}(1 \pm \gamma_5)f\,,$$

and similarly for f'. Substituting into (6-1) yields

$$\mathcal{L}_o(f,f') = i\left[\bar{f}_L\gamma\,\frac{\partial}{\partial x}\,f_L + \bar{f}_R\gamma\,\frac{\partial}{\partial x}\,f_R + \bar{f}'_L\gamma\,\frac{\partial}{\partial x}\,f'_L + \bar{f}'_R\gamma\,\frac{\partial}{\partial x}\,f'_R\right]. \tag{6-3}$$

Again, we look at $\mathcal{L}^{cc}$, eq. (5-1), and observe that the left-handed fermions (f_L and f'_L) seem to be intimately related to each other; they go into each other by sending off a W. For example, in (5-1)

$$\bar{\ell}_j\,\gamma^\lambda(1-\gamma_5)\nu_i = \frac{1}{2}\,\bar{\ell}_j(1+\gamma_5)\gamma^\lambda(1-\gamma_5)\nu_i = 2\bar{\ell}_{jL}\gamma^\lambda\nu_{iL}\,.$$

Therefore it is natural to write

$$\mathcal{L}_o(f,f') = i\,\left\{(\bar{f}_L,\bar{f}'_L)\gamma\,\frac{\partial}{\partial x}\begin{pmatrix} f_L \\ f'_L \end{pmatrix} + \bar{f}_R\,\gamma\,\frac{\partial}{\partial x}\,f_R + \bar{f}'_R\,\gamma\,\frac{\partial}{\partial x}\,f'_R\right\}. \tag{6-4}$$

Now we shall impose the Local Gauge Principle. Here, we require that neither the "direction" of $\begin{pmatrix} f_L \\ f'_L \end{pmatrix}$, in a weak isospin space spanned by f_L and f'_L, nor the phases of the three fields

$$\psi_1 \equiv \begin{pmatrix} f_L \\ f'_L \end{pmatrix}\,,\quad \psi_2 \equiv f_R\,,\quad \psi_3 \equiv f'_R \tag{6-5}$$

----

* See section 12 for generalization to several f's and f''s.

should be measurable. The gauge group is, therefore, SU(2) x U(1) and $\mathcal{L}_0(f,f')$ must be modified to be invariant under

$$\psi_j(x) \rightarrow \left\{\exp\left(i\ \bar{\alpha}(x)\cdot\frac{\bar{\sigma}}{2}\right)\exp\left(i\ \beta_j(x)\right)\right\}\psi_j(x)\ . \qquad (6\text{-}6)$$

$$\underset{SU(2)}{\uparrow} \qquad\qquad \underset{U(1)}{\uparrow}$$

Here $\bar{\alpha}(x)$ are the three (real but otherwise arbitrary) parameters of the group SU(2); $\bar{\sigma}$ are the Pauli matrices,

$$\left[\frac{\sigma_j}{2}, \frac{\sigma_k}{2}\right] = i\ \varepsilon_{jkm}\left[\frac{\sigma_m}{2}\right] .$$

$\frac{\bar{\sigma}}{2}$ represent the three generators of SU(2); $\bar{\sigma}\ \psi_2 = \bar{\sigma}\ \psi_3 = 0$, because $\psi_{2,3}$ are singlets. Finally $\beta_j(x)$, j = 1,2,3 are arbitrary real functions.

$\mathcal{L}_0$, as it stands, is not invariant under (6-6). From section 4, we know that we must introduce three vector bosons (one for each generator) for the SU(2) group and one additional vector boson for the U(1) group. Thus we replace

$$\mathcal{L}_0 = i\sum_j \bar{\psi}_j\ \gamma\frac{\partial}{\partial x}\psi_j$$

with

$$\mathcal{L}_1 = i\sum_{j=1}^{3}\bar{\psi}_j\ \gamma^\mu\left[\frac{\partial}{\partial x^\mu} - i\ g\frac{\bar{\sigma}}{2}\cdot\bar{W}_\mu - i\ g'y_j\ B_\mu\right]\psi_j\ . \qquad (6\text{-}7)$$

Here $\bar{W}^\mu = (W^{1,\mu}, W^{2,\mu}, W^{3,\mu})$; g and y_j are four constants, and the redundant constant g' has been introduced for convenience. Note that the W's couple only to the doublet ψ_1 and $\bar{\sigma}\ \psi_{2,3} = 0$. By construction (6-7) gives the correct form for the charged currents (see eq. (5-1) up to mixings in the hadronic sector. Eq. (6-7) is the analog of (4-4). We now ask, what is the analog of (4-5a)? How should the fields W and B transform in order to ensure us invariance under SU(2) x U(1), eq. (6-6). For the U(1) part we copy the procedure in section 4. We infer, by comparing equs. (6-7) and (4-8), that we just have to replace

$$e \rightarrow g'\ ,\quad Q_j \rightarrow y_j\ ,\quad \Lambda_j(x) \rightarrow \beta_j(x)$$

whereby we have invariance provided $\beta_j(x) = y_j\beta(x)$ and the gauge transformation of the B field is according to

$$B^\mu \rightarrow B^\mu + \frac{1}{g'}\ \frac{\partial\beta(x)}{\partial x_\mu}\ . \qquad (6\text{-}8)$$

Here $\beta(x)$ is an arbitrary real function. It remains to add the kinetic term

$$\mathcal{L}_o(B) = -\frac{1}{4} B^{\mu\nu}B_{\mu\nu} , \quad B_{\mu\nu} = \partial_\mu B_\nu - \partial_\nu B_\mu \tag{6-9}$$

and the "U(1) problem" is solved.

We now focus our attention on the SU(2) part, which only affects ψ_1. The SU(2) group is a nonabelian Lie group*, i.e., the commutator of generators (here represented by σ's) is nonvanishing and equals a linear combination of the generators. To repeat, the question is how should W transform so that $\mathcal{L}_1$ is invariant under $\psi_1 \to \exp(i\,\bar{\alpha}(x) \cdot \frac{\bar{\sigma}}{2})\psi_1$? The answer to this question was given by Yang and Mills (Phys.Rev. 96,191(1954)). The reader may check that for an infinitesimal transformation (small $\bar{\alpha}(x)$) one gets

$$\bar{W}^\mu \to \bar{W}^\mu + \frac{1}{g}\frac{\partial\bar{\alpha}}{\partial x_\mu} - \bar{\alpha}\times\bar{W}^\mu \tag{6-10}$$

i.e.,

$$W^j_\mu \to W^j_\mu + \frac{1}{g}\frac{\partial\alpha^j}{\partial x^\mu} - \varepsilon_{jkm}\,\alpha^k\, W^m_\mu \quad , \qquad j = 1,2,3$$

Note, since $\bar{W}$ and $\bar{\alpha}$ are vectors, eq. (6-10) expresses the most general behaviour expected, viz., W gets translated as well as rotated.

We now must supply W with a kinetic energy term which should remain invariant under (6-10). Again, here the analog of (6-9) does not work. Indeed, the requirement of invariance imposes self-interaction for W (the reader should check this)

$$\mathcal{L}_o(W) = -\frac{1}{4}\sum_{j=1}^{3} W^j_{\mu\nu}\, W^{\mu\nu,j} = -\frac{1}{4}\,\bar{W}_{\mu\nu} \cdot \bar{W}^{\mu\nu} , \tag{6-11}$$

where

$$W^j_{\mu\nu} = \partial_\mu W^j_\nu - \partial_\nu W^j_\mu + g\,\varepsilon_{jkm}\, W^k_\mu\, W^m_\nu \tag{6-12}$$

The self-interaction diagrams are

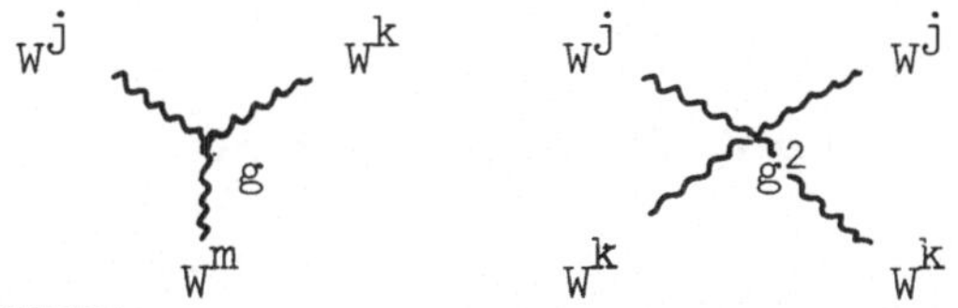

---

* Sophus Lie (1842-1899) was a mathematician from the most beautiful fjord area in Norway called Sognefjorden. I wish the interested reader will someday have the chance to visit that area.

Let us summarize: the interaction Lagrangian

$$\mathcal{L}_1 = i \sum_j \bar{\psi}_j(x)\gamma^\mu \left[\frac{\partial}{\partial x^\mu} - i\, g\, \frac{\bar{\sigma}}{2} \bullet \bar{W}_\mu - i\, g' y_j B_\mu\right] \psi_j(x) \qquad (6\text{-}13)$$

is invariant under the gauge transformation

$$\begin{aligned} \psi_j(x) &\to \left\{\exp\left(i\, \bar{\alpha} \bullet \frac{\bar{\sigma}}{2}\right) \exp\left(i\, y_j \beta(x)\right)\right\}\psi_j(x) \\ B^\mu &\to B^\mu + \frac{1}{g'} \frac{\partial \beta(x)}{\partial x_\mu} \\ \bar{W}^\mu &\to \bar{W}^\mu + \frac{1}{g} \frac{\partial \bar{\alpha}}{\partial x_\mu} - \bar{\alpha} x \bar{W}^\mu \,, \end{aligned} \qquad (6\text{-}14)$$

$$\psi_1 = \begin{pmatrix} f_L \\ f'_L \end{pmatrix}, \quad \psi_2 = f_R, \; \psi_3 = f'_R \;.$$

Furthermore, we learned how to add kinetic terms for the B and W fields. However, these particles must be massless, because a mass term $\frac{1}{2} M_B^2\, B_\mu\, B^\mu$, etc. violates the gauge principle, eq. (6-14).

7. UNIFICATION OF WEAK AND ELECTROMAGNETIC INTERACTIONS IN THE STANDARD MODEL

In the last section we constructed an SU(2) x U(1) model which has the correct form for the charged currents (up to mixings). The charged current interactions are mediated via $W^\pm$. From (6-13)

$$\begin{aligned} \mathcal{L}^{cc} &= g\, \bar{\psi}_1\, \gamma^\mu\, \frac{\sigma_1 W^1_\mu + \sigma_2\, W^2_\mu}{2}\, \psi_1 \\ &= \frac{g}{2} \left(\bar{f}, \bar{f}'\right)\left(\frac{1+\gamma_5}{2}\right)\gamma^\mu \begin{pmatrix} 0 & W^1_\mu - iW^2_\mu \\ W^1_\mu + iW^2_\mu & 0 \end{pmatrix} \frac{1-\gamma_5}{2} \begin{pmatrix} f \\ f' \end{pmatrix} = \\ &= \frac{g}{2\sqrt{2}} \left\{\bar{f}\, \gamma^\mu(1-\gamma_5) f'\, W^-_\mu + \text{h.c.}\right\} \,, \end{aligned} \qquad (7\text{-}1)$$

$$W^\pm_\mu = \frac{1}{\sqrt{2}} \left(W^1_\mu \pm i\, W^2_\mu\right)$$

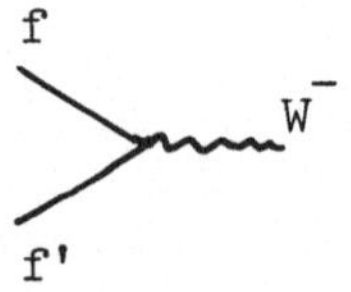

$\mathcal{L}^{cc}$ above does not describe the real world, because W must be massless giving long-range forces, whereas, we know that weak interactions are short-ranged. Later on we shall see that we can have both massive W's and the Lagrangian described above provided we introduce the mass via the so-called Higgs mechanism. Let us assume that we had done so. Then, a comparison with the local V - A Lagrangian for μ-decay yields

$$\frac{g^2}{8M_W^2} = \frac{G}{\sqrt{2}} \tag{7-2}$$

where M_W is the mass of the charged W and G is the Fermi constant.

Now we turn to the neutral currents, which result from $\mathcal{L}_1$, eq. (6-13). These are mediated by B_μ and W_μ^3. Since these objects are massless and as yet unphysical, we may form two orthogonal linear combinations of them denoted by A_μ and Z_μ

$$\begin{vmatrix} B_\mu \\ \\ W_\mu^3 \end{vmatrix} = \underbrace{\begin{vmatrix} a_{B\gamma} & a_{BZ} \\ \\ a_{W\gamma} & a_{WZ} \end{vmatrix}}_{\mathcal{R}} \cdot \begin{vmatrix} A_\mu \\ \\ Z_\mu \end{vmatrix} \tag{7-3}$$

where $\mathcal{R}$ is an orthogonal matrix with elements $a_{B\gamma}$, etc. We shall show later that via the Higgs mechanism, mentioned above, A_μ and Z_μ could be made to represent the physical particles, where A_μ (Z_μ) is a massless (heavy) particle. We wish to identify A_μ with the photon field, therefore, we must require that the terms in $\mathcal{L}_1$, eq. (6-13), containing A_μ must add up to give the electromagnetic Lagrangian, $\mathcal{L}^{em}$,

$$\mathcal{L}^{em} = i \sum_{j=1}^{3} \bar{\psi}_j \gamma^\mu(-ieQ_j)\psi_j A_\mu = e\left\{Q_f \bar{f}\gamma^\mu f + Q_{f'} \bar{f}'\gamma^\mu f'\right\} A_\mu \tag{7-4}$$

Here Q_j is the (matrix representation) of the electric charge, viz.,

$$Q_1 = \begin{pmatrix} Q_f & 0 \\ \\ 0 & Q_{f'} \end{pmatrix} = \frac{\sigma_3}{2} + \left(Q_f - \frac{1}{2}\right) , \quad Q_2 = Q_f , \quad Q_3 = Q_{f'} . \tag{7-5}$$

The three relations in (7-5) are summarized in the operator form

$$Q = \frac{\sigma_3}{2} + Y , \tag{7-6}$$

where

$$Q\psi_j = Q_j\psi_j, \quad Y\psi_j = Y_j\psi_j$$
$$Y_1 = Q_f - \frac{1}{2}, \; Y_2 = Q_f , \; Y_3 = Q_{f'} . \tag{7-6a}$$

Thus

$$\mathcal{L}^{em} = e \sum_j \bar{\psi}_j \gamma^\mu Q \psi_j A_\mu \equiv e \, j^{\mu,em} A_\mu . \tag{7-7}$$

Now we calculate the term in $\mathcal{L}$ which contains A^μ. Using eqs. (6-13) and (7-3), we find

$$\mathcal{L}^{em} = \sum_j \bar{\psi}_j \gamma^\mu \left[g \, a_{W\gamma} \frac{\sigma_3}{2} + g' \, a_{B\gamma} \, y_J \right] \psi_j A^\mu . \tag{7-8}$$

Comparing (7-8) with (7-7) yields

$$eQ_j = g \, a_{W\gamma} \frac{\sigma_3}{2} + g' \, a_{B\gamma} \, y_j , \tag{7-9}$$

$$e\left(\frac{\sigma_3}{2} + Y_j\right) = g \, a_{W\gamma} \frac{\sigma_3}{2} + g' \, a_{B\gamma} y_j \tag{7-9a}$$

The relation (7-9) is valid provided

$$g \, a_{W\gamma} = e \quad \text{and} \quad e \, Y_j = g' \, a_{B\gamma} \, y_j . \tag{7-10}$$

Remembering that the (redundant) constant g' may be choosen at will, we put

$$g \, a_{W\gamma} = e = g' \, a_{B\gamma} \tag{7-11}$$

$$y_j = Y_j \tag{7-12}$$

By orthogonality of the matrix , $(a_{W\gamma})^2 + (a_{B\gamma})^2 = 1$. Thus

$$e^2\left(\frac{1}{g^2} + \frac{1}{g'^2}\right) = 1 . \tag{7-13}$$

Putting $a_{W\gamma} = \sin\theta_W$ we have the famous relations

$$g = \frac{e}{\sin\theta_W} \quad , \quad g' = \frac{e}{\cos\theta_W} \quad , \tag{7-14}$$

where θ_W is usually called the Weinberg angle. Let us summarize: the SU(2) x U(1) model contains two neutral mediators W^3_μ and B_μ, from which, by a rotation, we may form the fields A_μ and Z_μ. The model contains four "coupling constants", viz. g, $g'y_j$ and a parameter, θ_W. Identifying A_μ with the photon field fixes these constants, viz., $g = \frac{e}{\sin\theta_W}$, $g' = \frac{e}{\cos\theta_W}$, $y_1 = Q_f - \frac{1}{2}$, $y_2 = Q_f$ and $y_3 = Q_{f'}$. Thus the model introduces one new parameter, the angle θ_W, which is empirically determined from the neutral current data. Note that the U(1) group is rather unpleasant; only for a very specific (and incomprehensible) choice of the phases $\beta_j(x) = y_j\beta(x)$, with y_j as given above, do the particles f and f' obtain their observed charges. The hope is that the SU(2) x U(1) model is a substructure in a more elaborate scheme, where the origin of the eigenvalues y_j and the quantization of charge is explained. The SU(5) model of weak, electromagnetic and strong interactions (see section 14) provides a good example of how such an elaborate scheme might look like.

The SU(2) x U(1) model unifies weak and electromagnetic interactions in the sence that A_μ and Z_μ are intimately related to each other (eq. 7-3) and to $W^\pm_\mu$. The weak and electromagnetic couplings are of the same order, viz. $g = \frac{e}{\sin\theta_W}$. Thus the apparent huge differences in the strength of these interactions are due to the differences in the masses, $M_\gamma = 0$,

$$M_W = \left(\frac{\sqrt{2}\, g^2}{8G}\right)^{\frac{1}{2}} = \left(\frac{\pi\alpha}{\sqrt{2}\, G \sin^2\theta_W}\right)^{\frac{1}{2}} = \frac{38 \text{ GeV}}{|\sin\theta_W|} .$$

We shall see later that $M_Z = \frac{M_W}{\cos\theta_W}$, if the Higgs mechanism is done à la Weinberg (Phys.Rev.Letters 19,1264(1967). Thus W and Z are beyond the reach of the present accelerators.

8. THE WEAK NEUTRAL CURRENT COUPLINGS IN THE STANDARD MODEL

We go back to the interaction Lagrangian of the standard model and, using eq. (7-3), determine the term proportional to Z_μ

$$\mathcal{L}^{n.c.} = \sum_j \bar{\psi}_j \gamma^\mu \left[g\, a_{WZ} \frac{\sigma_3}{2} + g'\, a_{BZ}\, y_j \right] \psi_j \, Z_\mu \tag{8-1}$$

Using $g\, a_{W\gamma} = g'\, a_{B\gamma} = e$ and $Q = \frac{\sigma_3}{2} + y$, obtained in section 7 we find

$$\mathcal{L}^{n.c.} = \frac{e}{a_{W\gamma}\, a_{B\gamma}} \sum_j \bar{\psi}_j\, \gamma^\mu \left[a_{B\gamma} \cdot a_{WZ} \frac{\sigma_3}{2} + a_{W\gamma} \cdot a_{BZ} \left(Q - \frac{\sigma_3}{2} \right) \right] \psi_j Z_\mu$$

$$= \frac{e}{a_{W\gamma} \cdot a_{B\gamma}} \sum_j \bar{\psi}_j\, \gamma^\mu \left[\det(\mathcal{R}) \cdot \frac{\sigma_3}{2} + a_{W\gamma}\, a_{BZ}\, Q \right] \psi_j Z_\mu \tag{8-2}$$

Since $\mathcal{R}$ is an orthogonal matrix and we had introduced $\sin\theta_W = a_{W\gamma}$, $\cos\theta_W = a_{B\gamma}$, we must have $\det(\mathcal{R}) = 1$, and $a_{BZ} = -\sin\theta_W$. Therefore,

$$\mathcal{L}^{n.c.} = \frac{e}{\sin\theta_W \cos\theta_W} \left\{ \bar{\psi}_1 \gamma_\mu \frac{\sigma_3}{2} \psi_1 - \sin^2\theta_W\, J_\mu^{em} \right\} Z^\mu \equiv J_\mu^{\,n.c.} \cdot Z^\mu \tag{8-3}$$

where $\psi_1 = \begin{pmatrix} f \\ f' \end{pmatrix}_L$; J_μ^{em} is the electromagnetic current, defined

in eq. (7-7), and $J_\mu^{n.c.}$ is the weak neutral current. We may rewrite J_μ^{em} as

$$J_\mu^{em} = Q_f \bar{f} \gamma_\mu f + Q_{f'} \bar{f}' \gamma_\mu f' = Q_f (\bar{f}_L \gamma_\mu f_L + \bar{f}_R \gamma_\mu f_R) + \ldots$$

Thus

$$J_\mu^{n.c.} = \frac{e}{\sin\theta_W \cos\theta_W} \Big\{ a_L^f\, \bar{f}_L \gamma_\mu f_L + a_R^f\, \bar{f}_R \gamma_\mu f_R + a_L^{f'}\, \bar{f}'_L \gamma_\mu f'_L + a_R^{f'}\, \bar{f}'_R \gamma_\mu f'_R \Big\} \tag{8-4}$$

where the left and right coupling $a_{L,R}$ are defined by

$$a_L = I_{3L} - Q\sin^2\theta_W$$
$$a_R = -Q\sin^2\theta_W \tag{8-5}$$

Here Q is the charge of the fermion involved and I_{3L} is the third component of its weak isospin ($I_{3L} = \frac{1}{2}$ for f and $I_{3L} = -\frac{1}{2}$ for f'). Although (8-4) was derived for a hypothetical couple of fermions (f,f') with $Q_f = Q_{f'}+1$, the relation (8-4) is valid also when there are several such pairs, provided they are all treated on the same footing. [We shall return to this question later on (see section 12)]. Accepting this fact, we may immediately write down the neutral current couplings for any pair of elementary fermions. For example

$$\text{(i)}\quad f,f' = \nu_e, e \Rightarrow I_{3L}^{\nu_e} = \frac{1}{2} = -I_{3L}^{e}\ ,\quad Q_e = -1\ ,\quad Q_\nu = 0$$

$$a_L^{\nu_e} = \frac{1}{2}\ ,\ a_R^{\nu_e} = 0\ ,\ a_L^{e} = -\frac{1}{2} + \sin^2\Theta_W\ ,\ a_R^{e} = \sin^2\Theta_W \tag{8-6}$$

$$\text{(ii)}\quad f,f' = u,d \Rightarrow Q_u = \frac{2}{3}\ ,\ Q_d = -\frac{1}{3}\ ,\ I_3^u = -I_3^d = \frac{1}{2}$$

$$a_L^u = \frac{1}{2} - \frac{2}{3}\sin^2\Theta_W\ ,\ a_R^u = -\frac{2}{3}\sin^2\Theta_W\ ,\ a_L^d = -\frac{1}{2} + \frac{1}{3}\sin^2\Theta_W,$$

$$a_R^d = \frac{1}{3}\sin^2\Theta_W\ . \tag{8-7}$$

Furthermore all sequential neutrinos (negative leptons) have the same couplings as ν_e (e). Similarly, in the hadronic sector, all charged 2/3 quarks have the same couplings as the up quark, etc.

We may derive a local four-fermion interaction Lagrangian from $\mathcal{L}^{n.c.}$, eq. (8-3), provided the Z is very heavy

$$\mathcal{L}^{n.c.} = -8\frac{G}{\sqrt{2}}\left\{a_L^j\,\bar{f}_j\,\gamma_\mu\,L\,f_j + a_R^j\,\bar{f}_j\,\gamma_\mu\,R\,f_j\right\}$$

$$\times\left\{a_L^k\,\bar{f}_k\,\gamma_\mu\,L\,f_k + a_R^k\,\bar{f}_k\,\gamma_\mu\,R\,f_k\right\}\ , \tag{8-8}$$

$$L = \frac{1}{2}(1-\gamma_5)\ ,\quad R = \frac{1}{2}(1+\gamma_5)\ .$$

The relation (8-8) describes the neutral current interactions of any two fermions f_j and f_k , $f_j \neq f_k$. For example to get the neutrino-up quark interactions we put $f_j = \nu$, $f_k = u$, $a_L^j = a_L^\nu$, $a_L^k = a_L^u$, etc. Note that we have used the Weinberg relation $M_W = \cos\Theta_W \cdot M_Z$. For $f_j = f_k$ the factor 2 in front of G must be removed.

9. SPONTANEOUS SYMMETRY BREAKING (SSB)

In the previous sections we discussed two examples of local gauge theories, namely QED and the SU(2) x U(1) model. Indeed the unobservability of the local phases of the fields led in a natural way to the existence of interactions, mediated via massless spin one (gauge) bosons. This masslessness of mediators, necessitated by the gauge principle, is just fine for QED but catastrophic for weak interactions. We know from experiments that the weak mediators ($W^{\pm}$, Z^{o}, etc.), if they exist at all, are certainly heavier than 10 GeV.

At this point one might take the attitude that gauge theories, in spite of their beauty, Ward-identities which ensure renormalizability, etc. are relevant for QED but not for weak interactions. Within the last few years, only a small group of brave physicists have dared to work along this line. Unconventional models for weak interactions have been constructed, but so far no viable alternative to gauge theories has been found. The mainstream of activity has gone in the direction of taking for granted the gauge principle in a world where there are no masses. In the real world, the gauge world, the gauge symmetry is, by assumption "spontaneously" broken whereby the masses are created.

Below, we give a short survey of the SSB as it is employed in particle physics.

The phenomenon of SSB is known to occur in several branches of physics. Usually systems exhibiting SSB have an infinite number of degrees of freedom. Since field theory has also an infinite number of degrees of freedom, it might well be that the theory of constituents of matter is a spontaneously broken one.

Let us consider a system, described by a classical Lagrangian $\mathcal{L}$ which is invariant under a continuous group of transformations G. We examine the ground states of the system. If the system possesses a unique ground (vacuum) state (which must be invariant under G) we have a situation in which the symmetry properties of $\mathcal{L}$ and the vacuum are the same. The relevant theory is a normal one. However, it might happen that the system has several ground states, which transform into each other under G. Then, if by some reason, one of the ground states is singled out as the physical ground state of the system (the others being unphysical) the symmetry is lost and the relevant theory is said to be spontaneously broken.

It is known that the spontaneous breaking of a <u>global</u> symmetry leads to existence of massless particles (excitations) called Goldstone bosons. Very fortunately, our gauge theories are based on invariance under a <u>local</u> group of transformations. When the local invariance is broken, instead of getting additional massless bosons, we get massive gauge bosons. We shall now describe the SSB mechanism in a little bit more detail.

9.1 The Spontaneous Breakdown of a Global Symmetry

A nice example of the SSB phenomenon is provided by the Goldstone model. Let us consider the Lagrangian

$$\mathcal{L} = + \frac{\partial \varphi^*}{\partial x_\mu} \frac{\partial \varphi}{\partial x^\mu} - m^2 \varphi^* \varphi - h(\varphi^* \varphi)^2 \equiv T - V ,$$

$$V = + m^2 \varphi^* \varphi + h(\varphi^* \varphi)^2 , \tag{9-1}$$

$$\varphi = \frac{1}{\sqrt{2}} (\varphi_1 + i\varphi_2) ,$$

where φ_i, $i = 1,2$ are real scalar classical fields and m^2, h are constants. We shall only consider the case $h > 0$, because otherwise $V(|\varphi|) \to -\infty$ as $|\varphi| \to \infty$, i.e., the Hamiltonian has no minimum and there is no ground state. Suppose that $m^2 \neq 0$, then there are two distinct possibilities. If $m^2 > 0$, the Lagrangian (9-1) would describe a self-interacting charged scalar field. For $m^2 < 0$, we have an example of a spontaneously broken theory. Perhaps nature utilizes both possibilities. We now look for the ground states, i.e., for those values of φ which minimize the potential

$$\frac{\partial V}{\partial |\varphi|} = 0 \tag{9-2}$$

For $m^2 > 0$ we find $V \geq 0$, the minimum of V corresponds to $|\varphi| = \varphi_1 = \varphi_2 = 0$. Thus there is a unique vacuum. Note that the Lagrangian (9-1) is invariant under the global U(1) group

$$\varphi(x) \to e^{i\Lambda}\varphi(x) \quad , \quad \Lambda = \text{real constant}. \tag{9-3}$$

The unique vacuum $\varphi_1 = \varphi_2 = 0$ is also invariant under U(1). Now we focus our attention on the case $m^2 < 0$. Put $\mu^2 = -m^2$. The potential $V = -\mu^2|\varphi|^2 + h|\varphi|^4$ is minimum for

$$2|\varphi|^2 = \varphi_1^2 + \varphi_2^2 = \frac{\mu^2}{h} \equiv v^2 \tag{9-4}$$

Thus the system has an infinite number of ground states, which 'lie' on a circle in the $\varphi_1 - \varphi_2$ plane. Under the action of the group, eq. (9-3), the ground states transform into each other.

Suppose now that we choose one of the ground states, for example $\varphi_o \equiv (\varphi_1 = v, \varphi_2 = 0)$ as the physical ground state then the symmetry is spontaneously broken, because the state φ_o is not invariant under the transformation (9-3). To see the structure of the theory, we expand the potential near the vacuum state, where the theory is expected to be stable. Put

$$\varphi_1' = \varphi_1 - v \quad , \quad \varphi_2' = \varphi_2$$

$$V(\varphi') = V_o + \sum_{i=1,2} \left(\frac{\partial V}{\partial \varphi_i}\right)_0 \varphi_i' + \frac{1}{2} \sum_{i,j=1,2} \left(\frac{\partial^2 V}{\partial \varphi_i \partial \varphi_j}\right)_o \varphi_i' \varphi_j' + \ldots \tag{9-5}$$

where the index o indicates that the quantities are evaluated at the point φ_o. We have

$$\frac{\partial V}{\partial \varphi_i} = \left[-\mu^2 + h(\varphi_1^2 + \varphi_2^2)\right] \varphi_i \quad , \quad \left(\frac{\partial V}{\partial \varphi_i}\right)_o = 0$$

$$\frac{\partial^2 V}{\partial \varphi_i \partial \varphi_j} = \left[-\mu^2 + h(\varphi_1^2 + \varphi_2^2)\right]\delta_{ij} + 2h\varphi_i\varphi_j, \quad \left(\frac{\partial^2 V}{\partial \varphi_i \partial \varphi_j}\right)_0 = 2h(\varphi_i\varphi_j)_o \tag{9-6}$$

Thus the mass matrix, defined by

$$m_{ij}^2 = \left(\frac{\partial^2 V}{\partial \varphi_i \partial \varphi_j}\right)_0 = 2h(\varphi_i\varphi_j)_o \tag{9-7}$$

$$\varphi_o = (\varphi_1 = v \, , \varphi_2 = 0)$$

has only one non-zero element, $m_{11}^2 = 2h\ v^2 = \frac{2h\mu^2}{h} = 2\mu^2$, while $m_{12}^2 = m_{21}^2 = m_{22}^2 = 0$. Therefore the 'particle' φ_2' is massless

$$V(\varphi') = \mu^2 \varphi_1'^2 + \text{interaction terms}$$

$$\mathcal{L}(\varphi') = \frac{1}{2} \sum_i \frac{\partial \varphi_i'}{\partial x_\mu} \frac{\partial \varphi_i'}{\partial x^\mu} - \mu^2 \varphi_1'^2 + \text{interaction terms} \tag{9-8}$$

The spontaneous breakdown of the U(1) symmetry has led to apparition of a massless (Goldstone) boson, represented by the field φ_2'. The field φ_1' has the mass $\sqrt{2}\,\mu$. The reader may wonder what happens if she would choose a different vacuum, say $\varphi_o' \equiv (\varphi_1 = \frac{v}{\sqrt{2}}, \ \varphi_2 = \frac{v}{\sqrt{2}})$? She could repeat the above analysis by introducing $\varphi_i' = \varphi_i - \frac{v}{\sqrt{2}}$. Then from $V(\varphi') = \frac{1}{2} m_{ij}^2 \varphi_i' \varphi_j' +$ interaction terms, $m_{ij}^2 = 2h(\varphi_i \varphi_j)_o$, she would find that the mass matrix is not diagonal

$$V(\varphi') = hv^2(\varphi_1'^2 + \varphi_2'^2 + 2\varphi_1'\varphi_2') + \ldots = hv^2(\varphi_1' + \varphi_2')^2 +$$
$$+ 0\cdot(\varphi_1' - \varphi_2')^2 + \ldots$$

Thus the linear combination $\frac{1}{\sqrt{2}}(\varphi_1' + \varphi_2')$ has acquired the mass $\sqrt{2}\mu$ while the combination $\frac{1}{\sqrt{2}}(\varphi_1' - \varphi_2')$ represents a massless (Goldstone) boson. This solution is obtainable from the earlier one by a rotation of 45 degrees in the $\varphi_1 - \varphi_2$ plane. Clearly, this solution and the former one are physically equivalent.

The arguments in this section can be easily generalized. Suppose we have a Lagrangian, $\mathcal{L}(\varphi_i, \frac{\partial\varphi_i}{\partial x}) = T - V(\varphi_i)$, $i = 1,\ldots,N$ Here φ_i are real scalar fields. Assume that $\mathcal{L}$ is invariant under a local group of transformations G. Let the elements of the group be represented by $\exp(i\,\bar{\alpha}\cdot\bar{T})$, where α^j are real constants and T^j, $j=1,\ldots, n$ are the matrices representing the generators

$$\varphi = \begin{pmatrix}\varphi_1\\ \vdots\\ \varphi_N\end{pmatrix}, \quad \varphi \to \exp(i\,\bar{\alpha}\cdot\bar{T})\varphi, \quad \delta\varphi = i\,\bar{\alpha}\cdot\bar{T}\,\varphi, \quad \text{small } \bar{\alpha} \tag{9-9}$$

$$T^j = N \times N \text{ matrices}, \quad j = 1,\ldots, n$$

Look for the vacuum state by using $\frac{\partial V(\varphi)}{\partial\varphi_j} = 0$. Suppose that there are several vacuua. These will transform into each other under the group. Choose a particular vacuum

$\varphi_o = \begin{pmatrix}\varphi_1\\ \vdots\\ \varphi_N\end{pmatrix}_o$, and disregard the rest. Under an infinitesimal

action of the group the vacuum is changed by the amount $\delta\varphi_o = i\,\bar{\alpha}\cdot\bar{T}\,\varphi_o$. We know that the vacuum is then non-invariant under some of the operations in the group, namely those for which $T^k \varphi_o \neq 0$. Suppose there are n' such generators, $k=1,\ldots, n'$, $1 \leq n' \leq n$. Then the remaining generators leave the vacuum invariant, $T^k \varphi_o = 0$, $k = n' + 1,\ldots, n$. We shall now show that to each generator which breaks the symmetry corresponds a Goldstone boson. Introduce $\varphi' = \varphi - \varphi_o$, and expand as before

$$V = V_o + \sum_j \left(\frac{\partial V}{\partial \varphi_j}\right)_o \varphi'_j + \frac{1}{2}\sum \left(\frac{\partial^2 V}{\partial \varphi_i \partial \varphi_j}\right)_o \varphi'_i \varphi'_j + \text{interaction terms},$$

$$\left(\frac{\partial V}{\partial \varphi_j}\right)_o = 0 \quad , \quad m^2_{ij} = \left(\frac{\partial^2 V}{\partial \varphi_i \partial \varphi_j}\right)_o \quad . \tag{9-10}$$

The potential is invariant under G implying

$$\delta V = \frac{\delta V}{\partial \varphi_j} \delta \varphi_j = 0 \quad ,$$

i.e.,

$$\frac{\partial V}{\partial \varphi_j} T^r_{jk} \varphi_k = 0 \qquad r = 1, \ldots, n \tag{9-11}$$

Take the derivative $\frac{\partial}{\partial \varphi_\ell}$ of (9.11)

$$\frac{\partial^2 V}{\partial \varphi_\ell \partial \varphi_j} T^r_{jk} \varphi_k + \frac{\partial V}{\partial \varphi_j} T^r_{j\ell} = 0 \tag{9-12}$$

At $\varphi = \varphi_o$, using (9-10), we find

$$m^2{}_{\ell j} T^r_{jk} (\varphi_k)_o = 0 \qquad r = 1, \ldots, n \tag{9-13}$$

By our assumption $T^r \varphi_o \neq 0$ for $r = 1, \ldots, n'$. Thus $T^r \varphi_o$ span a n' - dimensional subspace in the N dimensional space spanned by the φ's, and therefore, $m^2{}_{\ell j}$ has n' zero eigenvalues, or in other words there are n' Goldstone bosons in the theory.

Let us apply this formalism to the special case considered above

$$\varphi \to (\varphi + i\Lambda\varphi)$$

$$\delta\varphi = i\Lambda\varphi$$

$$\delta\varphi_1 + i\delta\varphi_2 = i\Lambda(\varphi_1 + i\varphi_2)$$

$$\delta\varphi_1 = -\Lambda\varphi_2 \ , \quad \delta\varphi_2 = \Lambda\varphi_1$$

Thus the generator of the U(1) group in the $\begin{pmatrix}\varphi_1 \\ \varphi_2\end{pmatrix}$ space has the form

$$T = \begin{pmatrix} 0 & -1 \\ 1 & 0 \end{pmatrix}$$ and

$$m^2_{ij}\ T_{jk}(\varphi_k)_0 = 0$$

$$\begin{pmatrix} m^2_{11} & m^2_{12} \\ m^2_{21} & m^2_{22} \end{pmatrix} \begin{pmatrix} 0 & -1 \\ 1 & 0 \end{pmatrix} \begin{pmatrix} v \\ 0 \end{pmatrix} = 0 \ .$$

From here we obtain $m^2_{12} = m^2_{22} = 0$, i.e., the φ_2' is massless, in accordance with previous results.

When a local symmetry is spontaneously broken the situation becomes quite different, as we shall discuss below.

In quantized field theory the SSB is assumed to be realized by letting the field operators acquire vacuum expectation values $\phi_i = \phi_i' + v_i$, $\langle 0|\phi_i'|0\rangle = 0$, $\langle 0|\phi_i|0\rangle = v_i$. The c-numbers v_i are chosen such as to minimize the classical potential.

10. SPONTANEOUS BREAKING OF A LOCAL SYMMETRY

We start with the same Lagrangian as in the previous section (eq. (9-1)) and require invariance under the local phase transformation $\varphi(x) \to e^{i\Lambda(x)}\varphi(x)$, where Λ is real. From our previous considerations (section 4) we know that we must introduce a massless vector field B_μ and replace the derivative ∂_μ by the covariant derivative $\partial_\mu - igB_\mu$, where g is a constant

$$\mathcal{L} = \left(\frac{\partial\varphi^*}{\partial x^\mu} + igB_\mu\varphi^*\right)\left(\frac{\partial\varphi}{\partial x_\mu} - igB^\mu\varphi\right) - V(\varphi) - \frac{1}{4}B_{\mu\nu}B^{\mu\nu} \ ,$$

$$V(\varphi) = m^2\varphi^*\varphi + h(\varphi^*\varphi)^2 \qquad (10\text{-}1)$$

$$B_{\mu\nu} = \frac{\partial B_\nu}{\partial x^\mu} - \frac{\partial B_\mu}{\partial x^\nu} \ , \quad \varphi = \frac{1}{\sqrt{2}}(\varphi_1 + i\varphi_2)$$

$\mathcal{L}$ is invariant under the joint gauge transformations

$$\varphi \to \exp(i\Lambda(x))\,\varphi(x) \ , \quad B_\mu \to B_\mu + \frac{1}{g}\frac{\partial\Lambda(x)}{\partial x^\mu} \ . \qquad (10\text{-}2)$$

Again h is taken to be positive. For $m^2 > 0$, $\mathcal{L}$ is the QED Lagrangian for a self-interacting charged spin zero particle, if we identify B^μ with the photon field. The vacuum of the φ-sector corresponds to $\varphi_1 = \varphi_2 = 0$ and is invariant under (10-2).

As before we turn to the interesting case $m^2 = -\mu^2 < 0$, and assume that the ground state is defined by the minimum of the potential $V(\varphi)$, $\frac{\partial V(\varphi)}{\partial \varphi_j} = 0$, whereby we obtain the relation (9-4), $\varphi_1^2 + \varphi_2^2 = v^2$. Among these infinite number of vacuua we select one, e.g. $\varphi_1 = v$, $\varphi_2 = 0$. Now the symmetry is spontaneously broken. Expanding near the physical vacuum, $\varphi_1' = \varphi_1 - v$, $\varphi_2' = \varphi_2$, we find

$$\mathcal{L} = \frac{1}{2}\left|\frac{\partial(\varphi_1' + i\varphi_2')}{\partial x_\mu} - igB^\mu(\varphi_1' + v + i\varphi_2')\right|^2 - \mu^2\varphi_1'^2 + 0 \cdot \varphi_2'^2 -$$

$$- \frac{1}{4} B_{\mu\nu}B^{\mu\nu} + \text{interactions terms involving } \varphi_1' \text{ and } \varphi_2'. \tag{10-3}$$

The Lagrangian (10-3) is indeed remarkable. It seems that we have a massive scalar φ_1' (with mass $\sqrt{2}\mu$), a massless one φ_2' and a massive vector particle B. Note that there is a term $\frac{1}{2} g^2v^2B^\mu B_\mu$, telling us that the B field has acquired a mass equal to $|gv|$. However this cannot be true because to begin with we had four degrees of freedom, one for each scalar field and two for the massless B. Now we have five, because a massive vector field has three spin states, $S_z = 0, \pm 1$. Thus there is a redundant degree of freedom in (10-3). Indeed the massless φ_2' is superfluous and can be "gauged away". To see this, we go back to our original Lagrangian, eq. (10-1) and perform a gauge transformation
$\varphi = e^{i\Lambda}\hat{\varphi}$ and $B = \hat{B} + \frac{1}{g}\frac{\partial \Lambda}{\partial x}$. The Lagrangian is of course invariant. We may now choose Λ such that $\hat{\varphi}_2 = 0$, viz.,

$$\varphi = e^{i\Lambda}(\hat{\varphi}_1 + i\hat{\varphi}_2)/\sqrt{2} \quad , \quad \hat{\varphi}_2 = 0$$

$$\varphi = \frac{1}{\sqrt{2}}(1 + i\Lambda + ..)\hat{\varphi}_1 \ ,$$

The exponential $e^{i\Lambda}$ disappears and the scalar field becomes real.

The spontaneous symmetry breaking can be done as before, viz. $\hat{\varphi}_1^2 = v^2$, for vacuum. Put $\hat{\varphi}_1 = \varphi_1 + v$, we find

$$\mathcal{L} = \frac{1}{2}\left|\frac{\partial \varphi_1}{\partial x_\rho} - igB_\rho(\varphi_1 + v)\right|^2 - \mu^2\varphi_1^2 - \frac{1}{4}B_{\rho\sigma}B^{\rho\sigma} -$$

$$- \frac{\mu^2}{v}\varphi_1^3 - \frac{\mu^2}{4v^2}\varphi_1^4$$

$$= \frac{-1}{4}B_{\rho\sigma}B^{\rho\sigma} + \frac{1}{2}v^2g^2B_\rho B^\rho + (g^2v\varphi_1 + \frac{1}{2}g^2\varphi_1^2)B_\rho B^\rho +$$

$$+\frac{1}{2}\left[\frac{\partial\varphi_1}{\partial x_\rho}\frac{\partial\varphi_1}{\partial x^\rho} - 2\mu^2\varphi_1^2\right] - \frac{\mu^2}{v}\varphi_1^3 - \frac{\mu^2}{4v^2}\varphi_1^4 \quad . \qquad (10\text{-}4)$$

Note that gauging away $\hat{\varphi}_1$ instead of $\hat{\varphi}_2$ leads to an unphysical . From (10-4), $M_B = |gv|$, $M_{\varphi_1} = \sqrt{2}\mu$, where M refers to the mass. $\mathcal{L}$ describes a massive vector particle interacting in a very specific way with a real scalar field. In the quantized theory the prescription is as given as the end of the last section. The particle φ_1 is usually referred to as a Higgs particle and the mechanism described above is an example of the Higgs, Kibble, etc. mechanism.

It is widely believed that all masses are generated via the Higgs mechanism, the reason being two-fold. Firstly it is appealing to imagine that a beautiful massless world built upon the local gauge doctrine suddenly undergoes a spontaneous breaking whereby all masses are created. If the reader does not buy this argument the second one can be given, i.e., no alternative has been found so far. Ever since the heroic discovery, by t'Hooft, that the SSB does not spoil the renormalizability, such theories are much cherished.

For each specific gauge theory, a spontaneous breaking scheme has to be cleverly invented in order to achieve the purposes in mind. We shall give some examples later on.

It is fair to say that the Higgs sector of gauge theories usually does not match at all the beauty of the rest. The Higgs sector introduces new parameters, sometimes many of them. Let us hope that some day we will understand the origin of the masses.

11. SPONTANEOUS BREAKING SCHEME FOR THE STANDARD SU(2) x U(1) MODEL

The standard model, described in sections 6-8, has initially 4 gauge bosons $W^\pm$, W^o and B. The breaking mechanism should be devised such that $W^\pm$ and (the linear combination of W^o and B corresponding to) Z^o become massive, whereas the photon remains massless. Thus we need to introduce at least 4 real scalar fields φ_i. These "Higgses" must transform as multiplets under SU(2) x U(1), so as to couple to the W's and B. We must make sure that the Higgs potential $V(\varphi)$ allows degenerate vacuua; then we would choose one of them as the physical vacuum and let $\varphi_i = \varphi_i' + v_i$, where $v_i = \langle 0|\varphi_i|0\rangle$. Since we believe in the conservation of the electric charge, we can only allow neutral scalars acquire vacuum expectation values. Thus $Q^\gamma v = 0$, where Q^γ is the charge operator and $v = \begin{pmatrix} v_1 \\ v_2 \\ \vdots \end{pmatrix}$. The group SU(2) x U(1) has four generators

T^k, $k = 1,2,3$, and y, where $Q = T^3 + y$. Clearly v should be invariant only under Q, then 3 "symmetries" are broken and the corresponding Goldstone bosons may be gauged away to yield three massive bosons. The four Higgses are introduced elegantly, à la Weinberg, in a doublet

$$\varphi = \begin{pmatrix} \varphi^+ \\ \varphi^o \end{pmatrix}, \quad \varphi^+ = \frac{1}{\sqrt{2}}(\varphi_1 + i\varphi_2), \; \varphi^o = \frac{1}{\sqrt{2}}(\varphi_3 + i\varphi_4) , \tag{11-1}$$

where φ_i are real. The Higgs Lagrangian now reads

$$\mathcal{L} = \left|\left(\frac{\partial}{\partial x} - ig\,\frac{\bar{\sigma}}{2}\cdot\bar{W} - ig'yB\right)\varphi\right|^2 - V(\boldsymbol{\varphi}) \tag{11-2}$$

where the requirement of local gauge invariance under SU(2) x U(1) has given the specific couplings to W and B. We may write $V(\varphi) = -\mu^2\varphi^*\varphi + h(\varphi^*\varphi)^2$, whereby SSB takes place. The arguments in the previous section go through mutatis mutandis

$$v = \text{vacuum} = \begin{pmatrix} 0 \\ v^o \end{pmatrix} , \qquad \varphi = \varphi' + v$$

$$\left|\left(g\,\frac{\bar{\sigma}}{2}\cdot\bar{W} + g'yB\right)v\right|^2 = \frac{1}{2}\,m^2_{ij}\,W^i\,W^j \quad , \quad W^o = B \tag{11-3}$$

Here m^2_{ij} is the mass matrix. We check easily that the photon is massless, $T_3 \equiv \sigma_{3/2}$

$$A_\mu A^\mu\left|\left(\underbrace{g\,a_{W_3\gamma}}_{e}\,T_3 + \underbrace{g'a_{B\gamma}}_{e}y\right)v\right|^2 = A_\mu A^\mu\left|eQ^\gamma v\right|^2 = 0 . \tag{11-4}$$

The mass of the Z is obtained from

$$\left|\left(g\,a_{W_3Z}\,T_3 + g'a_{BZ}\,y\right)v\right|^2 Z_\mu Z^\mu = \frac{1}{2}\,M_Z^2\,Z_\mu\,Z^\mu . \tag{11-5}$$

Using $y = Q - T_3$, $Qv = 0$, we find

$$M_Z^2 = \frac{v^{o2}\,e^2}{2(a^2_{W_3\gamma}a^2_{B\gamma})} \tag{11-6}$$

Finally, the mass of the charged W's

$$\left| g \frac{\sigma_1 W^1 + \sigma_2 W^2}{2} v \right|^2 = \frac{g^2 v^{o2}}{4} (W_1 + iW_2)(W_1 - iW_2) = \frac{g^2 v^{o2}}{2} W^+ W ,$$

$$M_W^2 = \frac{g^2 v^{o2}}{2} = \frac{e^2 v^{o2}}{2(a_{W_3\gamma})^2} . \tag{11-7}$$

From (11-6) and (11-7) we obtain the Weinberg relation

$$M_W^2 = a_{B\gamma}^2 M_Z^2 = \cos^2\Theta \, M_Z^2 \tag{11-8}$$

Using the empirical relation $M_W = \frac{38 \text{ GeV}}{|\sin\Theta|}$, we find that for the present preferred value $\sin^2\Theta = \frac{1}{4}$, the mediators are awfully heavy $M_W \sim 76$ GeV, $M_Z \sim 88$ GeV.

The analysis of this section can be generalized. Clearly the photon remains massless as long as the electric charge is conserved, viz.

$$\left| \sum g_k T_k W_k v \right|^2 = \frac{1}{2} W_i W_j m_{ij}^2$$

$$W_k = a_{W_k\gamma} A + \dots$$

$$A^\mu A_\mu \left| \sum_j g_j a_{W_j\gamma} T_j v \right|^2 = \frac{1}{2} m_\gamma^2 A^\mu A_\mu$$

Here g, T and W are the coupling constants, generators and gauge bosons respectively. The sum over j involves only those terms where $a_{W_j\gamma} \neq 0$. For these we have the condition

$$g_j a_{W_j\gamma} = e , \quad \text{and} \quad Q^\gamma = \sum_j T_j . \quad \text{Thus } m_\gamma^2 \sim |Q^\gamma v|^2 = 0$$

because v is electrically neutral.

The Higgs doublet (11-1) is sufficient for giving masses to all fundamental fermions. Of course fermion mass terms $m\bar{\psi}\psi$ do not spoil renormalizability and could be added by hand. Nevertheless, it is more attractive if all masses should have the same origin, say Higgses. Suppose we want to give masses to up and down quarks. We have the multiplets $\begin{pmatrix} u \\ d \end{pmatrix}_L$, u_R, d_R, $\varphi = \begin{pmatrix} \varphi^+ \\ \varphi^o \end{pmatrix}$. Note that $\varphi^c = \begin{pmatrix} \varphi^{o*} \\ -\varphi^- \end{pmatrix}$ has the same transformation properties as $\begin{pmatrix} \varphi^+ \\ \varphi^o \end{pmatrix}$. The gauge invariant Higgs-fermion interaction reads

$$\mathcal{L} = c_d(\bar{u},\bar{d})_L \begin{pmatrix}\varphi^+\\ \varphi^0\end{pmatrix} d_R + c_u(\bar{u},\bar{d})_L \begin{pmatrix}\varphi^{*0}\\ \varphi^-\end{pmatrix} u_R + \text{h.c.} \qquad (11\text{-}9)$$

Upon spontaneous breakdown $\varphi^0 \to \varphi^0 + v^0$, and after gauging away $\varphi^\pm$ we find

$$\mathcal{L}_{mass} = (c_d v^0)\, \bar{d}_L d_R + (c_u v^0)\, \bar{u}_L u_R + \text{h.c.}$$

If the c's are real (this can always be arranged as we shall discuss later)

$$c_d v^0(\bar{d}_L d_R + \bar{d}_R d_L) = c_d v^0 \bar{d}d \equiv m_d \bar{d}d \ .$$

Thus

$$\mathcal{L} = m_d\bar{d}d + m_u\bar{u}u + \frac{m_d}{v^0}\,\bar{d}d\varphi^0 + \frac{m_u}{v^0}\,\bar{u}u\varphi^0 \qquad (11\text{-}10)$$

A very specific prediction of the spontaneously broken SU(2) x U(1) model is the existence of the scalar boson φ^0, which couples to each elementary fermion with a strength proportional to the mass of that fermion. Unfortunately, the vacuum expectation value v^0 is huge (~200 GeV), see eqs. (7-2) and (11-7). Therefore only heavy elementary fermions do care about Higgses. The massive vector bosons couple strongly to Higgses, see eq. (10-4).

12. MORE QUARKS AND LEPTONS, MASSES AND THE CABIBBO ANGLES

In this section we start by reminding ourselves that Nature seems to be repeating herself and so far we have failed to understand her reasons for doing so. In fact by now there are three distinct families of leptons and quarks, viz;

the fundamental family	ν_e, e, u_1, u_2, u_3, d_1, d_2, d_3
the second (superfluous?) family	ν_μ, μ, c_1, c_2, c_3, s_1, s_2, s_3
the third (super superfluous?) family	ν_τ, τ, t_1, t_2, t_3, b_1, b_2, b_3.

(12-1)

The families are also sometimes referred to as generations, however, there does not seem to exist any mother-daughter relationship among them. In (12-1), the indices 1, 2, 3 refer to the colour degrees of freedom. As discussed before, the evidence for ν_τ, although substantial, is still indirect and the truth (t) remains to be discovered.

In gauge theories, one normally follows Nature's own pattern: a repeated world including n $(\text{super})^n$ fluous families is simply described by repeating the theory constructed for just one family, for example the fundamental family. (This state of affairs is, of course, not satisfactory.) Here we shall assume that there are n families,

$$\nu_j, \quad \ell_j, \quad u_j, \quad d_j \qquad j = 1, .., n$$

$$Q = 0, \quad -1, \quad 2/3, \quad -1/3, \tag{12-2}$$

where we have suppressed the colour indices. In (12-2), for example $u_1 = u$, $u_2 = c$, $u_3 = t$, etc. Furthermore, we shall assume that the standard $SU(2) \times U(1)$ model is correct, the left multiplets are doublets and the right-handed ones singlets. In constructing the theory, we follow the steps in section 6, including of course all the elementary fermions in nature, (12-2). The kinetic energy term reads

$$\mathcal{L}_o = i \sum_{j=1}^{n} \left[\bar{\nu}_j \gamma \frac{\partial}{\partial x} \nu_j + \bar{\ell}_j \gamma \frac{\partial}{\partial x} \ell_j + \bar{u}_j \gamma \frac{\partial}{\partial x} u_j + \bar{d}_j \gamma \frac{\partial}{\partial x} d_j \right], \tag{12-3}$$

where the sum over the colour has not been explicitly exhibited. This relation is the analog of (6-1). The decomposition into left and right components may be done exactly as in (6-3). Since we have more than 2 elementary fermions, we do not know, a priori, how to choose the doublets, viz., there are infinite number of possibilities and the physics depends on the choice made. Of course, because of the charge assignments, we cannot put leptons and quarks in the same doublets (otherwise we would get W's with fractional charges).

We now restrict ourselves to the hadronic sector. The generalization to leptonic sector is trivial (see below). Due to our lack of knowledge, the most general choice for multiplets is to take (arbitrary) linear combinations

$$\psi_{jL} = \begin{pmatrix} u'_j \\ \\ d'_j \end{pmatrix}_L, \qquad u'_{jR}, \qquad d'_{jR}, \qquad j = 1,.., n \tag{12-4}$$

where

$$u'_{jL} = A^{(u,L)}_{jk} u_{kL} \qquad u'_{jR} = A^{(u,R)}_{jk} u_{kR} \tag{12-5}$$

$$d'_{jL} = A^{(d,L)}_{jk} d_{kL} \qquad d'_{jR} = A^{(d,R)}_{jk} d_{kR}, \quad j = 1,..,n.$$

Here the A's are n x n matrices. They have to be unitary in order to leave $\mathcal{L}_o$ invariant (we shall return to the origin of these matrices soon). Suppressing the indices, etc.

$$\sum_j \bar{u}_j u_j = \sum_j \bar{u}'_j u'_j \quad , \quad \sum_j \bar{d}_j d_j = \sum_j \bar{d}'_j d'_j \quad , \tag{12-6}$$

$$A^+ = A^{-1}, \text{ for each } A,$$

$$\mathcal{L}_o = i \sum_j \left\{ (\bar{u}'_j \, , \, \bar{d}'_j)_L \, \gamma \frac{\partial}{\partial x} \begin{pmatrix} u'_j \\ d'_j \end{pmatrix}_L + \bar{u}'_{jR} \, \gamma \frac{\partial}{\partial x} u'_{jR} + \bar{d}'_{jR} \gamma \frac{\partial}{\partial x} d'_{jR} \right\}. \tag{12-7}$$

Now we "gauge" the kinetic energy term, whereby the mediators $\vec{W}$ and B are born (just as in section 6). Thus (see eq. 6-7)

$$\gamma \frac{\partial}{\partial x} \rightarrow \gamma^\mu D_\mu = \gamma^\mu \left[\frac{\partial}{\partial x^\mu} - i \, g \frac{\bar{\sigma}}{2} \cdot \bar{W}_\mu - i \, g' y \, B_\mu \right]$$

is sandwiched between the fields in (12-4) just as in (6-7); remembering $\bar{\sigma} \, u'_{jR} = \bar{\sigma} \, d'_{jR} = 0$, $y = Q - \frac{\sigma_3}{2}$, Q is the charge matrix, $\bar{\sigma}$ = Pauli matrices, viz.

$$\mathcal{L} = i \sum_j \left\{ (\bar{u}'_j , \bar{d}'_j)_L \, \gamma_\lambda D^\lambda \begin{pmatrix} u'_j \\ d'_j \end{pmatrix}_L + \bar{u}'_{jR} \gamma_\lambda \, D^\lambda \, u'_{jR} + \bar{d}'_{jR} \gamma_\lambda \, D^\lambda \, d'_{jR} \right\} ,$$

$$D = \frac{\partial}{\partial x} - i \, g \frac{\bar{\sigma}}{2} \cdot \bar{W} - i \, g' \, y \, B \tag{12-8}$$

So far all the (matter as well as gauge) fields are massless. According to our assumptions, in the real world, the SU(2) x U(1) symmetry is spontaneously broken, whereby the masses are generated. The breaking mechanism is assumed to be à la Weinberg, i.e., due to a doublet of Higgses. Thus we may repeat the arguments in section 11. The quark - Higgs interaction (see eq. 11-9) before SSB reads

$$\mathcal{L}(\text{quark-Higgs}) = \sum_{j,k} \left\{ c_{jk} (\bar{u}'_j \, , \, \bar{d}'_j)_L \begin{pmatrix} \overset{*}{\varphi^o} \\ -\varphi^- \end{pmatrix} u'_{kR} + \tilde{c}_{jk} (\bar{u}'_j \, , \, \bar{d}'_j)_L \begin{pmatrix} \varphi^+ \\ \varphi^o \end{pmatrix} d'_{kR} \right\} + \; + \text{h.c.} \tag{12-9}$$

Here the c's and $\tilde{c}$'s are arbitrary constants. The SSB is achieved by letting $\varphi^0 \rightarrow \varphi^0 + v$ and dropping terms involving $\varphi^\pm$. Then

$$\mathcal{L}_{mass} = v \sum_{j,k} \left\{ \bar{u}'_{jL}\, c_{jk}\, u'_{kR} + \bar{d}'_{jk}\, \tilde{c}_{jk}\, d'_{kR} + h.c. \right\}$$

$$= \bar{u}'_L M' u'_R + \bar{d}'_L \tilde{M}' d'_R + \bar{u}'_R M'^{+} u'_L + \bar{d}'_R \tilde{M}'^{+} d'_L \qquad (12\text{-}10)$$

where

$$u'_{L,R} = \begin{pmatrix} u'_1 \\ u'_2 \\ \vdots \\ u'_n \end{pmatrix}_{L,R} , \quad d'_{L,R} = \begin{pmatrix} d'_1 \\ d'_2 \\ \vdots \\ d'_n \end{pmatrix}_{L,R} , \quad M'_{jk} = v\, c_{jk} , \quad \tilde{M}'_{jk} = v\, \tilde{c}_{jk}$$

Now we see clearly the significance of the matrices A in eq. (12-5). They originate from diagonalizing the mass matrix and getting rid of the pseudoscalar terms in (12-10). For $Q = 2/3$ sector, we write M' in "polar coordinates"

$$M' = m\, U , \qquad (12\text{-}11)$$

where m is hermitian and U is unitary. m is diagonalized by a transformation $m = S^{-1} D\, S$, where S is unitary, and D is diagonal and real,

$$\bar{u}'_L \mathbf{M'}\, u'_R + h.c. = \bar{u}'_L\, S^{-1} D\, S\, U\, u'_R + h.c.$$

We put

$$u_R = S\, U\, u'_R \quad , \quad u = \begin{pmatrix} u_1 \\ \vdots \\ u_n \end{pmatrix}$$

$$u_L = S\, u'_L \quad , \qquad (12\text{-}12)$$

$$\bar{u}'_L M' u'_R + h.c. = \bar{u}_L D\, u_R + \bar{u}_R D\, u_L = \bar{u}\, D\, u .$$

Thus the elements of D are the physical masses

$$D = \begin{pmatrix} m_u & & & \\ & m_c & & \\ & & m_t & \\ & & & \ddots \end{pmatrix} , \quad u = \begin{vmatrix} u_1 = u \\ u_2 = c \\ u_3 = t \\ \vdots \\ u_n \end{vmatrix} . \qquad (12\text{-}13)$$

Evidently, the charge -1/3 sector can be diagonalized in the same way. Comparing (12-12) with (12-5), yields

$$A^{(u,L)} = S^+ \ , \ A^{(u,R)} = U^+ S^+, \text{ etc.}$$

We now return to the interaction term in (12-8). As we have just seen the primed fields are not the physical ones and therefore we should express them in terms of the physical unprimed fields, by utilizing (12-5). It seems that we then get a very complicated expression. Fortunately, the generalized GIM - mechanism is operating here and there is no trace of the matrices A in the neutral current sector, viz.

$$\sum_j (\bar{u}'_j, \bar{d}'_j)_L \frac{\sigma_3}{2} \begin{pmatrix} u'_j \\ d'_j \end{pmatrix}_L = \frac{1}{2} \sum_j \left\{ \bar{u}'_{jL}\, u'_{jL} - \bar{d}'_{jL}\, d'_{jL} \right\} = \tag{12-14a}$$

$$= \frac{1}{2} \sum_j \left\{ \bar{u}_{jL}\, u_{jL} - \bar{d}_{jL}\, d_{jL} \right\} ,$$

$$\sum_j \bar{u}'_{jR}\, y\, u'_{jR} = \sum_j Q_u\, \bar{u}'_{jR}\, u'_{jR} = \frac{2}{3} \sum_j \bar{u}_{jR}\, u_{jR} \ , \tag{12-14b}$$

$$\sum_j \bar{d}'_{jR}\, y\, d'_{jR} = -\frac{1}{3} \sum_j \bar{d}_{jR}\, d_{jR} \tag{12-14c}$$

The neutral currents are thus 'naturally' flavour conserving: there are no strangeness-changing or charm-changing, etc. neutral currents, independent of the specific form of the matrices A.

All we had to do is to treat all Q = 2/3 quarks on an equal footing (and similarly for Q = -1/3 quarks). The above arguments apply just as well for leptons, i.e., there are no flavour-changing neutral currents (such as $\bar{\mu}e$, $\bar{\nu}_e \nu_\mu$, etc. The absence of such currents in the standard model is one of its attractive features. Strangeness-changing neutral currents are known to be highly suppressed (for example from $K^o_L \to \mu^+\mu^-$). Similarly there are good limits on charm-changing neutral currents from $D^0 - \bar{D}^0$ transition. In summary the neutral current sector of the standard model is as given in (8-4) and (8-5), $I_{3L} = \frac{1}{2}$ for neutrinos and charge 2/3 quarks, $I_{3L} = -\frac{1}{2}$ for ℓ^- and all charge - 1/3 quarks .

We now focus our attention on the charged current sector in eq. 12-8,

$$\mathcal{L}^{c.c.} = \frac{g}{2\sqrt{2}} \bar{u}'_j \gamma^\lambda (1 - \gamma_5) d'_j W^-_\lambda + \text{h.c.}$$

$$\mathcal{L} = \frac{g}{2\sqrt{2}} \bar{u}_j \gamma^\lambda (1 - \gamma_5) \left\{ \left(A^{(u,L)}\right)^+ A^{(d,L)} \right\}_{jk} d_k W_\lambda^- , \quad (12\text{-}15)$$

where we have used (12-5). Note that { } is a unitary n x n matrix. Because of our lack of information, all we know about U = { } is that it is an arbitrary unitary matrix. A general n x n unitary matrix has n^2 real parameters; of these 2n - 1 may be removed by redefining the fields u and d, viz.

$$\bar{u}\, U\, d = \bar{u}\, \Phi^{-1}(\theta_1, \ldots, \theta_n) \Phi(\theta_1, \ldots, \theta_n) U\, \Phi^{-1}(\varphi_1, \ldots, \varphi_n) \Phi(\varphi_1, \ldots, \varphi_n) d ,$$

$$\Phi(\theta_1, \ldots, \theta_n) = \begin{pmatrix} e^{i\theta_1} & & & \\ & \cdot & & \\ & & \cdot & \\ & & & \cdot \\ & & & & e^{i\theta_n} \end{pmatrix} \quad (12\text{-}16)$$

The reader may now write $U_{jk} = |U_{jk}| e^{i\zeta_{jk}}$ and convince herself that, by judicial choice of $(\theta) = (\theta_1, \ldots, \theta_n)$ and $(\varphi) = (\varphi_1, \ldots, \varphi_n)$, 2n - 1 of the parameters in $\Phi(\theta) U\, \Phi^{-1}(\varphi)$ can be removed. Of cource, in our $\mathcal{L}$, we must now redefine $\Phi(\varphi)d = (d)_{new}$, $\Phi(\theta)u = (u)_{new}$, but this does not change anything anywhere else. Thus the transformed matrix $\Phi(\theta)U\, \Phi^{-1}(\varphi)$, which we simply refer to as U, has $n^2 - (2n-1) = (n-1)^2$ real parameters.

13. THE FOUR QUARK AND SIX QUARK MODELS; CP - VIOLATION

Up to 1975 we had only four quark flavours (u, d, c, s) and four leptons (ν_e, e, ν_μ, μ). The charged current interactions were described by

$$\mathcal{L}^{cc} = \frac{g}{2\sqrt{2}} \bar{u}_j \gamma^\lambda (1 - \gamma_5) U_{jk} d_k W_\lambda^- + \text{h.c.} \quad , \quad (13\text{-}1)$$

where U is a 2 x 2 general unitary matrix. According to our discussions in the previous section U has just one parameter, which we take to be a rotation angle, viz.,

$$u = \begin{pmatrix} u \\ c \end{pmatrix}, \quad d = \begin{pmatrix} d \\ s \end{pmatrix}, \quad U = \begin{pmatrix} \cos\theta_c & \sin\theta_c \\ -\sin\theta_c & \cos\theta_c \end{pmatrix}, \quad (13\text{-}2)$$

where θ_c is the Cabibbo angle. We may rewrite $\mathcal{L}^{cc}$ as

$$\mathcal{L}^{cc} = \frac{g}{2\sqrt{2}} \left\{ \bar{u}_L \gamma^\lambda d_{cL} + \bar{c}_L \gamma^\lambda s_{cL} \right\} W_\lambda^- + \text{h.c.} ,$$

$$d_{cL} = \cos\theta_c \, d_L + \sin\theta_c \, s_L \tag{13-3}$$

$$s_{cL} = -\sin\theta_c \, d_L + \cos\theta_c \, s_L .$$

Thus, the doublets chosen in nature are

$$\begin{pmatrix} u \\ d_c \end{pmatrix}_L , \quad \begin{pmatrix} c \\ s_c \end{pmatrix}_L , \tag{13-4}$$

i.e., the most general ones one could have. Of course the single parameter θ could also be put in the u - c sector. If we start with the doublets in (13-4) and the right-handed singlets (u_R, c_R, d_R, s_R) and construct the standard SU(2) x U(1) model we could forget about all rotation matrices because all their effects are included in the single parameter θ_c which appears in the doublets (13-4).

The relations (13-4) summarize the celebrated GIM mechanism. In the standard model the GIM mechanism (which tells us that there are no flavour-changing neutral currents) is perfectly "natural". We "understand" why there should be a Cabibbo angle, but cannot compute its magnitude.

Why is there not a second Cabibbo-like angle in the leptonic sector? Why do we not get

$$\mathcal{L}^{cc} = \frac{g}{2\sqrt{2}} (\bar{\nu}_e, \bar{\nu}_\mu)\gamma^\lambda \begin{pmatrix} \cos\theta' & \sin\theta' \\ -\sin\theta' & \cos\theta' \end{pmatrix} \begin{pmatrix} e \\ \mu \end{pmatrix} W_\lambda^- + \text{h.c.}, \tag{13-5}$$

where θ' is a new angle? The answer is that if ν_e and ν_μ are massless such an angle is unobservable. Because, then we could identify

$$\begin{pmatrix} (\nu_e)_{\text{physical}} \\ (\nu_\mu)_{\text{physical}} \end{pmatrix} = \begin{pmatrix} \cos\theta'\nu_e - \sin\theta'\nu_\mu \\ +\sin\theta'\nu_e + \cos\theta'\nu_\mu \end{pmatrix} \tag{13-6}$$

whereby θ' is eliminated. By definition ν_e is the neutrino which is the partner of the electron, etc. For quarks we cannot do so because the d and s quarks are distinguishable (they have different "masses").

In conclusion in the standard model, with four quarks and four leptons, the GIM mechanism is natural; the lepton numbers L_e and L_μ are also automatically conserved provided the neutrinos are massless.

Since 1975 the third family of leptons and quarks have entered into the arena. We know that the GIM mechanism (with just one Cabibbo angle), which was so natural and attractive as well as in very good agreement with data must be at most only approximately valid.

We now examine the generalization of the standard model to the case of six quark flavours (u, c, t, d, s, b) and six leptons. Applying the results in section 12, we find that the neutral current sector is flavour conserving and given by (8-4) and (8-5), where $I_{3L} = \frac{1}{2}$ for all Q = 2/3 quarks and neutrinos and $I_{3L} = -\frac{1}{2}$ for all Q = -1/3 quarks and negative leptons. The charged current sector is the only place where new parameters (generalized Cabibbos) can enter, viz.,

$$\mathcal{L}^{cc} = \frac{g}{2\sqrt{2}} (\bar{u}, \bar{c}, \bar{t})\gamma^\lambda(1 - \gamma_5)U \begin{pmatrix} d \\ s \\ b \end{pmatrix} W_\lambda^- + \text{h.c.} \quad , \tag{13-7}$$

where we have used (12-5). Here U is a 3 x 3 unitary matrix and (according to our discussions in section 12) may be taken to have $(3-1)^2 = 4$ parameters. These parameters are normally taken as three rotation angles and a phase à la Kobayashi and Maskawa. It is easily seen that the Kobayashi - Maskawa choice corresponds to taking

$$U = \mathcal{R}_1(\theta_2) \cdot \mathcal{R}_3(\theta_1)\Phi(0,0,\delta)\mathcal{R}_1(\theta_3) \quad , \tag{13-8}$$

where $\mathcal{R}_j$ is a rotation matrix about the axis j

$$\mathcal{R}_1(\theta) = \begin{pmatrix} 1 & 0 & 0 \\ 0 & c & s \\ 0 & -s & c \end{pmatrix} \quad , \quad \mathcal{R}_3(\theta) = \begin{pmatrix} c & s & 0 \\ -s & c & 0 \\ 0 & 0 & 1 \end{pmatrix} \quad ,$$

$$c = \cos\theta \quad , \quad s = \sin\theta \quad , \quad \Phi(0,0,\delta) = \begin{pmatrix} 1 & 0 & 0 \\ 0 & 1 & 0 \\ 0 & 0 & e^{i\delta} \end{pmatrix} \tag{13-9}$$

Multiplying the matrix we have

$$U = \begin{pmatrix} c_1 & s_1c_3 & s_1s_3 \\ -s_1c_2 & c_1c_2c_3-s_2s_3e^{i\delta} & c_1c_2s_3+s_2c_3e^{i\delta} \\ s_1s_2 & -c_1s_2c_3-c_2s_3e^{i\delta} & -c_1s_2s_3+c_2c_3e^{i\delta} \end{pmatrix} \quad (13\text{-}10)$$

this is the famous Kobayashi - Maskawa generalization of the Cabibbo - GIM - matrix which described the world as long as we had only four quarks. Comparing (13-10) and (13-2) we find that $\theta_1 = \theta_c$ (the Cabibbo angle) but now we have two more arbitrary angles and a phase δ. None of these is calculable. The phase could possibly be the origin of CP - violation. Note that this phase is removable if any of the angles θ_1, θ_2 or θ_3 vanishes. For example, if $\theta_3 = 0$ we have $_1(\theta_3) = 1$ and then putting $(d)_{new} = \Phi(0,0,\delta)d$, $d = \begin{pmatrix} d \\ s \\ b \end{pmatrix}$, removes the phase. Similarly, using $\mathcal{R}_3(\theta_1)\Phi(0,0,\delta) = \Phi(0,0,\delta)\mathcal{R}_3(\theta_1)$, we see that for $\theta_2 = 0$ again the phase can be absorbed into $u = \begin{pmatrix} u \\ c \\ t \end{pmatrix}$. Finally if $\theta_1 = 0$, we get $U = \mathcal{R}_1(\theta_2)\Phi(0,0,\delta)\mathcal{R}_1(\theta_3)$. The reader can easily convince herself that U is of the form

$$U = \begin{pmatrix} 1 & 0 & 0 \\ 0 & & \\ 0 & & V \end{pmatrix} ,$$

where V is a 2 x 2 unitary matrix. Removing of phases in such a 2 x 2 matrix is rather trivial.

We now discuss very briefly the phenomenological consequences of the six quark model. How big are the angles θ_j, j = 1,2,3?

i) <u>Cabibbo universality</u>. Before knowing that there are more than four quarks we used to identify the strength of $u \to d$ transition with "$\cos\theta_c$" and $u \to s$ by "$\sin\theta_c$", see eq. (13-3). Furthermore the analysis of baryon decays, etc. gave always $\sin^2\theta_c + \cos^2\theta_c = 1$ to a good approximation. Nowadays we know better. From (13-10) we have that

transition	relative coupling	where to measure
$u \leftrightarrow d$	c_1	β-decay
$u \leftrightarrow s$	s_1c_3	hyperon decays

$c \leftrightarrow d$	$-s_1 c_2$	$\nu N \to \mu^- \mu^+ X$
$c \leftrightarrow s$	$c_1 c_2 c_3 - s_2 s_3 e^{i\delta}$, etc.	"

Since the Cabibbo universality worked well with just four quarks, i.e., "$\sin^2\theta_c$" + "$\cos^2\theta_c$" = $s_1^2 c_3^2 + c_1^2 \approx 1$ we may conclude that θ_3 cannot be so large. Furthermore, since the four quark model predicted that charmed particles should predominantly decay into strange particles and this prediction is experimentally verified s_2 must also be a small angle. Remembering that the Cabibbo angle is also small we find that the Kobayashi - Maskawa matrix, eq. (13-10) is of the form

$$U \approx \begin{pmatrix} 1 & s_1 & 0 \\ -s_1 & 1 & s_3 + s_2 e^{i\delta} \\ 0 & -s_2 - s_3 e^{i\delta} & 1 \end{pmatrix} \tag{13-11}$$

Now we consider the present knowledge on the angle in (13-10). The muon lifetime together with the ratio of the supperallowed β-transitions give $\cos\theta_c = c_1$. We find

$$\cos^2\theta_c = (0.972 \pm 0.004)(1 + \Delta_\mu - \Delta_\beta) \text{ ,} \tag{13-12}$$

where $\Delta_\mu - \Delta_\beta$ is due to radiative corrections. Theoretically $\Delta_\mu - \Delta_\beta = -0.021$ for $\sin^2\theta_W = \frac{1}{4}$. Thus

$$\begin{aligned} \sin^2\theta_c = \sin^2\theta_1 &= 1 - \cos^2\theta_c = \\ &= 0.028 \pm 0.004 \quad \text{for} \quad \Delta_\mu - \Delta_\beta = 0 \\ &= 0.047 \pm 0.004 \quad \text{"} \quad \Delta_\mu - \Delta_\beta = -0.021 \end{aligned} \tag{13-13}$$

However, from the semileptonic decays of the baryon octet we obtain

$$\text{"}\sin^2\theta_c\text{"} = \sin^2\theta_1 \cos^2\theta_3 = 0.052 \pm 0.002 \text{ ,} \tag{13-14}$$

where "$\sin^2\theta_c$" emphasizes that this quantity was improperly identified as $\sin^2\theta_c$. Putting together (13-13) and (13-14) we find $\theta_1 \equiv \theta_c \approx 13^o$ and $\theta_3 \lesssim 8^o$.

The t quark is presumably heavier than the b quark, otherwise it would have revealed itself in some of the experiments which have established the existence of the Υ-particle. Thus no direct information is so far available on the elements of the third row in (13-10) and (13-11). Information on the angle θ_2 may be obtained by, for example, comparing $b \to c\,\ell\,\bar{\nu}$ and $b \to u\,\ell\,\bar{\nu}$, $\ell = e, \mu, \tau$, which have respectively the couplings $s_3 + s_2 e^{i\delta}$ and $s_1 s_3$. There is also some theoretical information, on θ_2, coming from the $\Delta s = 2$ weak processes responsible for the mass difference between K_L^0 and K_S^0, which is expected to be due to

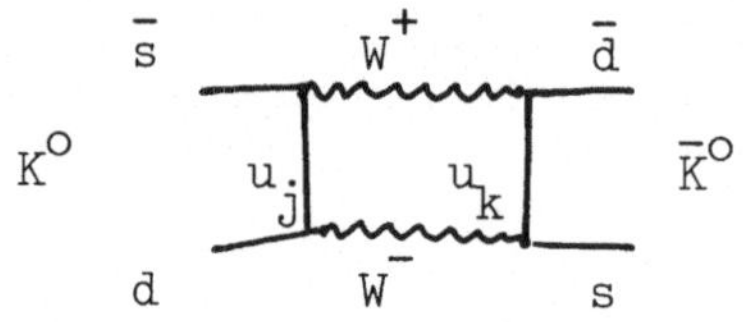

(13-15)

$$\Delta M \sim \sum_{j,k} \int \frac{d^4k\ \bar{v}_L\ \gamma^\mu\ (\not{k}+m_j)\ \gamma^\nu\ u_L \bar{u}_L \gamma_\nu (\not{k}+m_k)\ \gamma_\mu\ v_L}{(k^2-M_W^2)^2(k^2-m_j^2)(k^2-m_k^2)}\ F_{jk}$$

where u and v stand for the appropriate Dirac spinors; m denotes the mass. Note that the momenta of the quarks (inside the K^0 and $\bar{K}^0$) are small and thus have been neglected compared to the momentum flowing in the loop; furthermore the m's in the numerator drop out. F_{jk} is the product of the four coupling constants as the vertices in (13-15)

$$F_{jk} = a(u_j,s) \cdot a(d,u_j)\ a(u_k,s)\ a(d,u_k)\ ,$$
$$u_1 = u, \quad u_2 = c, \quad u_3 = t, \qquad (13\text{-}16)$$

The a's are the elements of the unitary matrix U discussed in section 12, viz. $U_{j2} = a(u_j,s)$, $a(d,u_j) = a^*(u_j,d) = U_{1j}^\dagger$, etc. For the four and six quark models these quantities were given in (13-2) and (13-10). Note that the leading term in (13-15), which is obtained by putting $m_j = m_k = 0$ vanishes because

$$a(u_j,s) \cdot a(d,u_j) = (U^+U)_{12} = 0\ .$$

For the four quark model, estimation of ΔM, gave an impressive prediction for m_c. Now that we have more than four quarks, the old prediction should not get spoiled. Estimation of (13-15) for the six quark model gives $\theta_2 < 27^0$ (see Ellis et al., Nucl Phys. B109, 213(76)) were the phenomenology of the model is discussed).

14. GRAND UNIFICATION

So far, in these lectures, we have discussed the conventional theory for weak and electromagnetic interaction. Nowadays, strong interactions are believed to be described by quantum chromodynamics (QCD), a nonabelien gauge theory based on the group SU(3); viz. three colours and eight mediators (gluons). This theory is not broken and the gluons are massless. In this section, we discuss the Georgi - Glashow model (Phys. Rev. Letters 32, 438(1974)) where the conventional SU(2) x U(1) model and chromodynamics are (grand) unified. Grand unification means that γ, $W^{\pm}$, Z^o, and the eight gluons should be the mediators (gauge bosons) of one and the same theory, where the latter is based on a simple or semisimple gauge groups, i.e., it contains only a single coupling constant. The SU(5) model by Georgi, and Glashow is the most economical scheme for unification of SU(2) x U(1) and (SU(3)) colour.

The SU(5) model does not explain the existence of the families 2 and 3 (see 12-1), so what we have to do is to write the theory for a hypothetical family

$$\nu_j', \ \ell_j', \ (u_j')_{1,2,3} \ , \ (d_j')_{1,2,3} \ , \tag{14-1}$$

where j denotes the family number: $Q(\nu') = 0$, $Q(\ell') = -1$, $Q(u') = 2/3$, $Q(d') = -1/3$; the indices 1,2,3 denote the colour. The primes on the fields emphasize that the quantities in (14-1) do not necessarily represent the physical particles; to get the physical fields, we have to diagonalize the mass matrix (see section 12). Once we have constructed the theory for the family j, we just take a sum over all families, whereby all families are democratically represented. In the following, we simplify the notation by suppressing the primes, etc. and replacing (14-1) by ν, e, $u_{1,2,3}$, $d_{1,2,3}$; we shall discuss mixings later on.

The SU(5) theory may be constructed (analogously to theories discussed before) by starting with the kinetic term

$$\mathcal{L}_o = i \sum_f (\bar{f}_L \gamma \frac{\partial}{\partial x} f_L + \bar{f}_R \gamma \frac{\partial}{\partial x} f_R) \ , \quad f = \nu, e, u, d \tag{14-2}$$

How should we put f in SU(5) multiplets? The left-handed ν_e and e may be put into a fundamental 5-plet of SU(5)

$$\psi_R = \begin{pmatrix} a_1 \\ a_2 \\ a_3 \\ e^+ \\ -\bar{\nu} \end{pmatrix}_R \qquad \begin{matrix} a_4 = e^+ = \psi_e^c \\ -a_5 = \bar{\nu} = \psi_\nu^c \end{matrix} \tag{14-3}$$

together with the flavour a (to be determined) in three colours. Here $\psi^c = c\,\bar{\psi}$, where c denotes the charge conjugation matrix.

The reason for choice (14-3) is clear, viz., $(e_R^+, -\bar{\nu}_R) \sim (\nu_L, e_L)$ are intimately connected and form a doublet under SU(2). In (14-3) these leptons are "unified" with a triplet of hadrons $a_{1,2,3}$. Requiring invariance under $\psi_R \to \exp(i\,\bar{\alpha}(x)\cdot\bar{T})\psi_R$ (where the T/2 are 5 x 5 matrices representing the 24 generators of SU(5)) gives 24 gauge bosons V^j, j = 1,.., 24. Thus, we get a term

$$\mathcal{L} = i\, \bar{\psi}_R \gamma \left(\frac{\partial}{\partial x} - i\, g_o\, \bar{T} \cdot \bar{V} \right) \psi_R + \ldots \qquad (14\text{-}4)$$

The matrices T may be chosen analogously as is done in SU(2) and SU(3). For example

$$T^j = \left(\begin{array}{ccc|cc} & \lambda^j & & 0 & 0 \\ & & & 0 & 0 \\ & & & 0 & 0 \\ \hline 0 & 0 & 0 & & \\ 0 & 0 & 0 & & 0 \end{array} \right), \quad j = 1,\ldots 8 \quad , \quad \lambda^j = \text{Gell-Mann matrices (SU(3))}$$

$$T^9 = \left(\begin{array}{ccc|cc} & & & 1 & 0 \\ & 0 & & 0 & 0 \\ & & & 0 & 0 \\ \hline 1 & 0 & 0 & 0 & 0 \\ 0 & 0 & 0 & 0 & 0 \end{array} \right), \quad T^{10} = \left(\begin{array}{ccc|cc} & & & -i & 0 \\ & 0 & & 0 & 0 \\ & & & 0 & 0 \\ \hline i & 0 & 0 & & \\ 0 & 0 & 0 & & 0 \end{array} \right), \ \ldots . \qquad (14\text{-}5)$$

We continue by putting 1 and ±i in the same pattern. It remains

$$T^{15} = \frac{1}{\sqrt{6}} \begin{pmatrix} 1 & & & & \\ & 1 & & & \\ & & 1 & & \\ & & & -3 & \\ & & & & 0 \end{pmatrix}, \quad T^{24} = \frac{1}{\sqrt{10}} \begin{pmatrix} 1 & & & & \\ & 1 & & & \\ & & 1 & & \\ & & & 1 & \\ & & & & -4 \end{pmatrix} \qquad (14\text{-}5a)$$

The properties of T's are $Tr[T^j] = 0$, $T^{j^+} = T^j$ so that the group elements $\exp(i\, \bar{\alpha}(x) \cdot \bar{T})$ are unitary and have determinant equal to unity. Furthermore $[T^j,T^k] = 2ic^{jkm}T^m$ defines the SU(5) Lie algebra with c^{jkm} = structure constants. Note that $Tr[T^jT^k] = 2\delta_{jk}$. Since our theory is supposed to include all interactions (except gravity), one of the gauge bosons, i.e., one linear combination of the V's in (14-4) should be the photon field. The generator coupling to photon is the charge operator, $Q = \Sigma c_j\, T_j$, where c_j are constants. Thus

$$Tr\ Q = 0 \qquad (14\text{-}6)$$

For the multiplet (14-3)

$$\begin{aligned} Tr\ Q &= 3Q_a + Q_\nu + Q_{e^+} = 3Q_a + 1 = 0, \\ Q_a &= \frac{-1}{3} \end{aligned} \qquad (14\text{-}7)$$

where we have used the fact that a_1, a_2 and a_3, except for colour, are identical objects. Relation (14-7) implies that a should be identified with the d quark

$$\psi_R = \begin{pmatrix} d_1 \\ d_2 \\ d_3 \\ e^+ \\ -\bar{\nu} \end{pmatrix}_R \qquad (14\text{-}8)$$

We see that (14-8) includes a doublet and three singlets under the weak isospin group. The nature of mediators in SU(5) model is easily established from (14-8), where any member can transform into any other by sending off a mediator. Thus we have $\bar{\nu}_R \leftrightarrow e_R^{\pm}$, i.e., $\nu_L \leftrightarrow e_L^-$ via $W^{\pm}$ mediators; $d_j \leftrightarrow d_k$ via gluons, etc. However, there are also lepton-quark transitions via new exotic mediators denoted by $X_j^{(\pm 4/3)}$, $Y_j^{(\pm\frac{1}{3})}$, j = 1,2,3

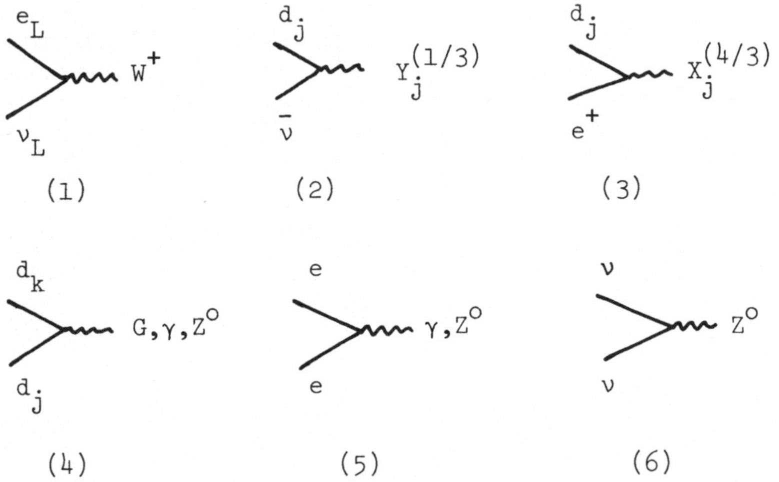

Thus neither the lepton number nor the baryon number is conserved. The twelwe exotics X and Y are sometimes referred to as lepto-quarks, although they are neither leptons nor quarks. They must be heavy because otherwise the stable matter (protons and neutrons) would evaporate into leptons, as we shall discuss below.

We have altogether 15 helicity states to account for if the neutrino is massless (ν_L, e_L, e_R, u_{jL}, u_{jR}, d_{jL}, d_{jR}). For a massive neutrino there is also a ν_R, i.e., altogether 16 states. In (14-8), 5 of these states are accounted for, leaving us with 10 (11) unaccounted ones if the neutrino is massless (massive). The remaining ten states can be easily placed into a 10 dimensional representation of SU(5), as we discuss shortly. For massive neutrinos we would need to introduce an additional singlet ν_R. This is ugly and, therefore, we conclude that SU(5) has at least two beautiful features, (i) the fractionally charged quarks are natural and (ii) the neutrino is massless.

From the five objects b_j, j = 1,.., 5 which transform as 5-plet of SU(5) we can form a ten-plet by taking the antisymmetric product

$$b_{jk} = \frac{1}{\sqrt{2}}(b_j\, b_k - b_k\, b_j) \ , \qquad , \ j,k = 1,..,5 \tag{14-9}$$

$$b = \begin{pmatrix} b_1 \\ b_2 \\ \vdots \\ b_5 \end{pmatrix} .$$

Evidently $b_{jk} = 0$ if j = k and $b_{jk} = -b_{kj}$. For j,k = 1,2,3 we have the antisymmetric product of two colour triplets. From SU(3) relation 3 x 3 = 6 + $\bar{3}$, where the six is symmetric and $\bar{3}$ antisymmetric, we see that b_{jk}, j,k = 1,2,3 represent a colour antitriplet. Furthermore these three objects are singlets under weak - e m SU(2) and have $Q = -\frac{2}{3}$. Thus b_{jk} may be identified with $\varepsilon_{jkm}\, \bar{u}_{mL}$. Similarly $b_{j4}(b_{j5})$, j = 1,2,3 represents a colour triplet which has charge $\frac{-1}{3} + 1 = \frac{2}{3}\left(\frac{-1}{3} + 0 = \frac{-1}{3}\right)$ and the third component of weak isospin $\frac{1}{2}(-\frac{1}{2})$. These can be identified with u_{jL} and d_{jL}. The only remaining state is b_{45} which is singlet under colour and weak isospin and has Q = +1, i.e., e_L^+. Thus

$$\psi_L = \frac{1}{\sqrt{2}} \begin{pmatrix} 0 & \bar{u}_3 & -\bar{u}_2 & u_1 & d_1 \\ -\bar{u}_3 & 0 & \bar{u}_1 & u_2 & d_2 \\ \bar{u}_2 & -\bar{u}_1 & 0 & u_3 & d_3 \\ -u_1 & -u_2 & -u_3 & 0 & e^+ \\ -d_1 & -d_2 & -d_3 & -e^+ & 0 \end{pmatrix}_L \equiv b_{jk} \tag{14-10}$$

Here $\bar{u}$ means $\psi_u^c = c\tilde{\bar{\psi}}_u$, i.e., the charge conjugated spinor.

The kinetic energy term, (14-2) can be expresses in terms of (14-8) and (14-10)

$$\begin{aligned} \mathcal{L}_o &= i\left\{\bar{a}_j\, \gamma \frac{\partial}{\partial x}\, a_j + \bar{b}_{k\ell}\, \gamma \frac{\partial}{\partial x}\, b_{k\ell}\right\}. \\ &= i\ \mathrm{Tr}\left[\bar{\psi}_R\, \gamma \frac{\partial}{\partial x}\, \psi_R + \bar{\psi}_L\, \gamma \frac{\partial}{\partial x}\, \psi_L\right] \rightarrow \mathcal{L}_1 \end{aligned} \tag{14-11}$$

$$\mathcal{L}_1 = i\ \mathrm{Tr}\left\{\bar{\psi}_R\, \gamma\left[\frac{\partial}{\partial x} - i\, g_o\, \bar{T}\cdot\bar{V}\right]\psi_R + \bar{\psi}_L\, \gamma\left[\frac{\partial}{\partial x} - i\, g_o\, \bar{T}'\cdot\bar{V}\right]\psi_L\right\}.$$

Here $2g_0$ is the grand unified coupling constant, $\bar{T}(\bar{T}')$ are the 24 matrices representing generators, for the 5 (10) dimensional representation. Furthermore

$$\bar{\psi}_R\ \bar{T}\cdot\bar{V}\ \psi_R = \bar{a}_j \sum_{\alpha=1}^{24} T^{\alpha}_{jk}\ V^{\alpha}\ a_k = \mathrm{Tr}(\bar{a}\ \bar{T}\cdot\bar{V}\ a)\ ,$$

$$\mathrm{Tr}(\bar{\psi}_L\ \bar{T}'\cdot\bar{V}\ \psi_L) = \bar{b}_{jk} \sum_{\alpha=1}^{24} (\acute{T}^{\alpha}\ V^{\alpha})_{jk,j'k'}\ b_{j'k'} \qquad (14\text{-}12)$$

Although the matrices representing generators look a bit frightening, for the left-handed 10-plet, they are quite simple because the b_{jk} decomposes into objects with simple transformation properties under colour and weak isospin. This ensures that, although some fermions are put into 5-plet and others into 10-plet, there is universality of strong, weak and electromagnetic interactions. For example the right-handed u's form a colour triplet - SU(2) singlet and are thus treated exactly as the right-handed d's, except for weak hypercharge; (u_{jL}, d_{jL}) form doublets under weak isospin and are treated as (ν, e), etc.

Let us look a bit into the 5-plet sector. There we have 4 diagonal generators T_3, T_8, T_{15} and T_{24}, see (14-5). T_3 and T_8 couple to gluons G_3 and G_8. T^{15} is proportional to the (right-handed electric charge operator, $Q_R = -\sqrt{\frac{2}{3}}\ T^{15}$, i.e., V^{15}_{μ} is the photon field A_{μ} and $g_0 = \sqrt{\frac{2}{3}}\ e$. If we are going to achieve unification, V^{24}_{μ} better be the Z_{μ} and $Q^Z_R = r\ T^{24}$, where r is a constant $\mathrm{Tr}[T^jT^k] = 2\ \delta_{jk}$, whereby $\mathrm{Tr}[Q^{\gamma}_R Q^Z_R] = 0$. This relation is normally not satisfied in the standard model $[Q^Z \sim (I_{3L} - Q \sin^2\theta_W)$, eq. (8-3)] as is seen by substituting I_{3L} and Q for the objects in (14-8) and remembering that $(e^+, -\bar{\nu})_R \rightarrow (\nu, e^-)_L$. We require

$$\mathrm{Tr}\left[Q^{\gamma}_R Q^Z_R\right] \sim \mathrm{Tr}(I_{3L}Q - \sin^2\theta_W Q^2) = 0\quad ,\quad Q = I_{3L} + Y$$

So $\sin^2\theta_W = \dfrac{\mathrm{Tr}(I^2_{3L})}{\mathrm{Tr}(Q^2)}$, where the sum goes over the members in (14-8),

i.e., $\sin^2\theta_W = \dfrac{2\cdot 1/4}{1 + 3\cdot\frac{1}{9}} = \dfrac{3}{8}$. Thus the unification hypothesis fixes the value of the angle θ_W to be about 38°, at the unification energy, which is going to be a very large energy. [At our energies θ_W is expected to be smaller.]

Now we consider the left-handed multiplet, (14-10). The generators $\bar{T}'$, in (14-11) are easily obtained by taking infinitesimal transformations (see 14-9)

$$b_j \to [(1 + i\,\bar{\alpha} \cdot \bar{T})b]_j$$

$$b_{jk} \to [(1 + i\,\bar{\alpha} \cdot \bar{T})b]_j [(1 + i\,\bar{\alpha} \cdot \bar{T})b]_k - j \leftrightarrow k$$

Comparison with (14-12) yields

$$T'_{jk,j'k'} = \delta_{jj'}\, T_{kk'} + \delta_{kk'}\, T_{jj'} \quad ,$$

$$\mathrm{Tr}(\bar{\psi}_L\, \bar{T}' \cdot \bar{V}\, \psi_L) = 2\, \bar{b}_{jk}\, \bar{T}_{kk'} \cdot \bar{V}\, b_{jk'} = 2\, \bar{b}_{jk}\, \bar{T}_{jj'}\, \bar{V}\, b_{j'k} \quad . \tag{14-13}$$

Again V^{15}_μ (V^{24}_μ) are respectively the photon (Z^o) fields. The reader can easily write down the various pieces of the interaction, e.g. the coupling to the Z^o of the right-handed electron is derived from

$$g_o\, \bar{\psi}_L\, \gamma_\mu\, T'^{15}\, \psi_L\, Z^\mu \Rightarrow \left\{\sqrt{\frac{2}{3}}\, e\, \left\{\bar{b}_{54}\, T^{15}_{44}\, b_{54} + \bar{b}_{45}\, T^{15}_{55}\, b_{45}\right\}\right\}$$

$$= -\frac{3}{\sqrt{10}} \sqrt{\frac{2}{3}}\, e\, \bar{\psi}^c_{eL}\, \gamma^\mu\, \psi^c_{eL}\, Z_\mu = \sqrt{\frac{3}{5}}\, e\, \bar{\psi}_{eR}\, \gamma^\mu\, \psi_{eR}\, Z^\mu \; .$$

This agrees with (8-3), where we find

$$\frac{e\, Z^\mu}{\sin\theta_W\, \cos\theta_W} (-\sin^2\theta_W) Q_e\, \bar{\psi}_{eR}\, \gamma^\mu\, \psi_{eR} = e \sqrt{\frac{3}{5}}\, \bar{\psi}_{eR}\, \gamma^\mu\, \psi_{eR} \; ,$$

Here we have used $\sin^2\theta_W = 3/8$.

15. MIXING ANGLES IN THE GEORGI-GLASHOW MODEL

In the previous section, we determined the structure of the couplings between the elementary fermions and the mediators. However, all our results are valid for the unphysical fermions (primes were suppressed) and massless intermediate bosons. We may now repeat the arguments in section 12 and rewrite the theory for the physical fermions, viz. $u'_L = A^{(u,L)}\, u_L$, $d'_L = A^{(d,L)} d_L$, etc., where as before,

$$u = \begin{pmatrix} u \\ c \\ t \\ \vdots \end{pmatrix} , \quad d = \begin{pmatrix} d \\ s \\ b \\ \vdots \end{pmatrix} , \quad \nu = \begin{pmatrix} \nu_e \\ \nu_\mu \\ \nu_\tau \\ \vdots \end{pmatrix} , \quad e = \begin{pmatrix} e \\ \mu \\ \tau \\ \vdots \end{pmatrix}$$

The six matrices $A^{(f,L)}$, $A^{(f,R)}$, $f = e,u,d$ (the neutrino is assumed to be massless) are all unitary. Furthermore, these matrices are independent of colour. We can now easily convince ourselves that in the interaction term

$$\mathcal{L} = g_o \, \mathrm{Tr}\left\{\bar{\psi}_R \gamma^\mu \bar{T} \cdot \bar{V}_\mu \, \psi_R + \bar{\psi}_L \gamma^\mu \cdot \bar{T}' \cdot \bar{V}_\mu \, \psi_L\right\} \tag{15-1}$$

the A's cancel out in the neutral current sectors (couplings to the photon, Z^o and the gluons G). The reason is that in the interactions terms (because of the γ^μ) the L and R cannot mix and, therefore, in neutral current interactions we get terms of the type $\bar{u}'_L\{\ldots\}u'_L = \bar{u}_L(A^{(u,L)})^+\{\ldots\}A^{(u,L)}u_L = \bar{u}_L\{\ldots\}u_L$, etc. However, in the charged currents (couplings to $W^\pm$, $X_m^{\pm(4/3)}$ and $Y_m^{\pm(\frac{1}{3})}$, $m = 1,2,3$) the A's show up, that is, there we find all kinds of Cabibbo angles. More explicitly, the charged currents from $\bar{\psi}_R(\bar{T} \cdot \bar{V})\psi_R$ are of the form

$$\overline{e'^c_R} \, \nu'^c_R \, W \sim \bar{\nu}'_L \, e'_L \, W \, , \; \bar{d}'_R \, \nu'^c_R \, Y \, , \; \bar{d}'_R \, e'^c_R \, X \quad . \tag{15-2}$$

From $\bar{\psi}_L(\bar{T}' \cdot \bar{V})\psi_L$, we get the charged currents of the form

$$\bar{u}'_L \, d'_L \, W \, , \; \overline{u'^c_L} \, d'_L \, Y \, , \; \overline{u'^c_L} \, u'_L \, X \, , \; \bar{u}'_L \, e'^c_L \, Y \, , \; \bar{d}'_L \, e'^c_L \, X \, . \tag{15-3}$$

In (15-2) and (15-3) we have left out the γ-matrices, colour indices, etc. We may now express the charged currents in terms of the physical states. For example

$$\bar{u}'_L \, d'_L \, W = \bar{u}_L \underbrace{\left[A^{(u,L)}\right]^+ A^{(d,L)}}_{U^{(u,d)}} d_L \, W = \bar{u}_L \, U^{(u,d)} d_L \, W, \tag{15-4}$$

where $U^{(u,d)}$ is a general n x n unitary matrix, n = number of families. For example, for three families U is the Kobayashi-Maskawa matrix, discussed in section 13. In the leptonic charged currents the corresponding U may be absorbed into the definition of the neutrino fields, exactly in the same fashion as in section 13.

$$\bar{\nu}'_L \, e'_L \, W = \bar{\nu}'_L \, A^{(e,L)} e_L \, W \equiv \bar{\nu}_L \, e_L \, W \quad . \tag{15-5}$$

Here

$$\nu_L \equiv \left[A^{(e,L)}\right]^+ \nu_L' \ ,$$

$$\nu_L\, e_L\, W = \left\{\bar{\nu}_{eL}\, e_L + \bar{\nu}_{\mu L}\, \mu_L + \bar{\nu}_{\tau L}\, \tau_L + \ldots\right\} \ . \qquad (15\text{-}6)$$

The remaining interactions in (15-2) and (15-3) may be treated analogously. Note that there are additional angles and phases in couplings to X and Y which are, in principle, completely different from the generalized Cabibbo angles and Kabayashi-Maskawa phases which are introduced to parameterize $U^{(u,d)}$. For example

$$\bar{d}_R'\, \nu_R'^{\,c}\, Y \sim \bar{\nu}_L'\, d_L'^{\,c}\, Y = \bar{\nu}_L \underbrace{\left(A^{(e,L)}\right)^+ \left(\overset{*}{A}{}^{(d,R)}\right)}_{U^{(\nu,d^c)}} d_L^c\, Y \ . \qquad (15\text{-}7)$$

Again the γ-matrices, etc, have been suppressed. The specific form of these new mixing matrices will depend on how one chooses to break the symmetry and generate the fermion masses. The mediators X and Y are expected to be very very heavy (10^{15}-10^{16} GeV). Thus it is going to be very tough to study the mixing angles and phases (such as CP-violation) in the X and Y charged currents.

The matrices A, as discussed in section 13, originate from diagonalization of the fermion mass matrix. We believe that the masses have to do with the spontaneous symmetry breaking à la Higgs. Below we describe, briefly, how the Higgs mechanism is done for the SU(5) model.

16. THE HIGGSES OF THE SU(5) MODEL

SU(5) has 24 mediators ($X_j^{\pm(4/3)}$, $Y_j^{\pm(1/3)}$, j = 1,2,3, $W^\pm$, Z, γ, G_i, i = 1,..,8). Clearly X and Y should be very heavy so as to suppress the violation of bayron number and lepton numbers. $W^\pm$ and Z^0 should turn out to be as à la Weinberg, (eq. 11-8), in order to be in accordance with experiments; the gluons and the photon should remain massless. Therefore the symmetry breaking should occur at two levels, viz.

(i) SU(5) → SU(3) x SU(2) x U(1)

	SU(3)	SU(2) x U(1)	
	colour	standard model	
massless:	G_i	$W^\pm$, Z, γ	(16-1)

ii) $SU(2) \times U(1) \rightarrow [U(1)]$.
standard charge

At the first stage one or more Higgses (which we have not introduced yet) should acquire a stupendous vacuum expectation value whereby X and Y get tremendous masses. Compared to these monsters W and Z are light as feather, i.e., their masses are due to a new class of Higgses which have acquired a much smaller vacuum expectation value (of the order of 100 GeV). After the second stage we have nine massless mediators (gluons and photon) and all symmetries except for colour and charge are broken.

The first stage of SSB is achieved by introducing 24 Higgses (in the adjoint representation, $5 \times \overset{*}{5} = 1 + 24$) φ^j, $j=1,2,\ldots,24$. The charge and colour quantum numbers of these spin zero objects are just the same as for the gauge bosons (see after eq. (14-8)). Thus we have 8 gluonic Higgses,

$$\varphi_G \equiv \varphi^j \ , \ j = 1,..,8 \quad \text{colour octet, electrically neutral.}$$

There are 12 Higgses with charges $\pm 4/3$, $\pm 1/3$. These form 2 colour triplets and two antitriplets

$$\varphi_X, \varphi_Y \equiv \varphi^j, \ j = 9\text{-}14, \ 16\text{-}21.$$

Similarly φ_W are two colour singlet Higgses with charges ± 1, viz. φ^j, $j=22,23$. Finally $\varphi_\gamma \equiv \varphi^{15}$, $\varphi_Z \equiv \varphi^{24}$ are colour singlets and electrically neutral. Clearly only φ_γ and φ_Z are permitted to acquire non-zero vacuum expectation values, because otherwise the conservation of electric charge or/and colour would be violated. We now copy the formalism in section 10 (adding more indices as needed). The kinetic term for the Higgses reads

$$\begin{aligned} \mathcal{L}_{KE} &= + \frac{1}{2} \sum_{j=1}^{24} \frac{\partial \varphi^j}{\partial x_\mu} \frac{\partial \varphi^j}{\partial x^\mu} = + \frac{1}{2} \sum_{j,k} \frac{\partial \varphi^k}{\partial x} \frac{\partial \varphi^j}{\partial x} \frac{1}{2} \mathrm{Tr}(T^j T^k) \\ &= + \mathrm{Tr} \frac{\partial \Phi}{\partial x} \frac{\partial \Phi}{\partial x} \quad , \end{aligned} \tag{16-2}$$

where

$$\Phi = \frac{1}{2} \sum_{j=1}^{24} \varphi^j T^j \quad . \tag{16-3}$$

The T are as given by (14-5) and (14-5a). Now we let φ_γ and φ_Z break the symmetry spontaneously

$$\Phi \to \Phi' + \langle\Phi\rangle \quad , \qquad\qquad (16\text{-}4)$$
$$\langle\Phi\rangle = \alpha\, T^{15} + \beta\, T^{24}$$

where α and β are as yet arbitrary constants. Of course, to do so we must add a potential to (16-2) and check that SSB can occur. We leave that task as an exercise to our reader or refer her to the literature quoted at the end of this article. As in section 10, $\mathcal{L}$ in (16-2) is not invariant under the local SU(5); we must replace

$$\frac{\partial\varphi^j}{\partial x^\mu} \to \frac{\partial\varphi^j}{\partial x^\mu} - i\, g_0 \sum_{s,r=1}^{24} \mathbb{T}^r_{js}\, V^r_\mu\, \varphi^s \quad . \qquad (16\text{-}5)$$

Here $\mathbb{T}^r$ are 24 matrices (24 by 24). They represent the generators in the 24 dimensional (adjoint) representation. It is easy to construct these matrices by taking infinitesimal transformations and remembering that φ^j transform as

$$\varphi^j \sim \bar{a}\, T^j\, a \quad ,$$

where a stands for a 5 (same quantum numbers as in (14-8) and T^j are given in (14-5), viz.

$$\bar{a}\, T^j\, a \to \bar{a}(1 - i\,\bar{\alpha}\cdot\bar{T})T^j(1 + i\,\bar{\alpha}\cdot\bar{T})a \to \qquad (16\text{-}6)$$
$$\bar{a}(T^j - i[\bar{\alpha}\cdot\bar{T},\, T^j]\, a \; .$$

We have to compare eq. (16-6) with

$$\varphi^j \to \exp(i\,\bar{\mathbb{T}}\cdot\bar{\alpha})_{jk}\,\varphi^k \approx (1 + i\,\bar{\mathbb{T}}\cdot\bar{\alpha})_{jk}\,\varphi^k \qquad (16\text{-}7)$$
$$= (\delta_{jk} + i\,\bar{\mathbb{T}}_{jk}\cdot\bar{\alpha})\varphi^k \quad .$$

Thus we find

$$\mathbb{T}^r_{jk} \sim \mathrm{Tr}\left([T^r, T^j]\, T^k\right) \sim i\, c^{rjk} \quad , \qquad (16\text{-}8)$$

where c^{rjk} are the SU(5) structure constants. This result is familiar from SU(3). [We apologize for being sloppy with factors of 2; T/2 are matrices representing the generators.]

Thus the vector boson mass matrix is obtained from

$$\frac{\partial\Phi}{\partial x} \to \frac{\partial\Phi}{\partial x} + c[\bar{T}\cdot\bar{V},\, \Phi] \quad , \qquad (16\text{-}9)$$

$$\mathrm{Tr}([\bar{T}\cdot\bar{V}, \langle\Phi\rangle][\bar{T}\cdot\bar{V}, \langle\Phi\rangle]) \sim (m^2)_{jk}\, V^j\, V^k \quad . \qquad (16\text{-}10)$$

Here c is a constant and the vacuum expectation value $<\Phi>$ is given in (16-4). We rewrite $<\Phi>$ in the form

$$<\Phi> = \begin{pmatrix} \begin{pmatrix} E \end{pmatrix} & \begin{pmatrix} 0 & 0 \\ 0 & 0 \\ 0 & 0 \end{pmatrix} \\ \begin{pmatrix} 0 & 0 & 0 \\ 0 & 0 & 0 \end{pmatrix} & \begin{pmatrix} F \end{pmatrix} \end{pmatrix} \tag{16-11}$$

where

$$F = a\, I_2 + b\, \sigma_3 \ , \qquad E = -\frac{2a}{3} I_3 \tag{16-12}$$

Here a and b are arbitrary constants; σ is the Pauli matrix and I_n is the unit n by n matrix. Note that $Tr(<\Phi>) = 0$. From eqs. (16-10), (16-11) and the definitions of T^j, we see that the vacuum is invariant under the colour SU(3), viz. the unit matrix E commutes with the colour generators T^j, j=1,..,8. Thus the gluons are massless. Similarly the photon and Z^0 are massless;

$$[T^j, <\Phi>] = 0 \ , \quad j = 15, 24 \ , \quad m_\gamma = m_Z = 0.$$

However for $b \neq 0$, in (16-12) the vacuum is noninvariant under the weak isospin and $W^\pm$ acquire masses. We avoid this by putting b = 0, without any good justification. Then the vacuum is only noninvariant under 12 operations in the group (generators corresponding to X's and Y's). The relevant generators are of the form (see (14-5))

$$T^j = \begin{pmatrix} \begin{pmatrix} 0 \end{pmatrix} & \begin{pmatrix} G^j \end{pmatrix} \\ \begin{pmatrix} G^{j+} \end{pmatrix} & \begin{pmatrix} 0 \end{pmatrix} \end{pmatrix} \tag{16-13}$$

Eqs. (16-10) - (16-12) yield

$$M_X^2 = M_Y^2 \sim g_0^2\, a^2 \tag{16-14}$$

where a is supposed to be tremendously large. Let us summarize: we introduced 24 Higgses and 12 gauge bosons (X_j, $\bar{X}_j$, Y_j and $\bar{Y}_j$) have acquired masses. Therefore half of the Higgses can be gauged away. In order to see who remains and who goes away, we go back to relation (9-13). Higgses corresponding to generator which break the invariance of vacuum must turn into Goldstone bosons, who get eaten up. The remaining Higgses are those corresponding to generators which leave the vacuum invariant. From (9-13) and (16-8) we find

$$\mathbb{T}^{r}_{jk} \langle\varphi\rangle^k \sim \mathrm{Tr}\left([T^r, \langle\varphi\rangle]T^j\right) \quad , \tag{16-15}$$

$$[T^r, \langle\varphi\rangle] = 0 \ , \ T = T_G, T_\gamma, T_Z, T_W \ .$$

Thus, the Higgses in the theory, up to now, are the φ_G (eight of them), φ_W (two of them), φ_γ and φ_Z.

The second step in the SSB pattern (16-1) is achieved by introducing a quintet of Higgses θ^j, j=1,..,5, in the fundamental representation. So θ^j, j=1,2,3 have charge -1/3 and form a colour triplet; θ^4 (θ^5) is colour singlet and has Q = 1 (Q = 0). Of course these objects have also antiparticles, so altogether we have 10 degrees of freedom. Only θ^5 may be given a nonzero vacuum expectation value

$$\theta = \begin{pmatrix} \theta^1 \\ \theta^2 \\ \theta^3 \\ \theta^4 \\ \theta^5 \end{pmatrix} \ , \quad \langle\theta\rangle = \begin{pmatrix} 0 \\ 0 \\ 0 \\ 0 \\ v \end{pmatrix} \tag{16-16}$$

$$\frac{\partial\theta^+}{\partial x}\frac{\partial\theta}{\partial x} \to \left|\left(\frac{\partial}{\partial x} - i\, g_0\, \bar{T}\cdot\bar{V}\right)\theta\right|^2$$

$$(m^2)_{jk}\, V^j V^k \sim g_0^2\, \langle\tilde{\theta}\rangle\, T^j\, T^k\, \langle\theta\rangle\, V^j V^k \quad . \tag{16-17}$$

Here v is a constant and the tilde stands for transposition. Clearly these new Higgses leave the photon and the gluons massless,

$$T^j \langle\theta\rangle = 0 \ , \quad j=1,\ldots,8, \text{ and } 15.$$

Furthermore the mass of the X is not altered, $T^X\langle\theta\rangle = 0$, however the Y are affected. We do not worry about this point because $|v| \ll |a|$. The $\langle\theta\rangle$ breaks the invariance under the weak isospin whereby W and Z become massive. In fact (θ^4, θ^5) are analogs of (φ^+, φ^0) of Weinberg and we obtain his relation, eq. (11-8), viz.

$$M_W^2 = K(0,\overset{*}{v})\ \sigma_1^2 \begin{pmatrix} 0 \\ v \end{pmatrix} = K|v|^2 \ ,$$

$$M_Z^2 = K\ \langle\tilde{\theta}\rangle\ (T^{24})^2 \langle\theta\rangle \ = \frac{8}{5}\,K|v|^2 \ ,$$

$$M_W^2/M_Z^2 = \frac{5}{8} = \cos^2\theta_W \ . \tag{16-18}$$

Again the angle is fixed at the unification energy. We have introduced 24 + 10 = 34 Higgses. Fifteen gauge bosons (all except the gluons and the photon) have become massive. Therefore there are 34 - 15 = 19 Higgses in the theory. The reader can easily convince herself that these consist of a neutral colour octet, three neutral colour singlets, two colour singlets with ±1 unit of charge, a colour triplet with charges -1/3 and the antiparticles of the latter. Although the Higgs sector of the SU(5) might seem outrageous, we must take it seriously because we lack an alternative for the generation of masses of the vector bosons.

The fermion masses are also believed to be due to the SSB mechanism. If the reader is familiar with the Young diagrams, she can easily see that

$$
\begin{aligned}
&\psi_R \sim 5 = \square\,, &&\bar{\psi}_R \sim 5^* = \begin{smallmatrix}\square\\\square\\\square\\\square\end{smallmatrix}\\
&\psi_L \sim 10 = \begin{smallmatrix}\square\\\square\end{smallmatrix}\,, &&\bar{\psi}_L \sim 10^* = \begin{smallmatrix}\square\\\square\\\square\end{smallmatrix}\\
&\phi \sim 24 = \begin{smallmatrix}\square\square\\\square\\\square\\\square\end{smallmatrix}\,, &&\\
&\theta \sim 5 = \cdot\cdot\,, &&\theta^* \sim 5^* = \cdots
\end{aligned}
\tag{16-19}
$$

Furthermore the charge conjugated spinors transform as

$$
(\psi_R)^c \sim 5^* \,,\quad (\psi_L)^c \sim 10^* \tag{16-20}
$$

Note also that $(\psi_L^c) = (\psi^c)_R$, etc. The Higgs-fermion invariant couplings connect left and right helicities. From (16-19) and (16-20) we find that

$$
\bar{\psi}_L\, \psi_R\, \theta,\ \text{h.c.}\,,\quad \overline{(\psi_L)^c}\, \psi_L\, \theta,\ \text{h.c.} \tag{16-21}
$$

with appropriate Clebsch-Gordan coefficients are the origin of masses. We have, in a short-hand notation, for example

$$
\begin{aligned}
\bar{\psi}_L\, \psi_R \langle\theta\rangle &\sim \sum_{j,k=1}^{5} \overline{b_{jk}}\, a_j \langle\theta^k\rangle \sim \sum_{j=1}^{5} \bar{b}_{5j}\, a_j\\
&\sim (\bar{d}_1\, d_1 + \bar{d}_2\, d_2 + \bar{d}_3\, d_3 + \bar{e}\, e)
\end{aligned}
\tag{16-22}
$$

where we have left out the uninteresting factors, and the primes denoting that the d and the e are not necessarily physical particles.

Similarly

$$\overline{\psi_L^c}\,\psi_L\,\theta \sim \sum_1^4 \varepsilon_{jkmn}\,\overline{b_{jk}^c}\,b_{mn} \sim (\bar{u}_1\,u_1 + \bar{u}_2\,u_2 + \bar{u}_3\,u_3) \quad (16\text{-}23)$$

17. THE PROTON/NEUTRON DECAY IN SU(5)

Let us recall that SU(5) has several nice features:

(i) It unifies the nongravitational forces.
(ii) The quantization of charge and fractional quark charges are naturally accounted for.
(iii) It has no room for a right-handed ν, unless the right-handed neutrino is, somewhat superficially, introduced as a singlet. So, we understand why the neutrino is massless.

Furthermore the leptons and quarks are unified, as they appear in the same multiplets. The reader might not like the 10 dimensional multiplets and the fact that the number of quark-lepton families is not explained. If she objects, we encourage her to produce a better theory.

What are the predictions of the model, in additional to the fact that there are 24 gauge bosons 19 Higgses, etc. as discussed before? First of all the model, by including the standard SU(2) x U(1) description of the weak and electromagnetic forces and SU(3) colour for strong interactions, reproduces the standard predictions of QCD, etc.

At our energies the weak, electromagnetic and strong interaction coupling constants are vastly different. The unification is expected to take place for energies well beyond all masses in the theory. The heaviest objects being the x and y gauge bosons, we have to go to energies $Q^2 >> M_x^2$, M_y^2 in order to have unifications. As we move down to lower values of Q^2, the strong, weak and electromagnetic couplings seperate and follow different trajectories. The renormalization of the coupling constant is due to higher order corrections (loops involving gauge bosons) and involves factors such as $\log(M_x^2/Q^2)$. One may ask how big should $M_x(=M_y)$ be in order to reproduce the presently measured values of the coupling constants at our energies. The answer is (D.A. Ross, Nucl. Phys. B140, 1(1978))

$$M_x = M_y \sim 5 \times 10^{15}\ \text{GeV !}$$

Thus the unification happens for energies much larger than

10^{16} GeV, presumably not far from the Planck value of 10^{19} GeV, where the gravitational interactions have already taken over!

We cannot hope to measure the trajectory of coupling constants up to such energies, however from

$$g = e/\sin\theta_W \quad , \quad g' = e/\cos\theta_W$$

the renormalization of g and g' imply that value of θ_W is also changing as function of Q^2. The quantity $\sin^2\theta_W$ which was 3/8 at the unification energy falls off slowly as we go down in Q^2. At our energies $\sin^2\theta_W \sim 0.2$ for three families of quarks and leptons (A. Buras et al. Nucl. Phys. B135, 66(1978)).

The most spectacular predictions of the SU(5) model (as well as several other grand unified models) is that the protons and neutrons can evaporate into "instable" matter (leptons and pions). The reason for proton decay is the existence of the bosons X and Y which violate the conservation of the baryon and lepton numbers.

The proton decay has been treated by A. Buras, loc. cit and by the author and F.J. Yndurain (to appear in Nucl. Phys. B). Here we follow the latter reference.

The dominant decay mechanism is that two quarks inside the nucleon annihilate, by exchanging either a X or Y boson. The third quark acts as spectator, viz.

(a)	$u + u \to e^+ + \bar{d}$	$P \to e^+ \pi^0$, etc.
	$\to \mu^+ + \bar{s}$	$P \to \mu^+ K^0$
(b)	$u + d \to e^+ + \bar{u}$	$P \to \pi^0 e^+$, $n \to \pi^- e^+$, ..
	$\to \mu^+ + \bar{c}$	forbidden by energy-momentum
(c)	$u + d \to \bar{\nu}_e + \bar{d}$	$P \to \pi^+ \bar{\nu}_e$, $n \to \pi^0 \bar{\nu}_e$, ..
	$\to \bar{\nu}_\mu + \bar{s}$	(small)
(d)	$d + d \to \bar{\nu}_e + \bar{u}$	$n \to \bar{\nu}_e \pi^0$, ..

The proton and neutron (baryon number violating) lifetimes are estimated to be

$$\tau_p \sim 10^{33} \text{ yrs} \quad , \quad \tau_n \sim 2 \times 10^{33} \text{ yrs.}$$

The errors are large due to present uncertainties in the input parameters. The lifetimes could be shorter or longer. Experimentally, the present best limit on the proton lifetime, $\tau_p \geq 2 \times 10^{30}$ yrs, is based on looking for muons (F. Reines and M.F. Crouch, Phys. Rev. Letters 32, 493(1974)). However, if the Georgi-Glashow model is correct, the branching ratio to muons is small, because the dominant channel (b) cannot produce muons.

Other measurements give limits which are more than an order of magnitude shorter than the quoted best limit.

Truly, the proton lifetime is much longer than the lifetime of our theories and our universe (if we believe the standard 10^{10} yrs). Measuring it is a great challenge.

ACKNOWLEDGEMENTS

I wish to thank Anne Grete Frodesen for reading these notes and Janny Asphaug for her efficient typing of the manuscript.

REFERENCES

An incomplete list of references on gauge theories and phenomenology includes

1. E.S. Abers and B.W. Lee, Phys. Reports 9C, 1 (1973).
2. M. Böhm and H. Joos, DESY Report 78/27 (in German); this article includes an abundant list of references.
3. H. Harari, Phys. Reports 42C, 235 (1978).
4. J. Iliopoulos, CERN Yellow Report 77-18, pp. 36-78.
5. J.C. Taylor, Gauge theories of weak interactions, Cambridge University Press (1976).

The SU(5) model and its consequences may be found in, for example,

6. H. Georgi and S.L. Glashow, Phys. Rev. Lett. 32, 438 (1974).
7. H. Georgi, H. Quinn and S. Weinberg, Phys. Rev. Lett. 33, 451 (1974).
8. A. Buras, J. Ellis, M. Gaillard and D. Nanapoulos, Nuclear Phys. B135, 66 (1978) and references therein.
9. C. Jarlskog, F.J. Yndurain, to be published in Nuclear Phys.
10. D.A. Ross, Nucl. Phys. B140, 1 (1978).

DRELL-YAN VERSUS EXPERIMENT

F. Vannucci

Lab. de Physique des Particules (LAPP)
Annecy, France

Since the experiment which first measured dimuon production in hadron collisions[1], several similar searches were started, and some of them proved amazingly successful[2,3]. The finding of new particles in the reaction

$$pN \to \ell^+\ell^- + X$$

explains the flourishing of this kind of study, but the understanding of the continuum (the "background" under the J/ψ or the Υ) now proves to be of primary interest by itself, and this presentation will be restricted to this aspect of dilepton production. As we will see, lepton pair measurements provide a way of testing various ideas about the distribution of quarks inside the hadrons, and new results from the CERN Intersecting Storage Rings (ISR) will be used to test, in particular, the prediction of scaling and the connection with lepton deep inelastic scattering.

The presentation will be as follows:

- description of an experimental set-up,
- the Drell-Yan model and its predictions,
- first symptoms of $q\bar{q}$ annihilation,
- test of scaling,
- extraction of quark distributions,
- problems of the basic model.

1. DESCRIPTION OF AN EXPERIMENT

Dilepton searches have been favoured by the relative ease of experimental detection. For instance, $\Gamma(J/\psi \rightarrow \pi^+\pi^-\pi^+\pi^-\pi^0)/\Gamma(J/\psi \rightarrow e^+e^-$ = 0.5, but in hadron collisions the J was found in the e^+e^- mode, and our only information about ψ decays comes from e^+e^- physics. This is explained by the fact that both electrons and muons have properties that are characteristic enough to allow extraction from an overwhelmi hadronic background.

As an example, and because original data from this experiment will be presented in the following, I will describe a set-up now in use at the CERN ISR whose aim is the measurement of dimuon production

Muons do not undergo strong interactions and can traverse many collision lengths of matter without being absorbed, and it is this property that allows them to be filtered out by absorbing the hadrons In order to measure the momentum, a magnetic field is needed, and these two functions of hadron absorber and magnetic field carrier are merged by using a set of magnetized iron toroids. The apparatus is shown in Fig. 1. There are two main sources of background in muon detection:

- hadron punch-through: in this experiment the filter represents a minimum of 15 collision lengths on the path of a particle.
- π or K decays giving a μ: the absorber starts 50 cm from the interaction point, thus drastically decreasing the probability of a deca

Magnetized iron, run at saturation, is the cheapest way of filli a large volume with a high magnetic field (1.8 T), and the whole detector gives a minimum B dl of 2.7 Tm over a solid angle of 8 sr. In polar angle the coverage is from 15° up to 110°. The drawback of thi technique comes from multiple scattering, and the resolution in mass is about 10%.

The tracks are detected in large drift chambers giving a spatial resolution of 0.7 mm over an area of 800 m^2. Time-of-flight techniqu allow the rejection of cosmic rays and also of stray particles accompanying the incoming beams.

The inner hole of the first yoke surrounding the interaction poi is filled with drift wire and delay line chambers, which help reconstruct the vertex and give information on the hadrons produced togeth with the muon pair.

The shielding proves effective: about 10^7 charged particles are produced each second but the trigger rate is only 0.5 per second. Th events which pass the analysis stream are very clean. This is shown

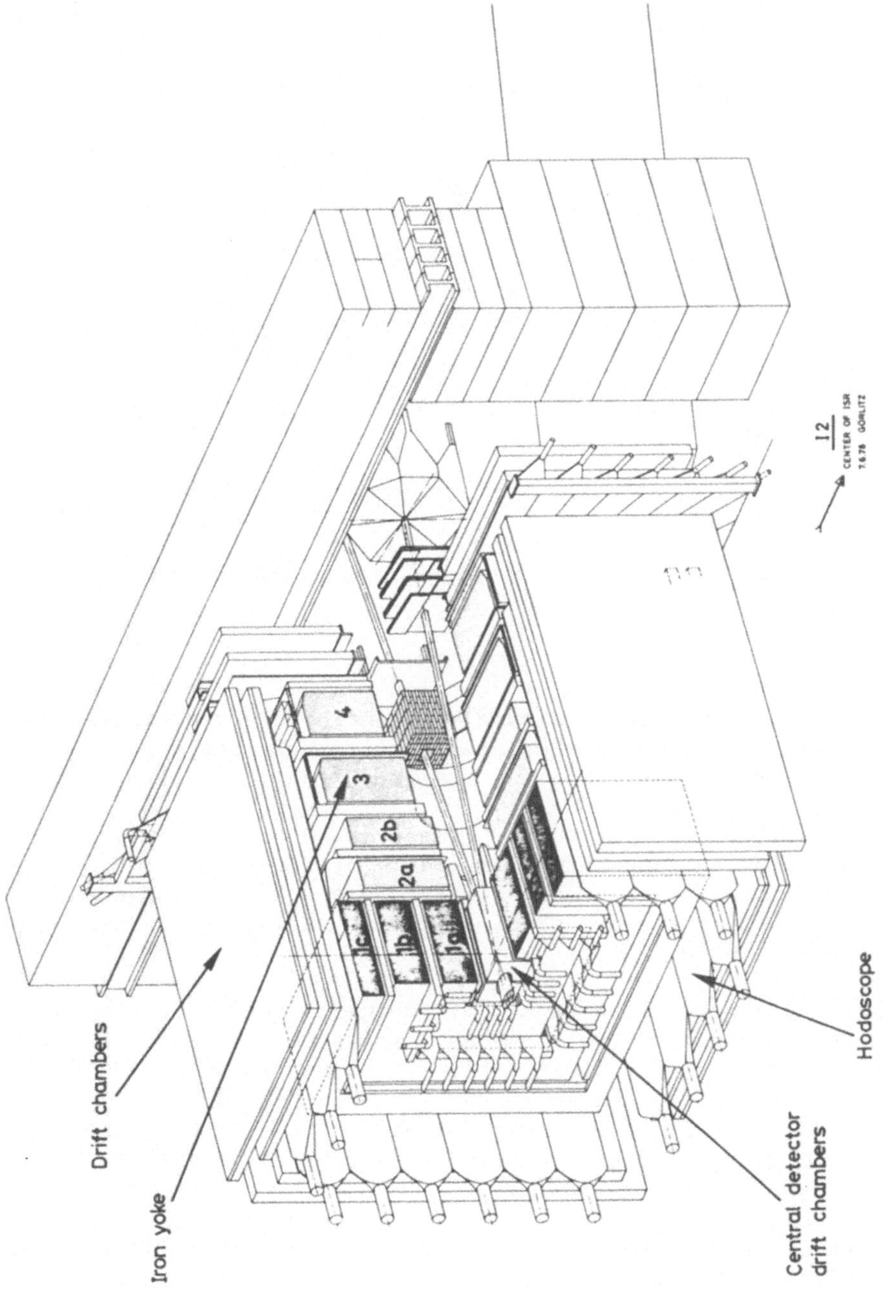

Fig. 1 : Detector of the experiment R209 at the ISR

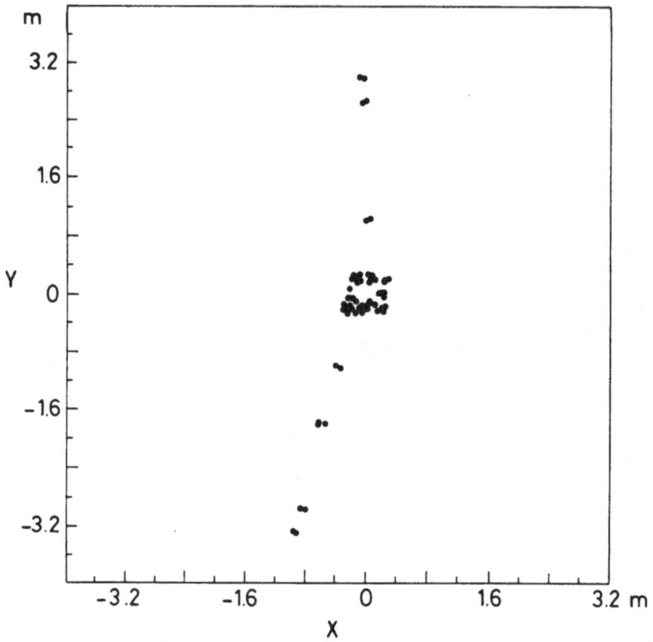

Figure 2. One dimuon event as seen in the detector.

in Fig. 2, where one sees a dimuon event with only two tracks beyond the first iron yoke, while the inner detector registers many hadron tracks.

The resulting plot of the $\mu^+\mu^-$ mass is seen in Fig. 3. This plot is based on an integrated luminosity of 2.6×10^{37} cm^{-2} s^{-1} and represents a total of about 600 events above 4 GeV/c^2. The backgrour level is checked by the amount of like-sign pairs; it is negligible above a mass of 6 GeV/c^2.

These data were taken at a c.m. energy $\sqrt{s}$ = 62 GeV. The mass plot can be compared to two previous results obtained at very differe energies. Together with these data, Fig. 3 displays the BNL data[1] taken at $\sqrt{s} \simeq 7$ GeV and those of the Columbia-FNAL-Stony Brook (CFS) Group[3] obtained at $\sqrt{s}$ = 27.4 GeV; it will be the task of the scaling stratagem to relate quantitatively these very different cross-sectic

When compared to a stationary target experiment, the ISR experiment loses a large factor in luminosity, but the ISR still remains by far the highest energy machine, and this is important for testing scaling. Furthermore, as will be seen, the ISR measures the continuu in a region where the FNAL or the CERN Super Proton Synchrotron (SPS) experiments are blind because of resonance production. Finally, the

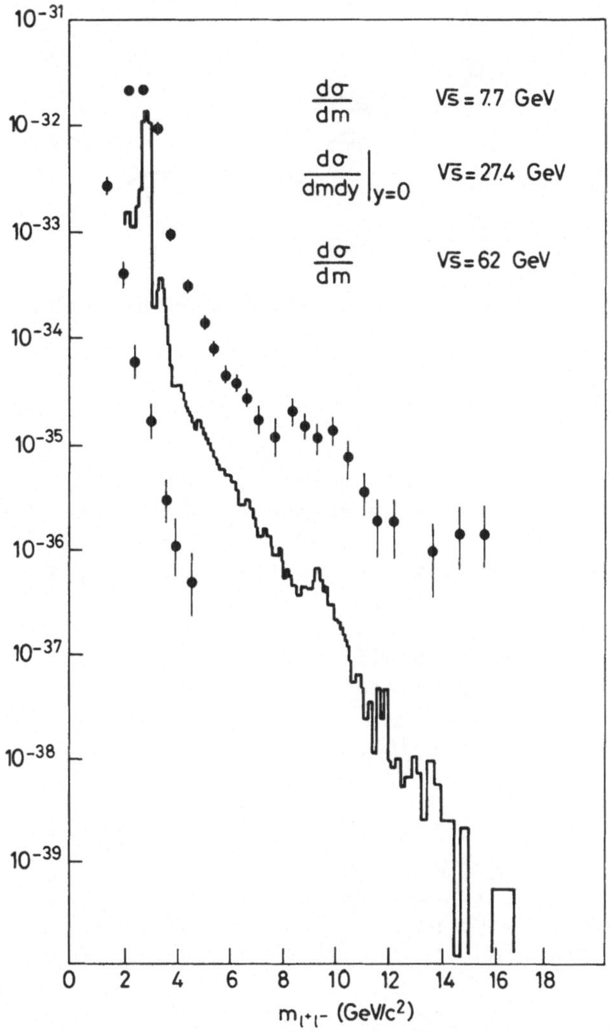

Figure 3. Three generations of dimuon events.

low luminosity ceases to be a handicap for the study of associated hadrons, about which virtually no information is yet available.

For a discussion of the experimental problems of dilepton searches, one can refer to Ref. 5.

2. THE CLASSICAL DRELL-YAN MODEL[6]

According to this model the massive lepton pair production

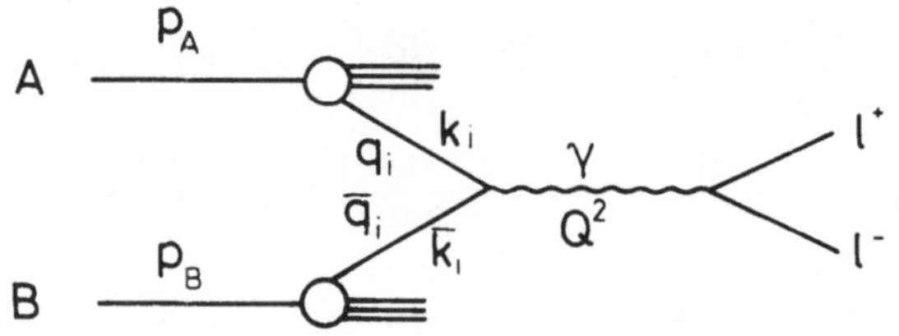

Figure 4. Schematic of the Drell-Yan mechanism.

$$A + B \to \ell^+ + \ell^- + X$$

takes place via the annihilation into a time-like photon of a quark of one incoming hadron with an antiquark of the other hadron. This is schematically shown in Fig. 4, where the kinematical variables are explained. It is essentially the inverse diagram of $e^+e^- \to q\bar{q} \to$ hadrons.

The cross-section can be written as the sum of elementary processes weighted by the probability of finding quarks and antiquarks which are allowed to annihilate with each other:

$$\sigma(A + B \to \ell^+\ell^- + X) = \iiint dx_i d\bar{x}_i \sum_i f_i^A(x_i)\bar{f}_i^B(\bar{x}_i) \times$$

$$\times\ \sigma(q_i + \bar{q}_i \to \ell^+\ell^-)\delta\left[Q^2 - (k_i + \bar{k}_i)^2\right] dQ^2 .$$

The sum is over all types i of quarks and antiquarks present in the initial hadrons with probabilities $f_i^A(x_i)$, $\bar{f}_i^B(\bar{x}_i)$. The variables x_i, $\bar{x}_i$ are the fraction of initial momentum carried by the constituents:

$$k_i = x_i P_A \ , \qquad \bar{k}_i = \bar{x}_i P_B \ .$$

The invariant mass squared of the photon, and thus of the dilepton, is

$$m^2 = Q^2 = (k_i + \bar{k}_i)^2$$

$$= (x_i P_A + \bar{x}_i P_B)^2$$

$$\simeq s\ x_i \bar{x}_i$$

in the limit of large s, where s is the total c.m. energy squared:

$$s = (P_A + P_B)^2 \ .$$

The model neglects any transverse momentum of the constituents. With this restriction the Feynman variable X for the lepton pair is simply

$$X = \frac{2P_L}{\sqrt{s}} = x_i - \bar{x}_i \ .$$

The model assumes a point-like coupling $\gamma q_i \bar{q}_i$; the elementary cross-section then reads:

$$\sigma(q_i + \bar{q}_i \rightarrow \ell^+\ell^-) = \frac{4\pi\alpha^2}{3Q^2} e_i^2 \ ,$$

e_i being the charge of the quark of type i. The total cross-section now takes the form

$$\frac{d\sigma}{dQ^2} = \frac{4\pi\alpha^2}{3Q^4} F(\tau)$$

$$\text{with } F(\tau) = \sum_i e_i \frac{\tau}{N_c} \int_0^1 dx_i \int_0^1 d\bar{x}_i \ \delta(x_i\bar{x}_i - \tau) f_i^A(x_i) \bar{f}_i^B(\bar{x}_i) \ ,$$

where N_c is the number of colour and τ is a new variable representing the "scaled" mass squared of the dilepton system:

$$\tau = \frac{m^2}{s} = x_i\bar{x}_i \ , \qquad 0 \leq \tau \leq 1 \ .$$

It is also useful to write down the double differential cross-section:

$$\frac{d^2\sigma}{dQ^2dX} = \frac{4\pi\alpha^2}{3N_cQ^4} \frac{\tau}{\sqrt{X^2 + 4\tau}} \sum_i e_i^2 f_i^A(x_i) f_i^B(\bar{x}_i)$$

$$x_{i,\bar{i}} = \frac{\sqrt{X^2 + 4\tau} \pm X}{2} \ .$$

By writing down the different flavours of quarks which can be excited in a given energy domain, one obtains

$$m^3 \frac{d\sigma}{dm dx} = \frac{8\pi\alpha^2}{3N_c} \frac{\tau}{\sqrt{X^2 + 4\tau}} \left\{ \frac{4}{9}(u^A\bar{u}^B + \bar{u}^Au^B) + \frac{1}{9}(d^A\bar{d}^B + \bar{d}^Ad^B) + \right.$$

$$\left. + \frac{1}{9}(s^A\bar{s}^B + \bar{s}^As^B) + \frac{4}{9}(c^A\bar{c}^B + \bar{c}^Ac^B) \right\} \ .$$

This model gives straightforward predictions which can be tested experimentally[7]:

- the cross-section should follow the scaling law $m^3(d\sigma/dm) = F(\tau)$ or, if X is measured, $m^3(d^2\sigma/dmdX) = G(\tau,X)$, F and G being universa functions of the dimensionless variable τ, independent of energy;
- if quarks have colour the probability of pairing a quark i with its antiquark of the same colour is reduced; then the cross-sectio can be expected to be inversely proportional to N_c;
- the quarks having spin ½ and point-like coupling to the photon, the angular distribution of one lepton in the dilepton c.m. system is $1 + \cos^2\theta$;
- the shape of the cross-section is determined by the quark distributions which are, in principle, measured in deep inelastic lepton scattering;
- the quarks having no transverse momenta, the intermediate photon is longitudinal to the beam, and the transverse momentum of the dilepton is expected to be small;
- finally the annihilation model predicts only two jets accompanying the lepton pair, in the direction of the initial hadrons.

3. FIRST SYMPTOMS OF $q\bar{q}$ ANNIHILATION

3.1 Target dependence

Because of shadowing effect, hadronic processes vary as $A^{2/3}$, where A is the atomic number of the target: the projectile sees the surface of the nucleus, and nucleons inside the nucleus act coherentl

The Drell-Yan process postulates an incoherent action of the different scatterers inside the nucleon. It can be expected that quarks inside different nucleons will also act incoherently, and the total cross-section, being the sum of incoherent elementary processes, will have a variation proportional to A.

Figure 5 shows the A dependence measured at Fermilab[8-10] with beryllium, carbon, copper, and tungsten targets. There is a clear transition from an $A^{2/3}$ dependence below a mass of 4-5 GeV/c to an A dependence above, indicative of a change of regime.

3.2 Beam dependence

The annihilation occurs between the quark of one hadron and the antiquark of the second hadron. The target consists of protons and neutrons for which antiquarks exist only in the sea; the projectile quark distribution offers more choice, and one expects the following relations, as a consequence of the valence-sea character of the quark and of their charges[11]:

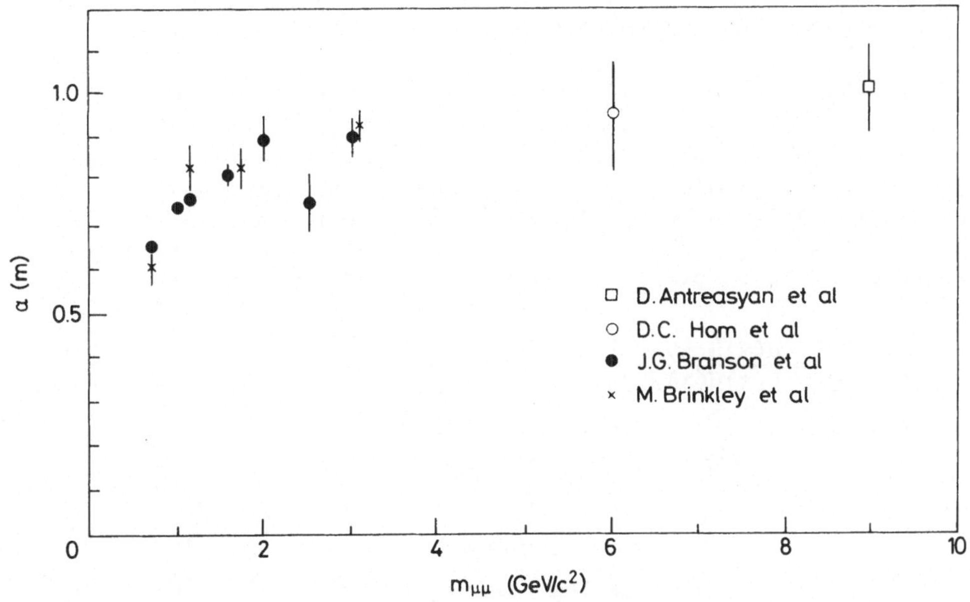

Fig. 5 : A dependence of dimuon production

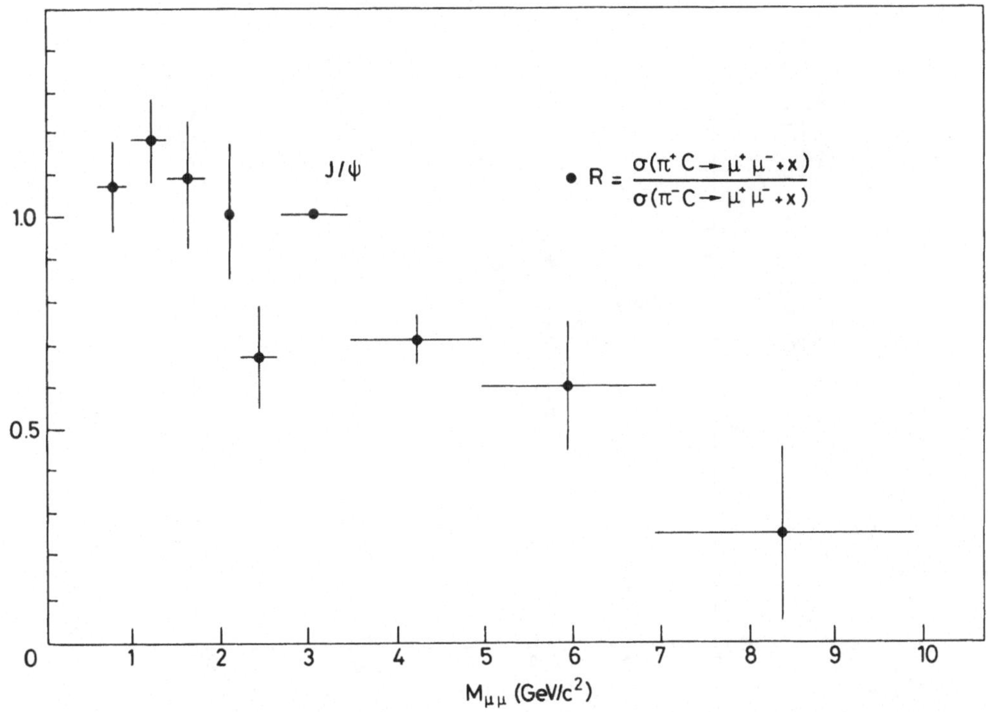

Fig. 6 : Ratio of $\mu^+\mu^-$ production cross-sections in π^+ and π^- induced reactions

$$\sigma(p\bar{p}) > \sigma(\bar{n}p) > \sigma(pp) > \sigma(np)$$

$$\sigma(\pi^-p) = \sigma(K^-p) > \sigma(\pi^+p) = \sigma(\bar{K}^0p) > \sigma(K^+p) = \sigma(K^0p) \ .$$

This last line holds if antiquarks in π and K have similar distributions. The ratio $\sigma(K^-N)/\sigma(\pi^-N)$ is found to be close to 1 [12], while $\sigma(\pi^+N)/\sigma(\pi^-N)$ is shown in Fig. 6 [10,13]. The π^+ has a $\bar{d}$ of charge $1/3$, while π^- has a $\bar{u}$ of charge $-2/3$; then, on an isoscalar target and if the reaction takes place via $q\bar{q}$ annihilation, one expects

$$\frac{\sigma(\pi^+N)}{\sigma(\pi^-N)} = \frac{d(x)\bar{d}(x)}{u(x)\bar{u}(x)} = \frac{1}{4} \ .$$

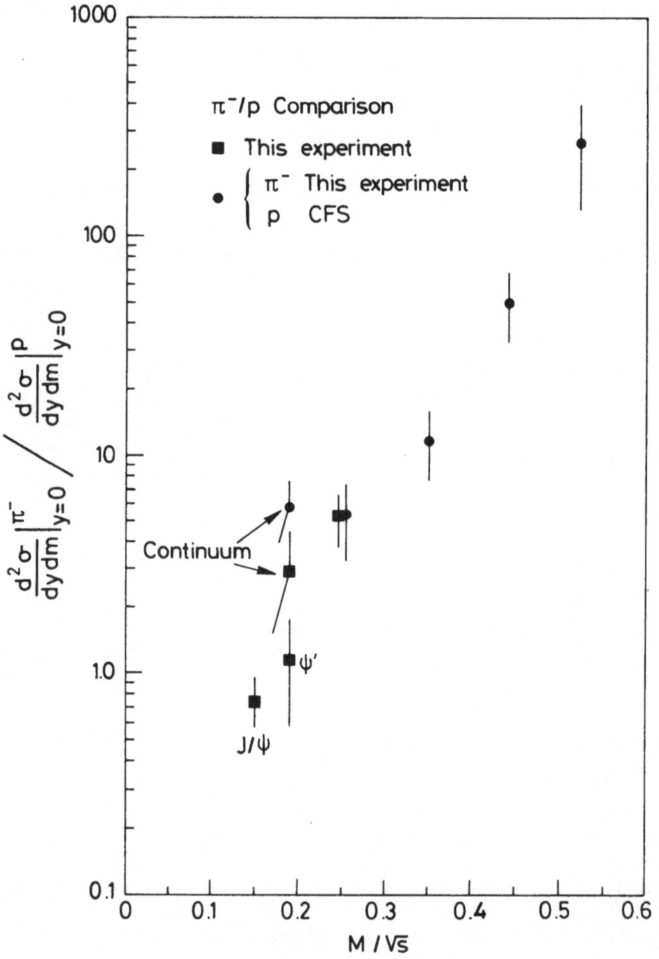

Figure 7. Ratio of $\mu^+\mu^-$ production cross-sections in π^- and p induced reactions.

This is true in the limit of no sea-quark contamination. As seen in Fig. 6 the experimental result indicates a transition from a ratio of 1 expected for hadronic production to a ratio of 0.25 for an 8 GeV/c^2 dimuon consistent with a $q\bar{q}$ annihilation process. The result for $\sigma(\pi^- N)/\sigma(pN)$ [13] is even more remarkable, as shown in Fig. 7. The ratio becomes much bigger than 1 (270 at a mass of 10 GeV/c^2), consistent with the existence of valence antiquarks in the π^-, while p has only sea antiquarks. This also corroborates the idea of distributions f(x) dying off much more rapidly for sea than for valence constituents.

There are other sum rules coming from simple charge counting:

$$\sigma(\pi^- p) - \sigma(K^+ p) = 4[\sigma(\pi^+ n) - \sigma(K^+ n)]$$

$$\sigma(\pi^- n) - \sigma(K^+ n) = 4[\sigma(\pi^+ p) - \sigma(K^+ p)] .$$

No data are as yet able to test those predictions.

3.3 Angular distribution

In the dimuon frame, the angular distribution of a final lepton is predicted to have the form

$$\frac{d\sigma}{d\Omega} = A(1 + \alpha \cos^2 \theta) .$$

If the quarks have no transverse momentum, the longitudinal direction is the direction of the initial hadrons.

New results from the CP II experiment at Fermilab[13] are shown in Fig. 8. While the distribution is flat for the J/ψ region, it is consistent with a form $1 + \alpha \cos^2 \theta$ with $\alpha = 1$ above the J/ψ. It has also the same shape below the J/ψ, although this region should not be dominated by the annihilation diagram!

These various observations all indicate very remarkably that lepton pairs at sufficiently high mass are not produced by normal hadronic interactions, and the annihilation mechanism seems to set in above a dilepton mass of $\sim$ 5 GeV/c^2 at Fermilab energies, corresponding to $\tau \sim 0.04$. If the transition occurred at a given τ, it should develop beyond a mass of 10 GeV/c^2 at the ISR. There are good reasons to believe that dimuon production is already accounted for by the Drell-Yan mechanism at lower masses. The transition then really seems related to a threshold mass or a threshold Q^2 below which a mechanism different from annihilation takes place independently of quark distributions unless we imagine some kind of coherent effect among quarks. This

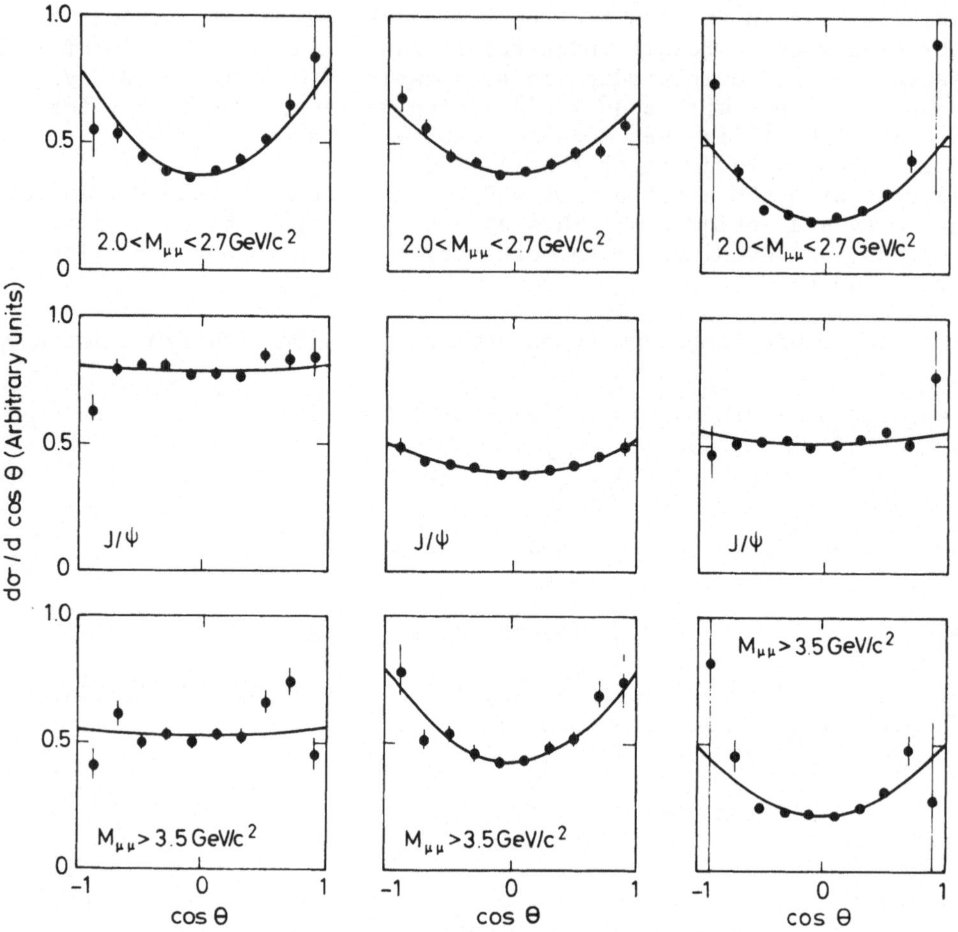

Figure 8. Angular distribution of one lepton in the dilepton c.m. frame. The reference axis of the sets of distributions are the recoil, the beam, and the target directions, respectively.

also indicates that to probe the very small τ region (the deeper sea) one has to go to higher energies, and not to smaller mass.

4. TESTS OF SCALING

The scaling law $m^3(d\sigma/dm) = F(\tau)$ is a basic prediction of the Drell-Yan model. However, it is more simply a consequence of dimensional analysis, a cross-section having dimension m^{-2} (in the system $\hbar = c = 1$). Scaling implies that the quark distributions are only functions of the dimensionless variable x, and are in particular independent of q^2. Scaling violations, on the other hand, could be generated by some quark form factors dependent on q^2.

Two experimental studies[9,14] of the scaling law have compared dimuon production for incident proton energies of 200, 300, and 400 GeV; $\sqrt{s}$ then varies from 20 GeV to 27.4 GeV. Figure 9a shows the mass spectra obtained for those three energies as a function of m, while Fig. 9b shows the same data plotted in a different way: the quantity $s\, d^2\sigma/d\sqrt{\tau}\, dy$ versus $\sqrt{\tau}$. The contrast between these two plots is a good illustration of the significance of the scaling test. Here, y is the rapidity variable,

$$y = \frac{1}{2} \log \frac{E + P_L}{E - P_L} .$$

Actually, if the experimental cross-section often measured is $d^2\sigma/dmdy$, the quantities which should rigorously scale are $m^3(d\sigma/dm)$ and $m^3(d\sigma/dmdx)$. To compare results from different experiments, we need a conversion formula between these different expressions. First we can write

$$\left.\frac{d^2\sigma}{dmdX}\right|_{X=0} = \frac{\sqrt{s}}{2m} \left.\frac{d^2\sigma}{dmdy}\right|_{y=0} .$$

The X variation of the cross-section has been measured in the range 3.5 to 4.5 GeV/c^2; it is found to be

$$\frac{d^2\sigma}{dmdX} = \left.\frac{d^2\sigma}{dmdX}\right|_{X=0} (1 - X)^{2.8} .$$

Note that the scaling test depends very sensitively on the exponent of this expression. We can then write the final correspondence formula:

$$\frac{d\sigma}{dm} = \frac{\sqrt{s}}{3.8\ m} \left.\frac{d^2\sigma}{dmdy}\right|_{y=0} .$$

With the help of this formula we can compare different experiments performed at energies over as wide a range as possible, and in particular we can try to relate the curves given in Fig. 4. Such an attempt has been made, comparing BNL data with Fermilab data[15]. The BNL data are essentially below a mass of 4 GeV/c^2, and the Drell-Yan regime is not expected to be dominant. In any case (precocious?) scaling is the conclusion of the study.

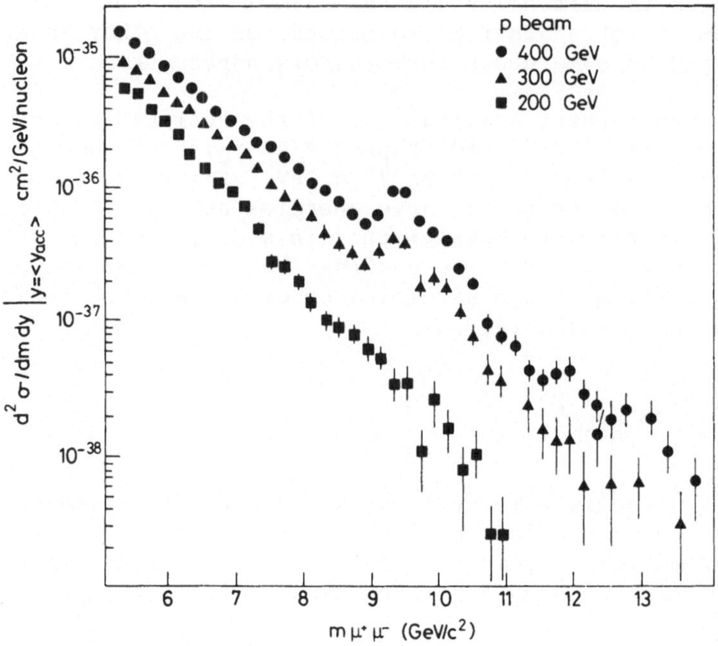

Figure 9a. Dimuon mass spectra with incident proton energies of 200, 300, and 400 GeV.

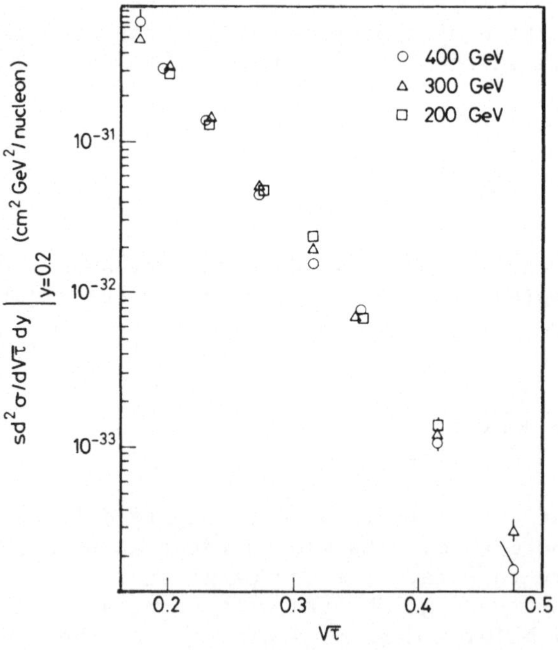

Figure 9b. $s\ d^2\sigma/d\sqrt{\tau}\ dy$ plotted versus $\sqrt{\tau}$ for the same data.

Here we will add a new energy region, namely $\sqrt{s}$ = 62 GeV, into the test. For the following comparison I have converted the CFS data at 400 GeV in order to plot the quantity $m^3(d\sigma/dm) = F(\tau)$. The resulting curve is shown in Fig. 10 as a function of $\sqrt{\tau}$. It represents an experimental measurement of $F(\tau)$, and it is clear that, in the range where it is measured ($0.2 < \sqrt{\tau} < 0.5$), $F(\tau)$ follows an exponential fall-off very nicely. The region around $\sqrt{\tau} = 0.35$ corresponds to the ϒ, and experimental points have been ignored. The line represents a straightforward fit:

$$F_1(\tau) = 13 \times 10^{-32} \exp(-18.6 \sqrt{\tau}) .$$

Now let us faithfully follow the rule, and scale this formula to the

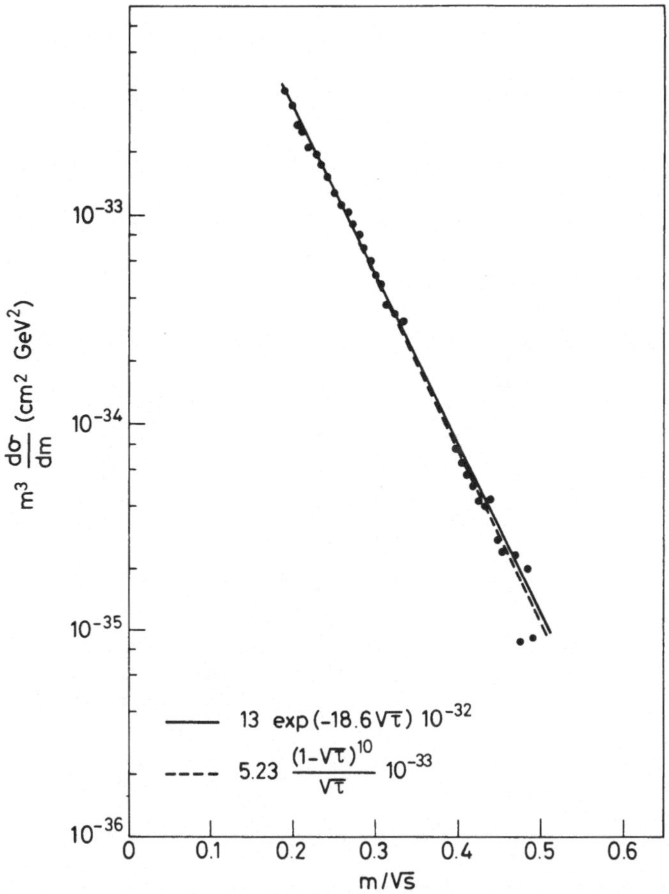

Figure 10. The quantity $m^3(d\sigma/dm)$ plotted versus $\sqrt{\tau}$ for the CFS data. The full line represents the fit: $F_1(\tau) = 13 \exp(-18.6\sqrt{\tau})\ 10^{-32}$. The dashed line represents the other fit: $F_2(\tau) = 5.23\ [(1 - \sqrt{\tau}\,^{10}/\sqrt{\tau}) \times 10^{-33}$.

ISR energy. Figure 11 shows the result. It is rather remarkable that such a simple procedure works so well considering that the cross-sections measured at Fermilab and at the ISR differ by a factor of 10 at a mass 5 GeV/c^2 and a factor of 100 at a mass 12 GeV/c^2. It is even more remarkable because a very different fit will give the same agreement. One can try a formula[16] diverging for $\tau = 0$:

$$F_2(\tau) = 5.23 \times 10^{-33}(1 - \sqrt{\tau})^{10}/\sqrt{\tau} \;,$$

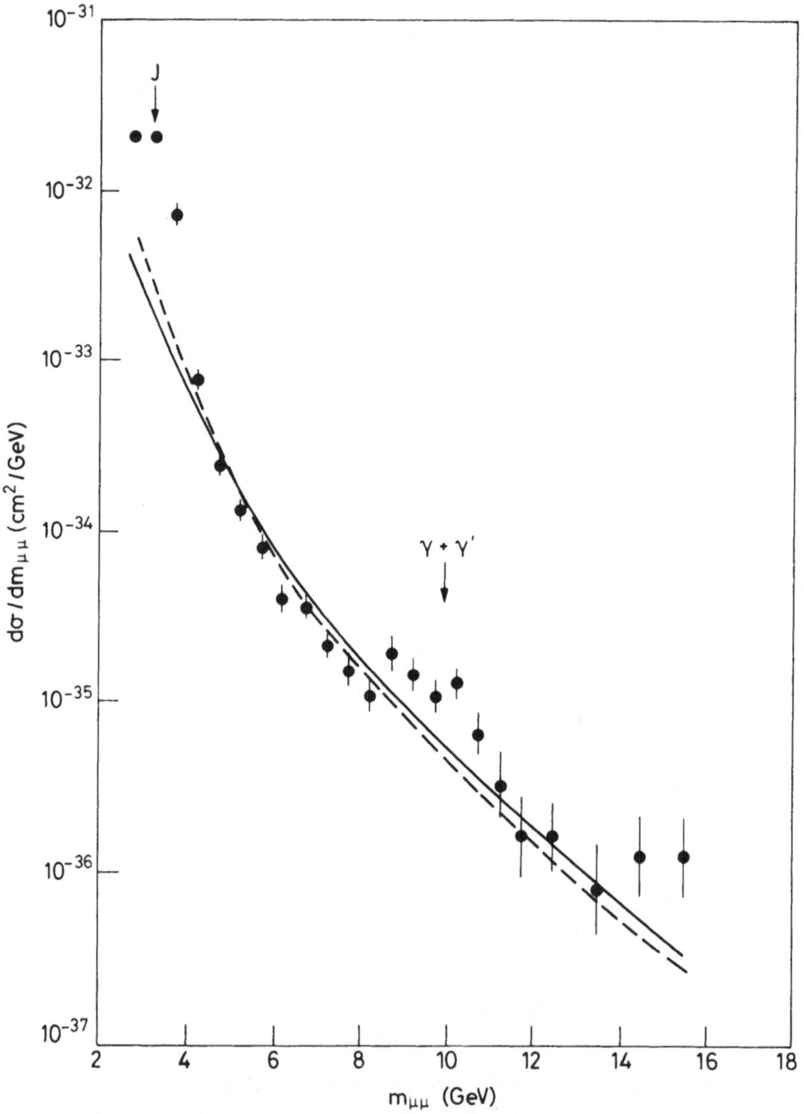

Figure 11. ISR mass spectrum with the scaling predictions based on $F_1(\tau)$ (full line) and $F_2(\tau)$ (dashed line).

and the result is shown by the dashed line, in both Figs. 10 and 11.

The insensitivity of the test can be understood by plotting the ISR result in the same way as we plotted the CFS data: $m^3(d\sigma/dm)$ versus $\sqrt{\tau}$. The two experiments do not overlap in $\sqrt{\tau}$: the CFS data map the function $F(\tau)$ for $\sqrt{\tau} > 0.2$, while the ISR data add points in the region $0.05 < \sqrt{\tau} < 0.15$. This indicates two facts: a real test of scaling must be done for the same $\sqrt{\tau}$ values, and this could be obtained with ISR data taken at energies 15 GeV × 15 GeV. On the other hand, the present ISR result gives unique information impossible to obtain at Fermilab, and we need even higher energies to continue mapping $F(\tau)$ at very small $\sqrt{\tau}$.

The conclusion of this scaling test is that any "smooth" extrapolation of $F(\tau)$ in the ISR range of τ works well.

In Fig. 12 I have summarized our present knowledge of $F(\tau)$, adding together the different measurements obtained so far, including the early BNL data after subtraction of the J contribution and some early ISR results[17]. I have also plotted a few points corresponding to low-mass dimuons ($1.5 < m < 2.5$ GeV/c^2) obtained with a 225 GeV proton beam[10] and the recent Omega result[12] at 40 GeV beam energy. The low-mass results coming from Fermilab show that, for the same τ, $m^3(d\sigma/dm)$ is up to an order of magnitude above the scaling prediction. This again stresses the point that the Drell-Yan mechanism is not applicable at low masses. To dramatize this effect, I have plotted in Fig. 14 the ratio $R_{DY} = (d\sigma/dm)/DY$, where DY is taken to equal to $m^{-3}F(\tau)$. Above a mass of 4 GeV/c^2, by construction, R_{DY} is compatible with 1 for measurements ranging from $\sqrt{s} = 20$ GeV to $\sqrt{s} = 62$ GeV, while below 3 GeV/c^2 the ratio R_{DY}, measured at three different energies, is consistently higher than 1. It is symptomatic that this change should occur for the same mass of the intermediate photon than is the case for the famous $R_{e^+e^-} = \sigma(e^+e^- \rightarrow \text{had.})/\sigma(e^+e^- \rightarrow \mu^+\mu^-)$ also plotted in Fig. 13. The ratio R_{DY} decreases while $R_{e^+e^-}$ increases at this threshold. It is tempting to link both effects to the same cause. The ratio R_{DY} is inversely proportional to the number of colour N_c, while $R_{e^+e^-}$ is proportional to N_c, and a colour thaw would qualitatively point towards the right direction. But this idea is probably heretical, and if the jump in $R_{e^+e^-}$ is associated with the charm threshold, the fall in R_{DY} does not yet seem to be well understood, and more low-mass experimental results are needed to confirm the trend.

5. EXTRACTION OF QUARK DISTRIBUTION

The scaling law is a consequence of the Drell-Yan model, but a more stringent prediction is given by the meaning of $F(\tau)$:

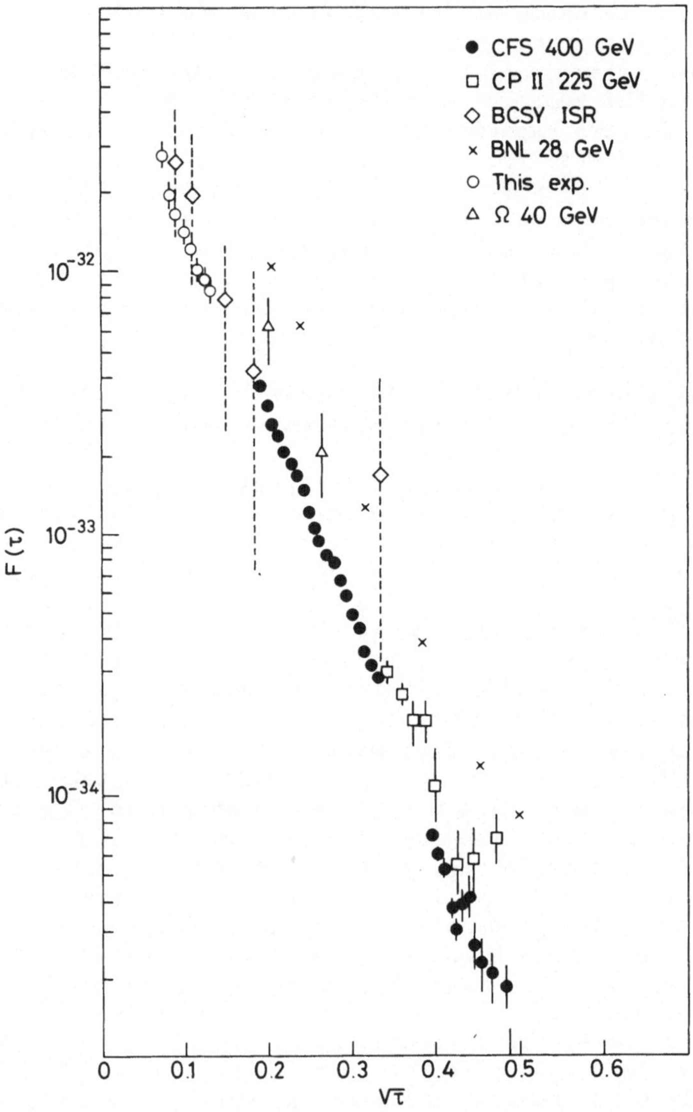

Figure 12. The experimental function F(τ) versus $\sqrt{\tau}$.

$$F(\tau) \propto \tau \sum e_i^2 \int_0^1 dx_i \int_0^1 d\bar{x}_i \; f_i^A(x_i) \; \bar{f}_i^B(\bar{x}_i) \; \delta(x_i \bar{x}_i - \tau) \;,$$

where f(x) are the quark distributions measured in deep inelastic scattering experiments. If the idea of constituents is at all relevant, one expects the same distributions to be present in the hadron, irrespective of the probe used to measure them.

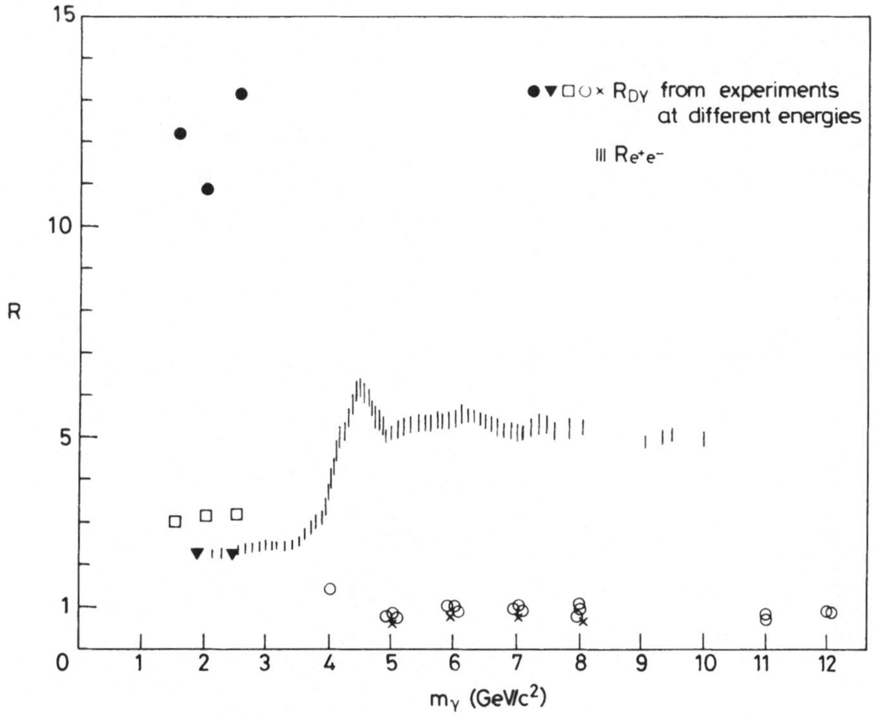

Figure 13. R_{DY} and $R_{e^+e^-}$ as a function of $m_{\mu\mu}$.

There is a total of eight distributions to be extracted: u(x), $\bar{u}(x)$, d(x), $\bar{d}(x)$, s(x), $\bar{s}(x)$, c(x), $\bar{c}(x)$.

The photon couples to all types of quarks, and deep inelastic electron or muon scattering gives the following relations on proton and neutron targets:

$$\nu W_2^{ep}(x) = \frac{4}{9} x(u + \bar{u}) + \frac{1}{9} x(d + \bar{d}) + \frac{1}{9} x(s + \bar{s}) + \frac{4}{9} x(c + \bar{c})$$

$$\nu W_2^{en}(x) = \frac{1}{9} x(u + \bar{u}) + \frac{4}{9} x(d + \bar{d}) + \frac{1}{9} x(s + \bar{s}) + \frac{4}{9} x(c + \bar{c}) \ .$$

On the other hand, $\nu(\bar{\nu})$ can only interact with d, $\bar{u}$, s, $\bar{c}$ ($\bar{d}$, u, $\bar{s}$, c) because of its left-handedness (right-handedness). This gives

$$\nu W_2^{\nu P}(x) = 2x(\bar{u} + d + s + \bar{c})$$

$$\nu W_2^{\bar{\nu} P}(x) = 2x(u + \bar{d} + \bar{s} + c)$$

$$\nu W_3^{\nu P}(x) = 2x(\bar{u} - d - s + \bar{c})$$

$$\nu W_3^{\bar{\nu} P}(x) = 2x(\bar{d} - u + c - \bar{s}) \ .$$

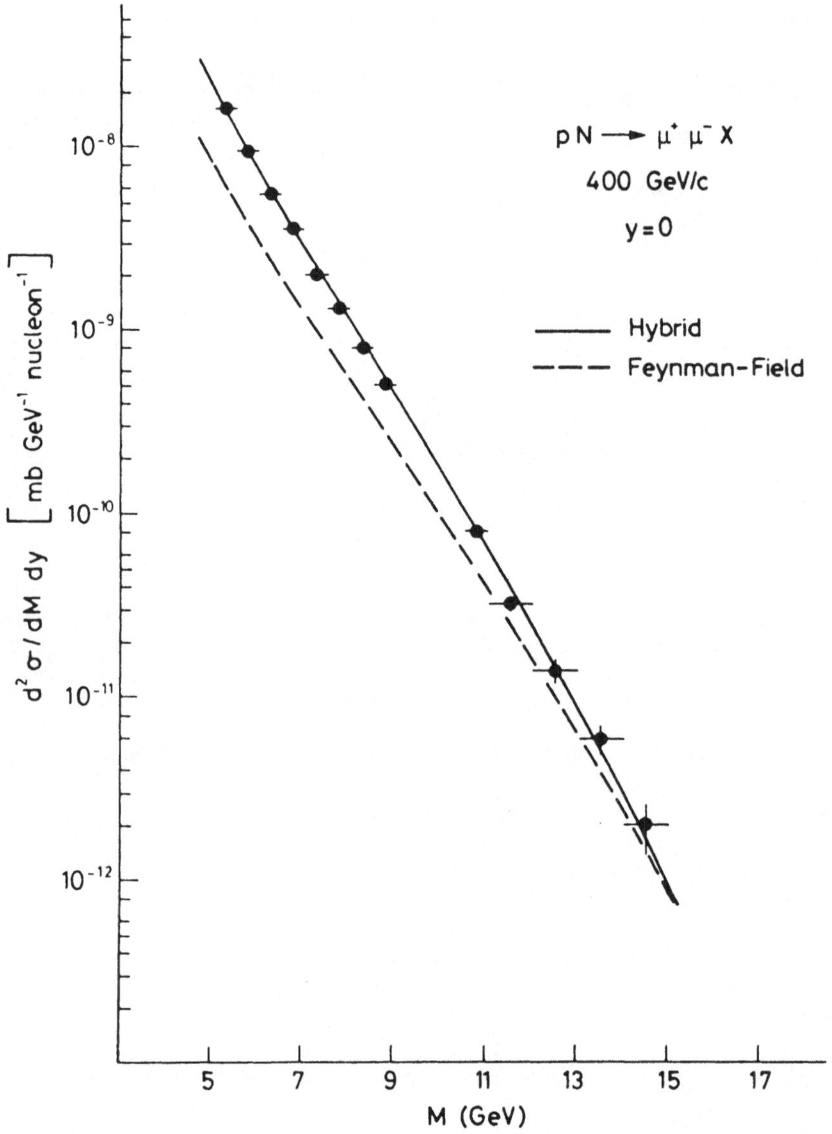

Figure 14. Fit of the CFS data with different sea-quark distributions.

W_2 and W_3 are the usual structure functions measured in lepton scattering.

If we neglect the charm contribution, or if we assume the sea to be SU(3) symmetric, those six relations are, in principle, sufficient to obtain all quark distributions. Field and Feynman[18] (FF) have extracted those distributions, and have found for the valence

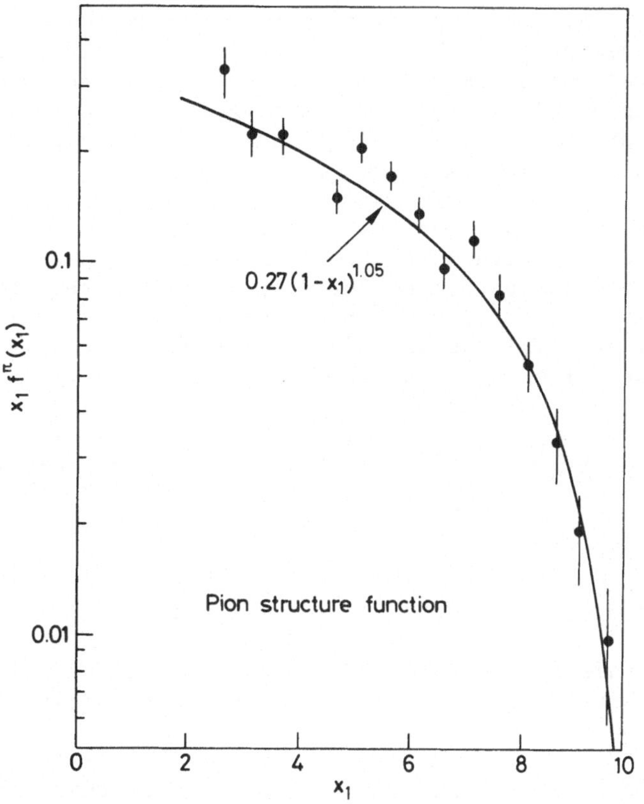

Figure 15. Valence quark distribution in pions as found in $\pi p \to \mu^+\mu^- + X$.

quarks:

$$xu(x) \sim (1 - x)^3 \quad \text{and} \quad xd(x) \sim (1 - x)^4$$

for large x, and for x near 0:

$$xu(x) \sim xd(x) \sim \sqrt{x}(1 - x)^2 \ .$$

The experimental results used so far are linear in the quark distributions and will be insensitive to the sea-quark content which is expected to be small. The advantage of dimuon production is that it gives a quantity that is quadratic in the quark distributions and directly proportional to the sea-quark content. This is where dilepton results can bring new information, and may prove to be the only way of extracting sea-quark distributions at relatively large X.

Berger[19] has fitted the CFS data with the FF distributions. As shown in Fig. 14, the FF prediction falls below the data by a factor of 3 for masses of ∿ 5 GeV/c. Then Berger uses the experimental curve to extract the sea-quark content, while keeping the valence-quark distributions of FF. He finds, assuming all sea quarks to have the same distribution, $\bar{x}S(\bar{x}) = 0.42(1 - \bar{x})^9$, and his fit reproduces the data very well, as is also shown in Fig. 15.

Another method has been used to extract the same information from the same data without the need of explicit valence-quark distributions. The idea is to feed the deep inelastic results directly into $(d^2\sigma/dmdy)_{y=0}$ [20] ($y = 0$ corresponds to $x_i = \bar{x}_i = x = \sqrt{\tau}$). Forgetting $c(x)$ and $\bar{c}(x)$ and assuming $\bar{u}(x) = \bar{d}(x) = s(x) = \bar{s}(x) \equiv S(x)$, we obtain the following relation:

$$\left.\frac{d^2\sigma}{dmdy}\right|_{y=0} = \frac{8\pi\alpha^2}{9m^3}\,\bar{x}S(\bar{x})\left[\nu W_2^{\ell P}(x) + \nu W_2^{\ell n}(x) - \frac{4}{3}\,\bar{x}S(\bar{x})\right]$$

The authors then find

$$\bar{x}S(\bar{x}) = 0.6(1 - x)^{10}, \qquad 0.2 < x < 0.5 .$$

Much flatter distributions have been obtained in neutrino scattering, but at much smaller $q^2 \sim 20\ \mathrm{GeV}^2$, and they are not in contradiction with the present results.

There is another kinematic region where further simplification occurs[21]. At small τ we have $x_i \approx X$, $\bar{x}_i \approx 0$; then:

$$\frac{d^2\sigma}{dmdX} = \frac{8\pi\alpha^2}{9m^3}\,\frac{1}{X}\,\bar{x}S(\bar{x})\,2\nu W_2^{ep}(X) .$$

The only dimuon data existing for such a comparison correspond to low masses[10], and an aover-all discrepancy of a factor of 3, alread discussed, is apparent. Here the ISR result could shed some light, as it corresponds to small τ; unfortunately, no x-dependence exists as yet from this experiment. Still, we can try the following consistency check:

The ratio of cross-sections for $\tau_1 = 0.0053$ and $\tau_2 = 0.0166$ is found to be 16:

$$\frac{d\sigma}{dm} = \frac{8\pi\alpha^2}{9m^3}\,\bar{x}S(\bar{x})\int dX\,\frac{2\nu W_2^{ep}(X)}{X} .$$

Using the sea-quark distribution previously quoted, we arrive at

$$\frac{d\sigma/dm(\tau_1)}{d\sigma/dm(\tau_2)} = \left(\frac{m_2}{m_1}\right)^3\left(\frac{1 - \bar{x}_1}{1 - \bar{x}_2}\right)^{10} = 16 .$$

Now it is interesting to notice that for small τ,

$$\Delta\tau = x_i\Delta\bar{x}_i + \bar{x}_i\Delta x_i \approx x_i\Delta\bar{x}_i \ ,$$

the variation of τ mainly reflects the variation in sea-quark content. Let us then assume that X varies little in the range explored in τ:

$$\tau \approx \langle x_i \rangle \, \bar{x}_i \ .$$

Then the relation between cross-sections give an average x_i value for the valence quark,

$$\langle x_i \rangle \approx 0.12 \ .$$

This was just a small exercise, but it is interesting to note that because of $\Delta\tau \approx x_i\Delta\bar{x}_i$ at small τ, the function $F(\tau)$, experimentally measured, offers a direct picture of the sea quark fall-off, and any new point at smaller τ is, for that reason, quite important:

$$F(\tau) \propto \bar{x}S(\bar{x}) \quad \text{for small } \tau \ .$$

A very important and unique contribution of dilepton production concerns the quark distribution in pions. This is the only way of getting such distributions, and new data[13] in $\pi p \to \mu^+\mu^- X$ make the method possible. The result (Fig. 15) shows that the valence-quark distribution in the π is substantially flatter than that of the proton.

6. PROBLEMS OF THE BASIC MODEL

Qualitative predictions of the Drell-Yan model agree very well with the data. It is a little surprising, considering that hadrons are supposed to be made of quarks, antiquarks, and gluons, to witness that the hadron-hadron interaction takes place almost totally through quark-antiquark annihilation. In QCD other interactions may occur; their diagrams are shown in Fig. 16. In first order they all give the same scaling law $m^3(d\sigma/dm) = F(\tau)$ [22] and their relative effect must be searched for elsewhere.

In the basic model the dimuon system gets no transverse momentum p_T, as the quarks are supposed to have no intrinsic p_T in the hadron. It seems obvious that the constituents carry some p_T because of their confinement in a region of small transverse size. One then expects an intrinsic p_T of the order of 300-500 MeV/c independent of the beam energy.

Experimental results[10,13] are shown in Fig. 17; p_T first increases with mass then levels off, but this plateau varies with beam

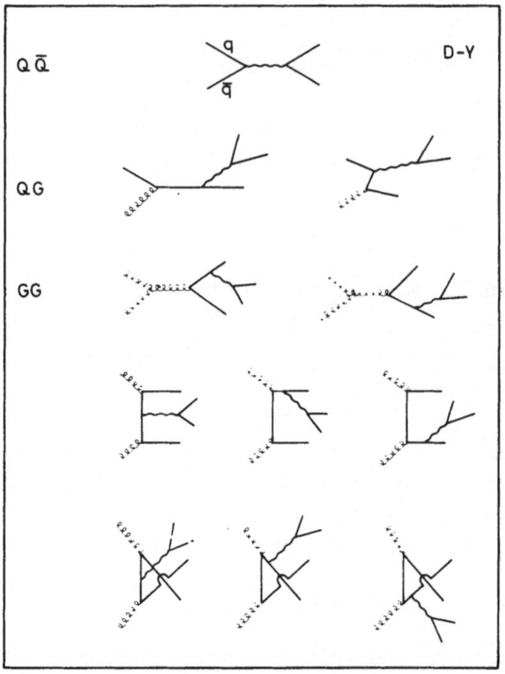

Figure 16. QCD constituent scattering diagrams for dimuon production.

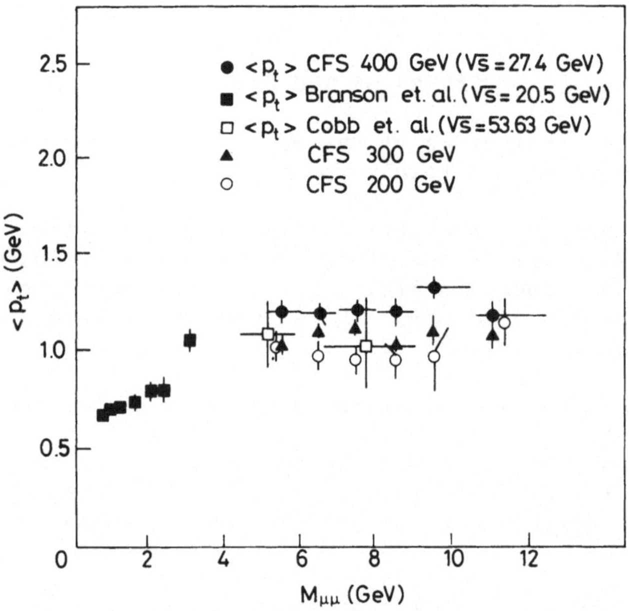

Figure 17. $<p_T>$ distribution as a function of the dimuon mass.

energy, in clear contradiction with the naïve prediction. Hard-scattering bremsstrahlung diagrams shown in Fig. 16 can give rise to transverse momentum for the dilepton even without assuming an intrinsic p_T to the constituents: the p_T of the intermediate photon is balanced by a jet of hadrons. Large p_T pairs seem to have an origin that is different from the pure annihilation diagram, and the detailed study of this new kinematical region will be necessary to better understand the dynamics of the quarks inside the hadrons.

Many theoretical papers[23] have appeared on this subject of surprisingly high p_T. Present judgement is that the p_T behaviour may be interpreted in a QCD framework, taking into account the full series of constituent scattering diagrams. This should generate q^2 dependences of the quark distributions and consequently a violation of scalling. The present results are not accurate enough to test

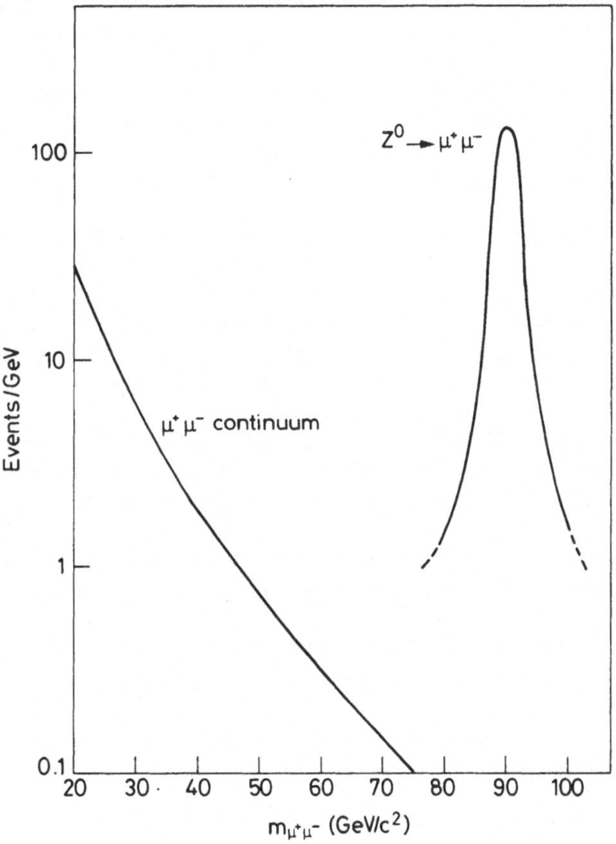

Figure 18. Expectations for dimuon production in 1000 hours of running at the $p\bar{p}$ collider. The Z^0 signal is shown as predicted for a mass of 90 GeV/c^2 and a width of 2 GeV/c^2.

scaling to better than 20%, and nobody would be surprised by such violation, considering the scaling violation in deep inelastic scattering which already forces us to include q^2-dependent structure functions. These two scaling violations are not independent, and if we insert the q^2 dependent structure functions measured in deep inelastic scattering into the dilepton function, for the same q^2, then the effects of higher-order graphs are automatically included[19], and the connection between the two phenomena is preserved. This obviously means that theorists give a recipe for going from a space-like photon to a time-like photon.

Despite these difficulties, dilepton production gives a new support to the idea of hadron consituents carrying the quantum number of the quarks. It also stresses the difference between valence quarks and sea quarks.

The future of dilepton production points, as usual, towards better data, especially at higher energies, but also towards a better understanding of a QCD formulation. This new formulation must retain the striking successes of the Drell-Yan model but also explain the discrepancies that we underlined. In the last two years the sensitivity of dilepton production has been improved by several orders of magnitude, and, if the trend continues, all questions still pending may find an answer very quickly.

In particular, within the next two years CERN will have a $p\bar{p}$ collider with a total c.m. energy of 540 GeV [24]. This represents a jump in $\sqrt{s}$ of almost a factor of 10 with respect to the ISR. It will be interesting to see if the scaling formula still holds. The result of this exercise is shown in Fig. 18: it shows the number of events detected in 1000 hours of running time in an apparatus which covers a 4π solid angle. I have taken into account a reduction in luminosity with respect to the ISR (it is expected to be 10^{30} cm^{-2} s^{-1} at the $p\bar{p}$ collider) and also the advantage of using $\bar{p}$ which offer their valence antiquarks[25]. The expected rate is dramatically low. Fortunately the intermediate vector boson Z^0 is expected to be produced with a cross-section 3-4 orders of magnitude above the continuum, thus giving semi-confortable rates, as is also seen in Fig. 18.

Will this curve be what experimentalists find at the $p\bar{p}$ collider? In less than 3 years we will know if this prediction comes true.

I wish to thank Professors Ting and Fries for the opportunity of preparing this study and Professor Jacob for reading the manuscript.

REFERENCES

1) J. Christenson et al., Phys Rev. Letters 25, 1523 (1970).

2) J.J. Aubert et al., Phys. Rev. Letters 33, 1404 (1974).

3) S.N. Herb et al., Phys. Rev. Letters 39, 252 (1977).

4) D. Antreasyan et al., Measurement of high-mass muon pairs at very high energies, 19th Internat. Conf. on High-Energy Physics, Tokyo, 1978.

5) Workshop on Future ISR Physics (ed. M. Jacob) ISR workshop 12-8, CERN, 1977.

6) S.D. Drell and T.M. Yan, Phys. Rev. Letters 25, 316 (1970).

7) T.M. Yan *in* Workshop on Future ISR Physics (ed. M. Jacob) ISR workshop 12-8, CERN, 1977.

8) D.C. Hom et al., Phys. Rev. Letters 37, 1274 (1976).

9) D. Antreasyan et al., Phys. Rev. Letters 37, 1451 (1976).

10) J.G. Branson et al., Phys. Rev. Letters 38, 457 and 1334 (1977).

11) F. Renard, Proceedings, Ecole d'été de Gif, Sept. 1977.

12) M.J. Corden et al., Phys. Letters 76B, 226 (1978).

13) K.J. Anderson et al., Hadronic production of high-mass muon pairs, 19th Internat. Conf. on High-Energy Physics, Tokyo, 1978.

14) J.K. Yoh et al., Phys. Rev. Letters 41, 684 (1978).

15) L.M. Lederman and B.G. Pope, Phys. Letters 66B, 486 (1977).

16) K. Kinoshita, H. Satz and D. Schildknecht, BI-TP 77/74 (1977).

17) C. Kourkoumelis et al., Electron pair production at the ISR, 19th Internat. Conf. on High-Energy Physics, Tokyo, 1978.

18) R.D. Field and R.P. Feynman, Phys. Rev. D 15, 2590 (1977).

19) E.L. Berger, report ANL-HEP-PR 78-12 (1978).

20) D.M. Kaplan et al., Phys. Rev. Letters 40, 435 (1978).

21) F.T. Dao et al., Phys. Rev. Letters 39, 1388 (1977).

22) C.T. Sachradja, preprint CERN TH-2416 (1977).

23) R.C. Hwa, report RL 78-044, and references therein.

24) C. Rubbia, P. McIntyre and D. Cline, Proc. Internat. Neutrino Conf., Aachen, 1976, p. 683.

25) R.F. Peierls et al., Phys. Rev. Letters 16, 1397 (1977).

PARTON MODELS

O. Nachtmann

Institut für Theoretische Physik der Universität HD
Heidelberg, Fed. Rep. of Germany
Max-Planck-Institut für Physik und Astrophysik,
München, Fed. Rep. of Germany

1. THE PARTON CONCEPT

In these lectures we will describe the concept and application of the naive parton model for hadrons invented in the late sixties [1]. This model has been extremely successful in comparison with experiment and we will mention some of its successes later on.

The idea to think of large objects as composites of smaller objects is hardly new in physics and usually has met with success. In the parton model large objects, hadrons, are again described as bound states of smaller objects, partons, In the naive version of the model partons are taken as pointlike in the same sense as electrons or muons are pointlike. This assumption alone does not, however, lead us very far. To give the model predicitive power a second assumption has to be introduced. We will assume that in certain "hard" processes partons can be treated as quasi free and use the impulse approximation. What is meant by a "hard" process can best be explained by giving a list of "hard" reactions. This list certainly should contain the following reactions

(i) Electron positron annihilation into hadrons (fig. 1)

$$e^+ + e^- \rightarrow X$$

with s, the center of mass energy squared tending to infinity.

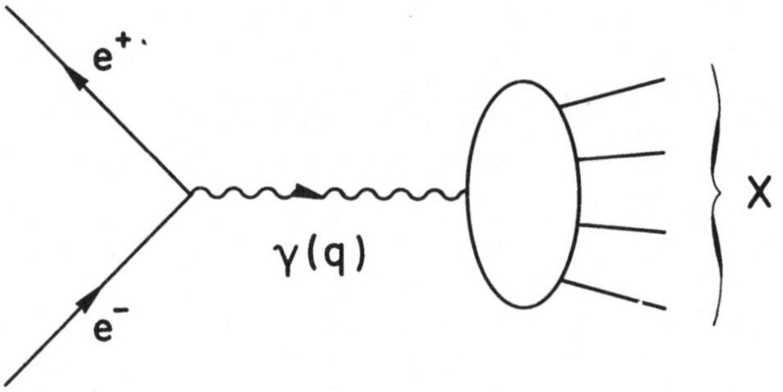

Fig. 1 e^+e^--Annihilation into hadrons.

(ii) Deep inelastic electron nucleon scattering (fig. 2)

$$e^+ + N \rightarrow e + X$$

with

$$\begin{aligned} Q^2 &= -q^2 \rightarrow \infty \\ \nu &= pq/M \rightarrow \infty \\ x &= Q^2/2M\nu \text{ fixed} \end{aligned} \tag{1.1}$$

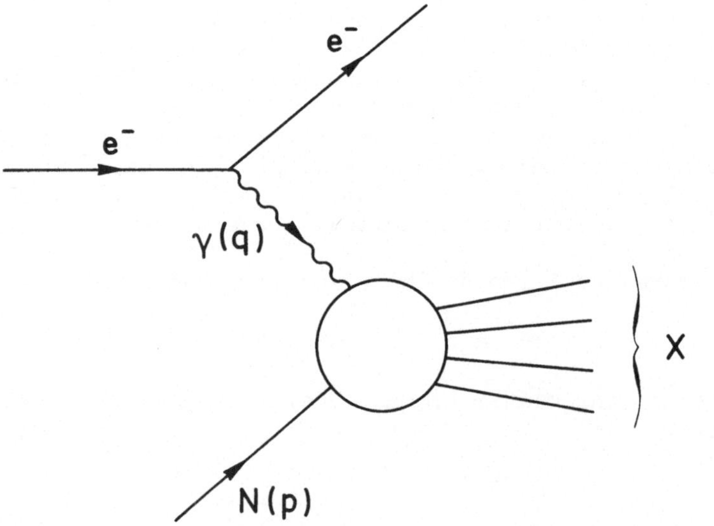

Fig. 2 Deep inelastic electron nucleon scattering.

(iii) The Drell Yan process [2] (fig. 3)

$$N_a + N_b \rightarrow \mu^+ + \mu^- + X$$

with

$$s = (p_1 + p_2)^2 \rightarrow \infty$$

$$m^2 = (k_1 + k_2)^2 \rightarrow \infty \qquad (1.2)$$

$$\tau = m^2/s \qquad \text{fixed.}$$

The naive model has been extremely successful in the description of these and other processes. The experimental analysis in this framework revealed the nature of charged and weakly coupled partons as we will review in chapter 3. Very soon after its invention, however, it became clear that the naive model also had its limits: it was shown not to be compatible with a realistic quantum field theory. The situation changed dramatically with the discovery of asymptotic freedom and the introduction of quantum chromodynamics [3]. In this theory the naive model survives as a zero order approximation. An extension of the naive parton model, developed by Kogut and Susskind [4], even allows to understand all the subtleties of renormalization and gliding coupling constant in the simple intuitive terms of larger objects built out of smaller ones.

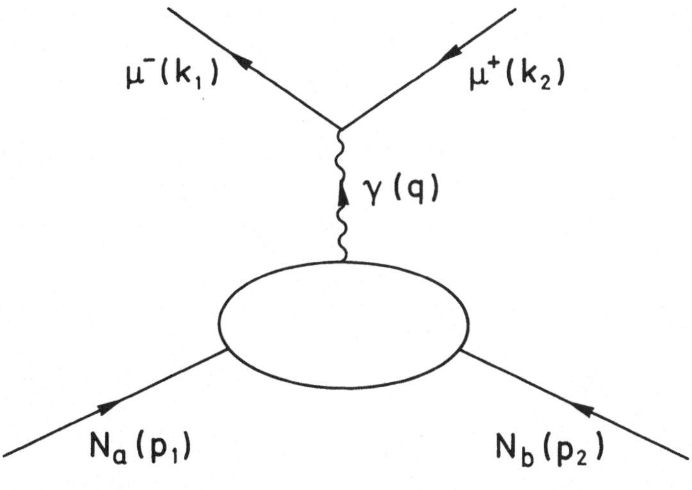

Fig. 3 Muon pair creation in nucleon nucleon collisions

2. THE NAIVE PARTON MODEL AND SOME OF ITS APPLICATIONS

In the parton model hadrons are considered as bound states of constituents whose internal motion can be described by a wave function. A Schrödinger type wave function is, however, not a relativistically invariant quantity. This can be considered a defect but actually turns out to be a blessing in disguise. We may hope that in special reference frames the form of the wave function simplifies. The general belief is that this happens in a reference frame where the nucleon moves very fast.

Without going into the more or less plausible justifications (ref. 1),5)) we will now state the dogmas of the parton model.

(i) A fast moving hadron looks like a jet of partons all moving more or less in the same direction as the parent hadron. The partons share the 3-momentum of the parent hadron.

(ii) For "hard" processes we can use the impulse approximation. The prescription for calculating reaction rates is the following: calculate the reaction rate for the basic free parton process and sum incoherently over all partons contributing.

To see how this works consider three simple processes. A technical hint. In the course of the calculations we find it very convenient to use parton and hadron states normalized to one in a finite normalization box of volume V.

2.1 Electron Positron Annihilation into Hadrons

The parton model is applied to this reaction for c.m. energy squared $s \to \infty$. The basic free parton process is the creation of a parton antiparton pair by a virtual photon (fig. 4). This process

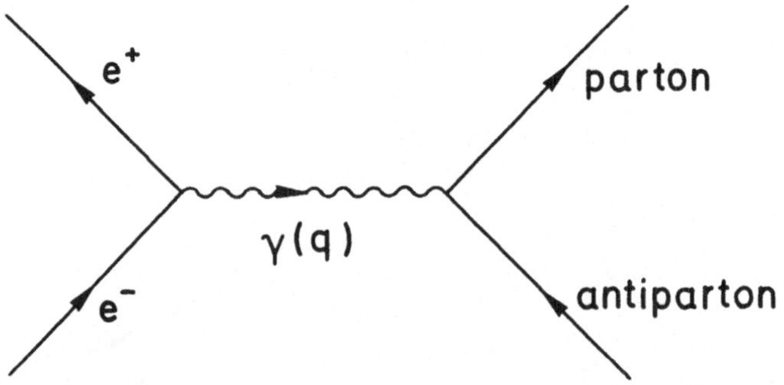

Fig. 4 The creation of a parton antiparton pair in e^+e^--annihilation

is completely analogous to μ-pair creation in e^+e^--annihilation (substitute muon for parton in fig. 4). For $s \to \infty$ we can neglect all masses. We find then very simply for a parton of spin 1/2 and charge e_i

$$\left.\frac{\sigma(e^+e^- \to \text{parton} + \text{antiparton})}{\sigma(e^+e^- \to \mu^+\mu^-)}\right|_{s\to\infty} = e_i^2 \qquad (2.1)$$

Applying the dogma (ii) of the parton model we get the cross section for $e^+e^- \to$ hadrons as incoherent sum of all parton contributions

$$\left.\frac{\sigma(e^+e^- \to \text{hadrons})}{\sigma(e^+e^- \to \mu^+\mu^-)}\right|_{s\to\infty} = \sum_i e_i^2 \qquad (2.2)$$

If we also consider spin 0 partons we find the more general result

$$\left.\frac{\sigma(e^+e^- \to \text{hadrons})}{\sigma(e^+e^- \to \mu^+\mu^-)}\right|_{s\to\infty} = \sum_{\text{spin } \frac{1}{2}} e_i^2 + \frac{1}{4} \sum_{\text{spin } 0} e_i^2 \qquad (2.3)$$

2.2 Deep Inelastic Electron Nucleon Scattering

Deep inelastic electron nucleon scattering (fig. 2) in the Bjorken limit (eq. (1.1) was the first testing ground for the parton model.

The essential part of the diagram in fig. 2 is the absorption of the virtual photon by the nucleon, the emission of the photon by the electron is simple QED. What experimentalists measure can, therefore, be regarded as the total absorption rate of a virtual photon on a nucleon.

For a nucleon at rest this absorption rate is given by the famous tensor

$$W_{\rho\lambda}(p,q) = \left(-g_{\rho\lambda} + \frac{q_\rho q_\lambda}{q^2}\right) W_1(\nu,Q^2)$$

$$+ \frac{1}{M^2}\left(p_\rho - \frac{(pq)q\rho}{q^2}\right)\left(p_\lambda - \frac{(pq)q\lambda}{q^2}\right) W_2(\nu,Q^2) \qquad (2.4)$$

To apply the parton model we have to consider the corresponding free parton process shown in fig. 5. The absorption rate for a fast moving parton of spin 1/2, charge e_i and 4-momentum p_i is easily found to be

$$\Gamma^{(i)}_{\rho\lambda}(p_i,q) = \frac{e_i^2}{(p_o)_i} \delta(2p_i q + q^2)$$

$$\{(-g_{\rho\lambda} + \frac{q_\rho q_\lambda}{q^2})(-\frac{q^2}{2}) \tag{2.5}$$

$$+ (p_{i\rho} - \frac{(p_i q)q_\rho}{q^2})(p_{i\lambda} - \frac{(p_i q)q_\lambda}{q^2})2\}$$

According to the dogma of the parton model we have to sum incoherently over all partons found in the nucleon. Denoting by $N_i(\xi)d\xi$ the number of partons of species i found in the nucleon with momentum fraction between ξ and $\xi + d\xi$ (transverse momenta are neglected) we find easily for the absorption rate on a fast moving nucleon

$$\frac{M}{p_o} W_{\rho\lambda}(p,q) = \sum_i \int_o^1 d\xi\, N_i(\xi)\, \Gamma^{(i)}_{\rho\lambda}(p_i,q)\Big|_{\vec{p}_i=\xi\vec{p}} \tag{2.6}$$

The Lorentz factor M/p_o in front of $W_{\rho\lambda}$ transforms the absorption rate for a nucleon at rest, $W_{\rho\lambda}$, into the absorption rate for a nucleon in flight. It is a consequence of the time dilation effect.

The final step is to evaluate this integral inserting in the

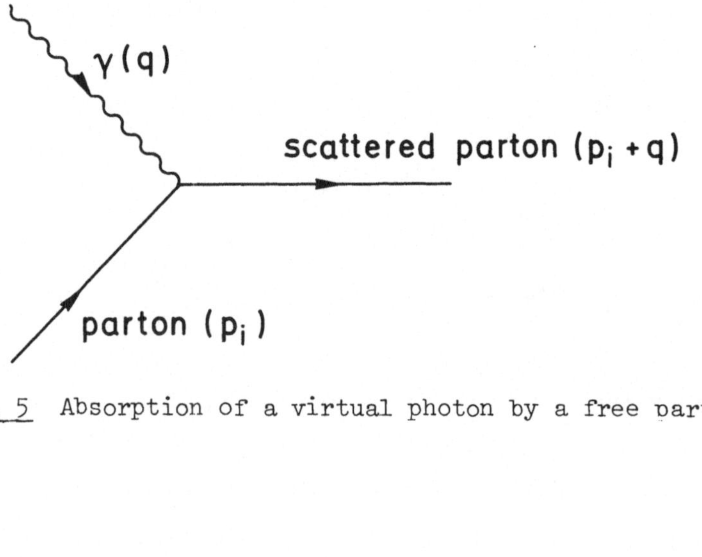

Fig. 5 Absorption of a virtual photon by a free parton.

spirit of a high energy approximation for the parton energy, neglecting masses $(p_o)_i \simeq \xi p_o$. This leads to the celebrated results

$$\begin{aligned} 2M\, W_1(\nu,Q^2) &= \sum_i e_i^2 N_i(x) \equiv F_1(x) \\ \nu\, W_2(\nu,Q^2) &= x \sum_i e_i^2 N_i(x) \equiv F_2(x) \\ x &= Q^2/2M\nu \end{aligned} \tag{2.7}$$

The naive parton model predicts Bjorken scaling for the structure functions. Our assumption of spin 1/2 charged partons moreover led to the relation

$$\frac{F_2(x) - x\,F_1(x)}{xF_1(x)} = 0 \tag{2.8}$$

Had we used spin 0 charged partons we would find

$$\frac{F_2(x) - x\,F_1(x)}{x\,F_1(x)} = \infty \tag{2.9}$$

Eqs. (2.8) and (2.9) relating the spin of partons to measurable quantities are due to Callan and Gross [6].

2.3 Muon Pair Creation in Hadron Hadron Collisions

This reaction is shown in fig. 3 for the collision of two nucleons N_a and N_b. The basic kinematic variables are defined in eq. (1.2). The basic parton process is

$$\text{parton} + \text{antiparton} \rightarrow \mu^+ + \mu^- \tag{2.10}$$

and the parton picture of the whole process is shown in fig. 6.

To calculate the basic parton process (eq. (2.10)) we assume again partons of spin 1/2, charge e_i and negligible mass.

We find then easily for the differential reaction rate of the parton process for two fast moving partons

$$\frac{d\Gamma^{(i)}}{dm^2}(\vec{p}_1^{\,i},\vec{p}_2^{\,i}) = \frac{1}{v(p_{1,o})^i(p_{2,o})^i}$$

$$\frac{2\pi\,\alpha^2}{3}\,e_i^2\;\delta((p_1^i+p_2^i)^2-m^2) \tag{2.11}$$

The reaction rate for the hadron hadron collision

$$N_a(p_1) + N_b(p_2) \quad\to\quad \mu^+ + \mu^- + X \tag{2.12}$$

is obtained as the incoherent sum of all parton contributions. Denoting by $N_i^a(\xi)$ and $\bar{N}_i^b(\xi)$ the distribution functions of parton spe species i in nucleon a and antiparton i in nucleon b we find for two fast moving nucleons

$$\frac{d\Gamma(\vec{p}_1,\vec{p}_2)}{dm^2} = \sum_i \int_o^1 d\xi_1 \int_o^1 d\xi_2\; N_i^a(\xi_1)$$

$$\bar{N}_i^b(\xi_2)\;\frac{d\Gamma^{(i)}}{dm^2}\,(\vec{p}_1^{\,(i)} = \xi_1\vec{p}_1,\;\vec{p}_2^{\,(i)} = \xi_2\vec{p}_2) \tag{2.13}$$

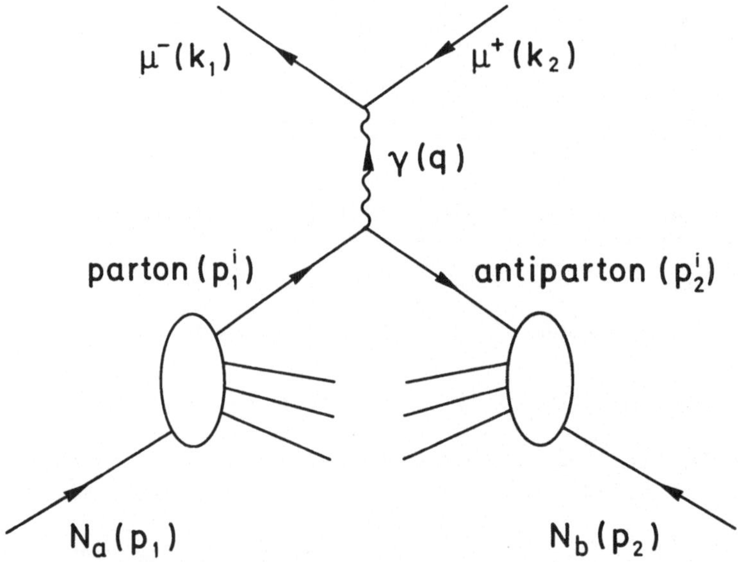

Fig. 6 Muon pair production in nucleon nucleon collisions in the parton model.

The reaction rate for two nucleons in flight is related to the cross section by some simple kinematic factors whose derivation is left as an exercise. For high energies we find

$$\frac{d\sigma}{dm^2}(N_a + N_b \to \mu^+ + \mu^- + X) = \frac{v\, p_1^o p_2^o}{(p_1 p_2)} \frac{d\Gamma}{dm^2}(\vec{p}_1, \vec{p}_2) \qquad (2.14)$$

Inserting eqs. (2.13) and (2.11) we obtain the final formula due to Drell and Yan [2].

$$\frac{d\sigma}{dm^2}(N_a + N_b \to \mu^+ + \mu^- + X)$$

$$= \frac{4\pi\alpha^2}{3} \frac{1}{m^4} \sum_i e_i^2 \int_0^1 d\xi_1 \int_0^1 d\xi_2 \qquad (2.15)$$

$$(\xi_1 N_i^a(\xi_1))\,(\xi_2 \bar{N}_i^b(\xi_2))\;\delta(\xi_1 \xi_2 - \frac{m^2}{s})$$

Of course, the antiparton could also come from nucleon a. This is taken into account if the sum over i runs over all partons and antipartons.

Eq. (2.15) is the basis of all modern analyses of our reaction.

This ends our discussion of application of the naive parton model. Many other processes have been treated in the literature. The reader who has worked through our examples should find it easy to apply the basic dogmas of the parton model in other cases.

3. THE NATURE OF PARTONS

The naive parton model had enormous success in comparison with experiment, starting with the verification of Bjorken scaling in deep inelastic electron nucleon scattering [7] and neutrino nucleon scattering [8]. The analysis of experiments in this framework furthermore led to a rather complete elucidation of the nature of partons. In the following we will retrace some crucial steps of this analysis.

3.1 The Spin of Charged Partons

Information on the spin of partons is contained in the Callan Gross relation eqs. (2.8), (2.9). The most recent experiments [9] give

$$\frac{F_2(x) - x\,F_1(x)}{x\,F_1(x)} = 0.25 \pm 0.10 \tag{3.1}$$

This number is small but not terribly close to the theoretical prediction zero for spin 1/2 partons (eq. (2.8)). It is perhaps fortunate that the experimental number used to be smaller, around 0.15, for a long time thus leading theorists to the conclusion that charged partons have spin 1/2 exclusively. The non-zero experimental result in eq. (3.1) is then blamed on non-asymptotic terms which should go away for high enough Q^2, where experiments have yet to be performed.

We may note that the result eq. (3.1) presents a problem as well for modern QCD calculations (cf. ref. 10 for a review).

In any case we will take the result eq. (3.1) from electron nucleon scattering as evidence that charged partons have spin 1/2. In charged current neutrino nucleon scattering experimentalists found [11]

$$\frac{F_2(x) - x\,F_1(x)}{x\,F_1(x)} = 0.15 \pm 0.10 \tag{3.2}$$

indicating that weakly interacting partons have spin 1/2 also.

3.2 The Flavour Quantum Numbers of Partons

In this subsection we will discuss the evidence for the charge and isospin assignment of partons, which comes mainly from a comparison of deep inelastic electron nucleon and charged current neutrino nucleon scattering. We denote by $F_2^{eN}(x)$ and $F_2^{\nu N}(x)$ the structure functions for the reactions

$$e + N \rightarrow e + X \tag{3.3}$$

$$\nu_\mu + N \rightarrow \mu^- + X \tag{3.4}$$

where N stands for an average over proton and neutron. Experimentally one finds

$$\frac{F_2^{\nu N}(x)}{F_2^{eN}(x)} \simeq 3.60 = \frac{18}{5} \tag{3.5}$$

for $x \gtrsim 0.2$ (cf. fig. 7 taken from H. Deden et al. [8]). This is a key number telling us that partons are fractionally charged quarks.

To see that the quark parton model leads to this result is easy. Take the simplest assumption for the structure of proton and neutron, i.e. that they are composed of the usual up and down quarks only.

$$\begin{aligned} p &= u\,u\,d \\ n &= d\,d\,u \end{aligned} \tag{3.6}$$

where u and d form an isodoublet with charges 2/3 and -1/3 respectively. Let us denote by $N_u(x)$ and $N_d(x)$ the distribution functions

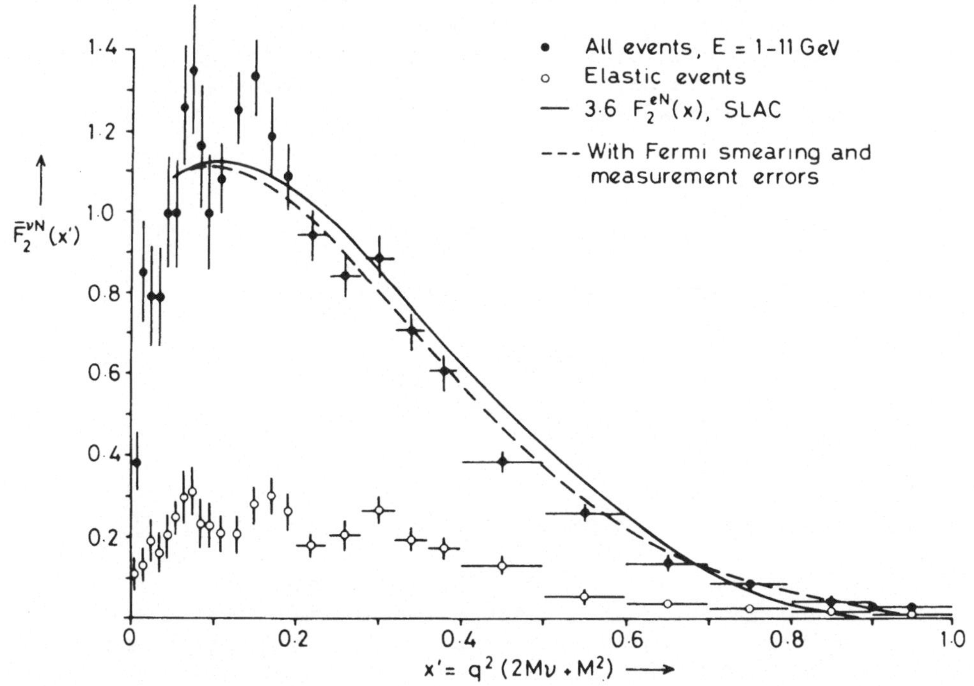

Fig. 7 The weak structure function of the nucleon compared to the electromagnetic (from H. Deden et al., ref. 8)).

of up and down quarks in the proton. The distribution functions for the neutron are obtained from those of the proton by a charge symmetry operation. An easy calculation using eq. (2.7) gives then

$$F_2^{eN}(x) = \frac{5}{18}\, x\, (N_u(x) + N_d(x)) \tag{3.7}$$

To compute the structure function for neutrino nucleon scattering (eq. (3.4)) we have to write down the charged weak current J^+ in the quark model. Neglecting the Cabibbo angle we have

$$J_\lambda^+ = \bar{u}\, \gamma_\lambda\, (1 - \gamma_5)\, d \tag{3.8}$$

which means that the basic parton weak scattering processes are

$$\begin{aligned} \nu_\mu + d &\rightarrow \mu^- + u \\ \nu_\mu + \bar{u} &\rightarrow \mu^- + \bar{d} \end{aligned} \tag{3.9}$$

Since we have neglected antiquarks in eq. (3.6) we find that in neutrino scattering (eq. (3.4)) we measure essentially the number of d-quarks in the nucleon. Since the number of d-quarks in the neutron equals the number of u-quarks in the proton we find with the conventional normalization

$$F_2^{\nu N}(x) = x\, (N_u(x) + N_d(x)) \tag{3.10}$$

This leads immediately to the theoretical prediction

$$\frac{F_2^{\nu N}(x)}{F_2^{eN}(x)} = \frac{18}{5} \tag{3.11}$$

in perfect agreement with experiment (eq. (3.5) and fig. 7).

Note that the prediction eq. (3.11) depends crucially on the fractional charges of quarks. Assume for a moment that u and d form an isodoublet with charges +1 and 0 like the proton and neutron. Then we would find

$$\frac{F_2^{\nu N}(x)}{F_2^{eN}(x)} = 2 \tag{3.12}$$

in contradiction with experiment (eq. (3.5)).

This simple argument can in fact be generalized considerably. It can be shown that allowing antipartons or higher isospin partons does not help to make integrally charged partons work (cf. ref. 5,12). The innocent looking experimental result, eq. (3.5) respectively fig. 7, leads unambiguously to the result that partons have fractional charges.

3.3 The Colour Quantum Number of Partons

Experiment tells us about another quantum number of partons, colour. Assume that the world consists of u, d, and s-quarks only and compute the cross section for e^+e^--annihilation into hadrons according to eq. (2.2). We find

$$\left.\frac{\sigma(e^+e^- \to \text{hadrons})}{\sigma(e^+e^- \to \mu^+\mu^-)}\right|_{s\to\infty} = \left(\frac{2}{3}\right)^2 + \left(-\frac{1}{3}\right)^2 + \left(-\frac{1}{3}\right)^2 = \frac{2}{3} \tag{3.13}$$

The experimental number is, however, close to 2. To get this right we can assume that quarks come in three varieties, labelled by a colour, e.g. red, green, blue. Then the result is

$$\left.\frac{\sigma(e^+e^- \to \text{hadrons})}{\sigma(e^+e^-) \to \mu^+\mu^-)}\right|_{s\to\infty} = \frac{2}{3}\cdot 3 = 2 \tag{3.14}$$

in good agreement with experiment. This tripling of quarks does not, by the way, change any results for deep inelastic scattering like eq. (3.11).

Another independent argument for colour comes from $\pi^0 \to 2\gamma$ decay. Without colour for quarks the lifetime of the π^0 would be nine times too long. Colour just supplies a factor 9 in this case and brings experiment and theory in agreement [13].

A third process where colour effects can be measured is the Drell Yan process. We return to eq. (2.15). If $N_1^a(\xi)$ etc. denote the quark distribution functions measured in deep inelastic scattering then colour introduces a suppression factor of 3 in eq.

(2.15). This comes about because a quark can only annihilate with an antiquark of the same colour. Unfortunately the antiquark distribution in the nucleon is not known well enough in order to test the presence of the colour factor in the Drell Yan process. Muon pair creation in proton antiproton collisions should. however, settle the question since the antiquark distribution in the antiproton equals the well-known quark distribution in the proton.

To summarize: colour is needed for explaining e^+e^--annihilation, π^0 decay and not in conflict with other experiments. On the contrary, outside of the parton model framework it is also very welcome in baryon spectroscopy, where it saves the Pauli principle for quarks.

3.4 Are There Other Partons Than Quarks?

To answer this question we consider the momentum balance of the constituents of a nucleon. If quarks are the only constituents of the nucleon we should find:

$$\int_0^1 dx \sum_{\text{quarks}} (N_q(x) + \bar{N}_q(x)) = 1 \qquad (3.15)$$

Experimentally one finds that quarks and antiquarks account only for about 50 percent of the nucleon momentum (cf. e.g. ref. 5)). This leaves half of the nucleon's momentum to be carried by partons which do not couple to electromagnetic and weak probes. Theorists invented the name gluons for these partons and made them also responsible for holding the quarks in the nucleon together. This binding must even be such that the constituent quarks do not come out, it must be confining, since no one ever saw a fractionally charged particle in the laboratory. On the other hand, the success of the naive parton model tells us that in "hard" processes the constituents act as quasi free. How nature manages to arrange things in this way is still a big puzzle.

To conclude this section we can say that experiments told us a lot about partons. It revealed to us the building blocks of hadrons/ fractionally charged quarks of spin 1/2 and some neutral objects, gluons. The natural theory of these building blocks turns out to be just the celebrated Quantum Chromodynamics.

REFERENCES

1. R.P. Feynman: Phys.Rev.Lett. 23, 1415 (1969). Photon-Hadron-Interactions (W.A. Benjamin, New York, 1972);
J.D. Bjorken and E.A. Paschos: Phys.Rev. 185, 1975 (1969);
S.D. Drell and T.M. Yan: Ann.Phys. 66, 578 (1971) and references cited therein;
P.V. Landshoff, J.C. Polkinghorne and R.D. Short: Nucl.Phys. B28, 225 (1971);
P.V. Landshoff and J.C. Polkinghorne: Nucl.Phys. B28, 240 (1971)

2. S.D. Drell and T.M. Yan: Phys.Rev.Lett. 25, 316 (1970)

3. G. t'Hooft: Remarks at a conference in Marseille (1972), unpublished;
D.J. Gross and F. Wilczek: Phys.Rev.Lett. 30, 1343 (1973);
H.D. Politzer: Phys.Rev.Lett. 30, 1346 (1973);
Several important advantages of QCD were pointed out by H. Fritzsch, M. Gell-Mann and H. Leutwyler: Phys.Lett. 47B, 365 (1973)

4. J. Kogut and L. Susskind: Phys.Rev. D9, 697, 3391 (1974)

5. O. Nachtmann: Interactions of Neutrinos with Hadrons. In: TEPP, M.K. Gaillard and M. Nicolic (eds.), (Institut National de Physique Nucléaire et de Physique des particules, Paris 1977)

6. C.G. Callan Jr. and D.J. Gross: Phys.Rev.Lett. 22, 156 (1969)

7. G. Miller et al.: Phys.Rev. D5, 528 (1972);
A. Bodek et al.: Phys.Rev.Lett. 30, 1o87 (1973)

8. T. Eichten et al.: Phys.Lett. 46B, 274 (1973);
H. Deden et al.: Nucl.Phys. B85, 269 (1975)

9. L.N. Hand: Elastic and Inelastic Electron and Muon Scattering. In: Proc. 1977 International Symposium on Lepton and Photon Interactions at High Energies, F. Gutbrod (ed.), DESY (1977)

10. O. Nachtmann: Deep Inelastic Lepton Scattering. In: Proc. 1977 International Symposium on Lepton and Photon Interactions at High Energies, F. Gutbrod (ed.), DESY (1977)

11. P.C. Bosetti et al.: Oxford University report 16/78 (1978)

12. H. Kühnelt and O. Nachtmann: Nucl.Phys. B88, 41 (1975)

13. S.L. Adler: Perturbation Theory Anomalies. In: Lectures on Elementary Particles and Quantum Field Theory (S. Deser, M. Grisaru and H. Pendleton (eds.), MIT Press, 1970)

THE LAST HEAVY LEPTON AND THE NEXT ONE

T.F. Walsh

Deutsches Elektronen-Synchrotron, DESY

Notkestr. 85, 2000 Hamburg 52

1. LEPTONIC WEAK INTERACTIONS

The present leptonic $SU_2 \times U_1$ weak interaction theory was formulated in 1967 (Ref. 1), ten years after the familiar V-A theory (Ref. 2). A decade later still, the low energy limit of this theory is now the successor to the V-A theory (Ref. 3). A brief explanation of why this is so (and why I will not discuss models) is in order. Most of us believe that charged current interactions are mediated by exchange of a massive charged vector boson (Ref. 1,2). In the low energy limit the V-A theory with eμ universality results. ((1) shows

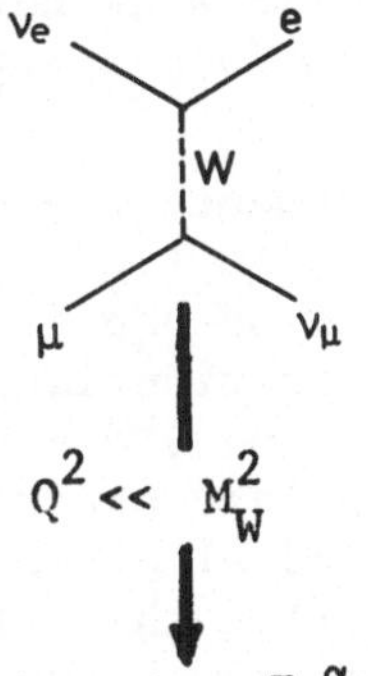

$$\frac{G_F}{\sqrt{2}} \bar{\mu} \gamma_\alpha (1+\gamma_5)\nu_\mu \; \bar{e}\gamma^\alpha (1+\gamma_5)\nu_e \qquad (1)$$

μ-decay in this theory; other processes are described similarly.) Neutrinos interact in a chirally pure way; they are left-handed (Ref. 2). Only left-handed electrons enter, too. Now we also know the low energy behaviour of neutral current processes (Refs. 3, 4,5). ((2) shows $\nu_\mu e \to \nu_\mu e$)

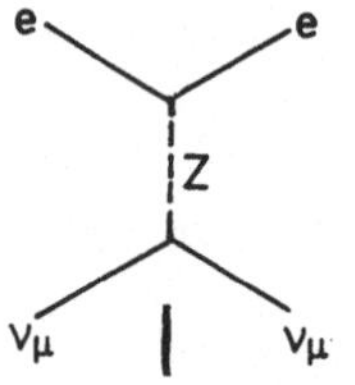

(2)

$$Q^2 << M_Z^2$$

$$\frac{G_F}{\sqrt{2}} \bar{\nu}_\mu \gamma_\alpha (1+\gamma_5) \nu_\mu \; \bar{e}(\gamma^\alpha g_V + \gamma^5 \gamma^\alpha g_A) e$$

$$g_V = -\frac{1}{2} + 2 \sin^2\theta_W ; \quad g_A = -1/2$$

The neutrino is still left handed, but now the electron is not. The factor -1/2 in (2) corresponds to the left-handed piece of the interactions; it is an SU_2 Clebsch-Gordon coefficient, the 3rd component of a "weak isospin". The extra vector interaction arises because the massive Z^0 and the massless photon are orthogonal mixtures of the 3rd component of an SU_2 triplet and a U_1 boson. Experimentally it happens the (Ref. 3)

$$\sin^2\theta_W \approx \frac{1}{4} \qquad (3)$$

so that the electron interaction is axial, and left and right handed components enter equally. (This is not so for quarks, because the extra vector interaction is proportional to the fermion charge.)

Recent neutral current experiments and their theoretical analysis convincingly support (2). In fig. 1,I show a plot of g_V versus g_A arrived at from analyses of $\nu_\mu e \to \nu_\mu e$ scattering, $\bar{\nu}_\mu e \to \bar{\nu}_\mu e$, $\bar{\nu}_e e \to \bar{\nu}_e e$ (which has contributions from both charged and neutral current interactions) and $e^- N \to e^- N$ for longitudinally polarized e^- (Ref. 6). In this last experiment a cross section different for right and left handed electrons is evidence for a weak interaction contribution. (Its theoretical analysis needs Z^0-quark couplings, which are also now known. The experimentally allowed region is stippled.)

The low energy charged and neutral current leptonic weak interaction is now essentially established. (Of course there is still room for unconventional effects at the 10-30 % level in some experiments.) Besides this, there is now a new heavy lepton τ which appears to behave exactly like e and μ except for its

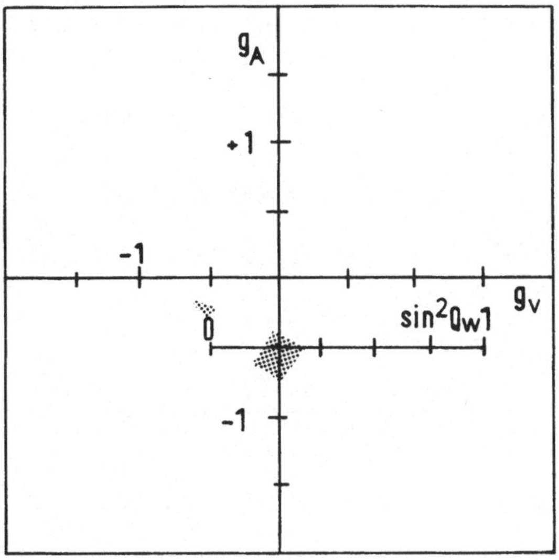

Fig. 1

large mass (Ref. 7). So eμ universality is replaced by eμτ universality. Models are not currently of much interest or relevance.

2. HADRONIC DECAYS OF

Hadronic decays of τ provide us access to the matrix element (Ref. 8)

$$< \text{HADRONS} \; |J_\alpha^{HAD}| 0 > \qquad (4)$$

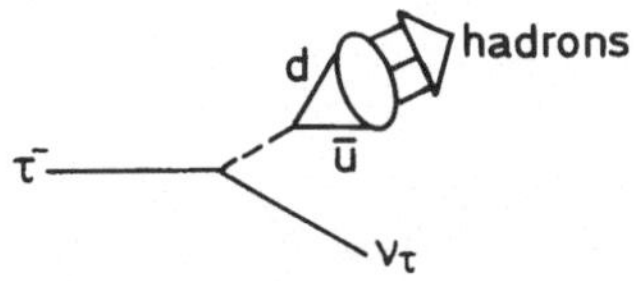

Since the quark couplings are known, this actually gives us access to the strong interaction through

$$\bar{u}d \rightarrow \text{HADRONS} \qquad (5)$$

for virtual confined quarks. Interesting issues include

(i) CVC

This is just the statement that the hadronic current behaves just as the vector quark current $\bar{u}\,\gamma_\alpha d$ (for zero Cabibbo angle). This and $e\mu\tau$ universality give $\tau^- \to \nu_\tau \rho^-$ from $\rho^0 \to e^+e^-$

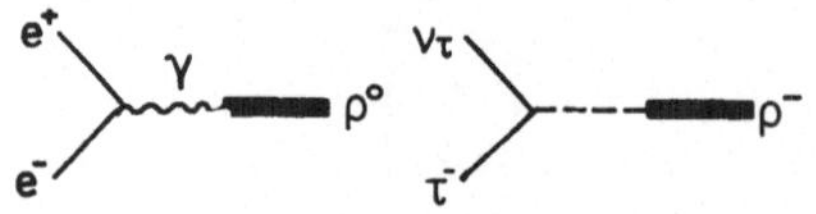

(6)

$$\frac{\Gamma(\tau \to \nu_\tau \rho^-)}{\Gamma(\tau \to \nu_\tau e\nu)} = \begin{cases} 1.2 \text{ THEORY (Ref. 8)} \\ 1.3 \pm .5 \text{ DASP (Ref. 7)} \end{cases}$$

For the 4π decay (Ref. 10),

$$\frac{\Gamma(\tau^- \to \nu_\tau (4\pi)^-)}{\Gamma(\tau^- \to \nu_\tau e\nu)} \simeq 0.56 \tag{7}$$

from $e^+e^- \to 4\pi$ data.

For the <u>axial</u> current, $\pi \to e\nu_e$ and $e\mu\tau$ universality give (Ref. 8)

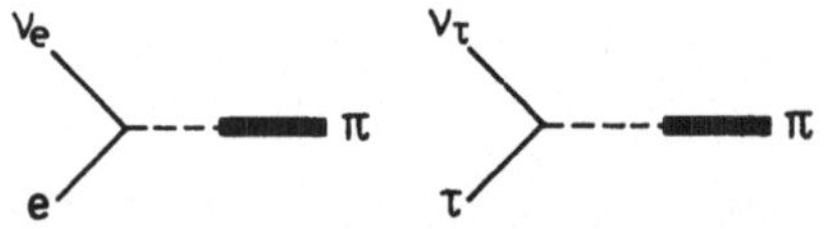

(8)

$$\frac{\Gamma(\tau \to \nu_\tau \pi)}{\Gamma(\tau \to \nu_\tau e\nu)} = \begin{cases} .55 \text{ THEORY} \\ .5\ (\pm\ 30\%) \text{ EXP. (ref. 7)} \end{cases}$$

This brings us to

(ii) The $J^{PC} = 1^{++}$ A_1 Meson and Current Algebra

The simplest theory for $\tau \to \nu_\tau \rho\pi$ is shown below

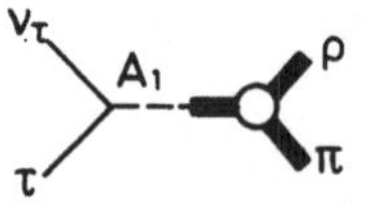

$$\tau \to \nu_\tau A_1 \searrow \rho\pi \tag{9}$$

where the coupling of the A_1 to the current is fixed by the Weinberg sum rules, and A_1 to ρ is fixed by current algebra. This simplest theory does not work. It gives an acceptable $\Gamma(\tau \to \nu_\tau A_1)$ but

$$\frac{\Gamma(\tau \to \nu_\tau A_1)}{\Gamma(\tau \to \nu_\tau e\nu)} = \begin{cases} .44 \text{ THEORY} \\ .56 \pm .2 \text{ EXPT}^{(7)} \end{cases} ; \quad \Gamma(A_1 \to \rho\pi) = 25 \text{ MeV} \tag{10}$$

much too small and A_1 width (Ref. 11).

A less simple theory (due to Geffen and Wilson (Ref. 12) deals with the matrix element

$$<\rho\pi|\, J_\alpha^{HAD}(0)\,|0> \;=\; \epsilon_\alpha F_o \;+\; \epsilon\cdot\pi \left\{ (\rho-\pi)_\alpha F_+ \;+ (\rho+\pi)_\alpha F_- \right\}$$

$$C_o + \frac{C_o^A}{(\rho+\pi)^2 - M_A^2} \;;\quad \frac{C_+^A}{(\rho+\pi)^2 - M_A^2} \;;\quad \frac{C_-^\pi}{(\rho+\pi)^2 - M_\pi^2} + \frac{C_-^A}{(\rho+\pi)^2 - M_A^2} \qquad (11)$$

(ϵ is the ρ polarization and particle momenta and labels are the same). The constants are fixed by current algebra and C_o allows for a possible low energy non resonant piece in F_o. (It seems unwise to add such constants to $F\pm$, as then the "cross section" $\sigma(\nu_e e \rightarrow \rho\pi)$ would be badly behaved above the A_1 peak.) The authors have a "no A_1" solution with C_o, a pion pole but no A_1 pole. They also have an A_1 solution (fig. 2).

Evidently, the A_1 is not yet seen in $\tau \rightarrow \nu_\tau A_1 \rightarrow \nu_\tau \rho\pi$. Once it is established, we will have a nice opportunity to test the decade old ideas of current algebra (Ref. 13).

(iii) Second Class Currents

Besides the standard weak current which transforms like the ρ, and A_1 (or π) mesons,

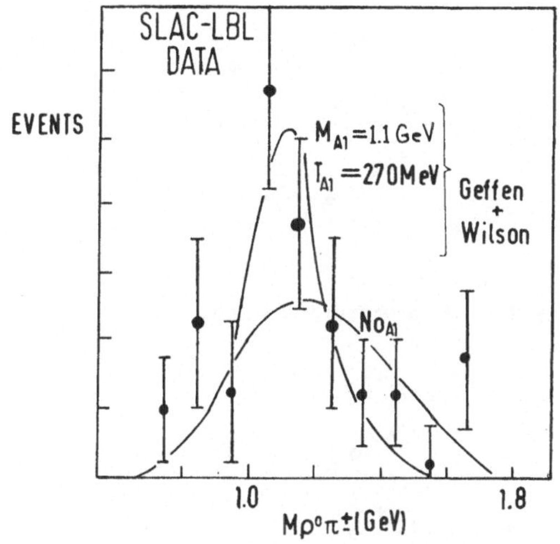

Fig. 2

current	C_n	P	$G = C_n(-)^I$	meson	
$\bar{u}\gamma_\alpha d$	-	-	+	ρ	(12)
$\bar{u}\gamma_\alpha\gamma_5 d$	+	+	-	$A_1(\pi)$	

One can imagine currents which have "wrong" G parity properties (e.g. a G = + axial piece), and which don't appear in the usual quark current,

"current"	C_n	P	G	meson	
$\sim \bar{u}(p_u-p_d)_\alpha d$	+	+	-	ρ(960)	(13)
$\sim i\bar{u}\sigma_{\alpha\beta}\dfrac{(p_u-p_d)^\beta}{m_u+m_d}d$	-	+	-	B(1250	

The former are "first class" currents and the latter "second class" (Ref. 9). It has been suggested that "second class" currents do not even exist. However, it may well be that small <u>effective</u> second class currents can appear in nature due to $m_u \neq m_d$ and the effects of quark confinement, or virtual gluon interactions of u and d not envisaged in (12) (Ref. 14).

Because of the interesting possibilities it would open up, one ought to look for processes forbidded for class currents, (Ref. 15)

$$\tau^- \to \nu_\tau \delta^-(960) \searrow \eta\pi^-,\ K^-K_s \tag{14}$$

$$\tau^- \to \nu_\tau B^-(1250) \searrow \omega\pi^-$$

and also their SU_3 rotated compatriots,

$$\tau^- \to \nu_\tau \kappa(0^{++}) \searrow K\pi \tag{15}$$

$$\tau^- \to \nu_\tau K_B(1^{+-}) \searrow K^*\pi, K\rho$$

(These are, of course, suppressed also by a Cabibbo factor $\sin^2\theta_c$.)

3. THE NEXT HEAVY LEPTON

Perhaps τ is the last heavy lepton. (Many people think so.) But it may be that there are more leptons (and quarks),

$$\begin{array}{cccc} u & d & \nu_e & e \\ c & s & \nu_\mu & \mu \\ t & b & \nu_\tau & \tau \\ g & h & \nu_\sigma & \sigma \\ \cdot & \cdot & \cdot & \cdot \\ \cdot & \cdot & \cdot & \cdot \\ \cdot & \cdot & \cdot & \cdot \end{array} \tag{16}$$

How would we recognize a next heavy lepton like σ ? The classic signatures are (Ref. 7)

$$e^+e^- \to \sigma^+ \sigma^-, \quad \sigma^- \to \mu^-(e^-)\nu\nu, \quad \sigma^+ \to e^+(\mu^+)\nu\nu \tag{17}$$

and

$$e^+e^- \to \sigma^+\sigma^-, \quad \sigma^- \to \text{anything}, \quad \sigma^+ \to \text{hard } e^+(\mu^+) + \nu\nu \tag{18}$$

There is no evidence for such a signal for $\sqrt{s} \leq 7.4$ GeV (Ref. 7).

(i) σ Decays

We have

$$\begin{array}{ll} \sigma^- \to \nu_\sigma W^- & \qquad \sigma^- \to \nu_\sigma W^- \\ \qquad\quad \downarrow & \qquad\qquad\quad \downarrow \\ \qquad \bar{\nu}_e e^- & \qquad d\bar{u} + O(\theta_c^2)s\bar{u} \\ \qquad \bar{\nu}_\mu \mu^- & \qquad s\bar{c} + O(\theta_c^2)d\bar{c} \\ \qquad \bar{\nu}_\tau \tau^- & \qquad b\bar{t} + O((\theta')^2)b\bar{c} \\ & \qquad\quad + O((\theta'')^2)b\bar{u} \end{array} \tag{19}$$

Estimates of m_b, m_t are $m_b \sim 4.5$ GeV, $m_t \gtrsim 8$ GeV (from the absence of a $pN \to \mu^+\mu^+ \ldots$ signal (Ref. 16). Since θ', $\theta'' < \theta_c$ (Ref. 17), no significant b production is expected, and no $b\bar{t}$ for $M_\sigma < 12.5$ GeV. So σ decays will be dominated by $\bar{\nu}_e e$, $\bar{\nu}_\mu \mu$, $\bar{\nu}_\tau \tau$, $d\bar{u}$, $s\bar{c}$. To estimate the branching ratio $\sigma \to \nu_\sigma e \nu$ or $\sigma \to \nu_\sigma \mu \nu$, it's necessary to calculate the V-A rate in (20)

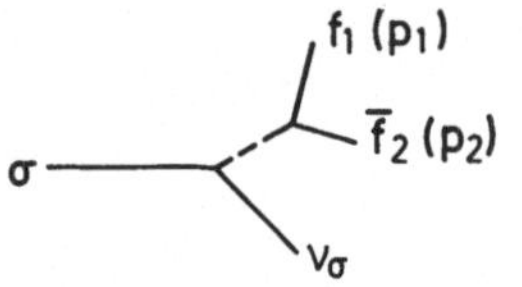

$$x = \frac{(p_1+p_2)^2}{M_\sigma^2} \tag{20}$$

The decay rate differential in x is (Ref. 18)

$$\frac{1}{\Gamma(\sigma \to \nu_\sigma e \nu)} \frac{d\Gamma}{dx} = 2u(x)(1-x)^2(1+2x) \tag{21}$$

where

$$n(\sqrt{Q^2}) \doteq \frac{\sigma(\nu_e e \to \Sigma f_1 \bar{f}_2)}{\sigma(\bar{\nu}_e e \to \bar{\nu}_\mu \mu)}$$

is the number of active fermion flavors in the charged weak current (3 per cent quark pair, 1 for a lepton pair). It is the analog of

$$R = \frac{\sigma(e^+e^- \to \Sigma f\bar{f})}{\sigma(e^+e^- \to \mu^+\mu^-)} \tag{22}$$

$n(\sqrt{Q^2})$ looks as follows

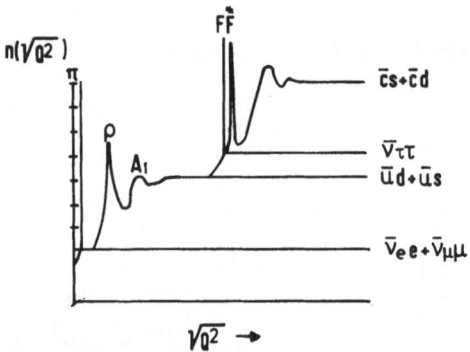

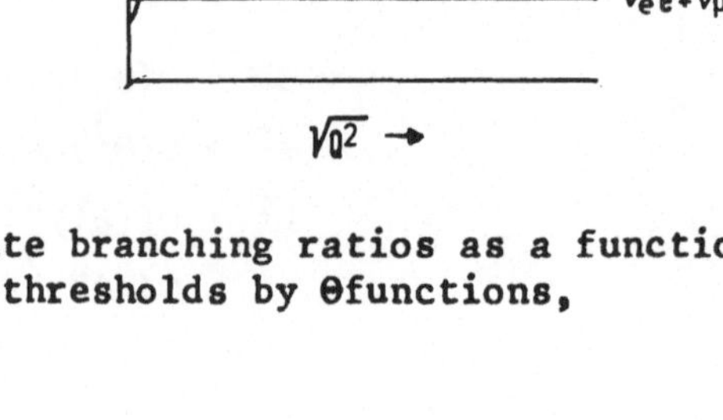

In order to estimate branching ratios as a function of σ mass, approximate these thresholds by θfunctions,

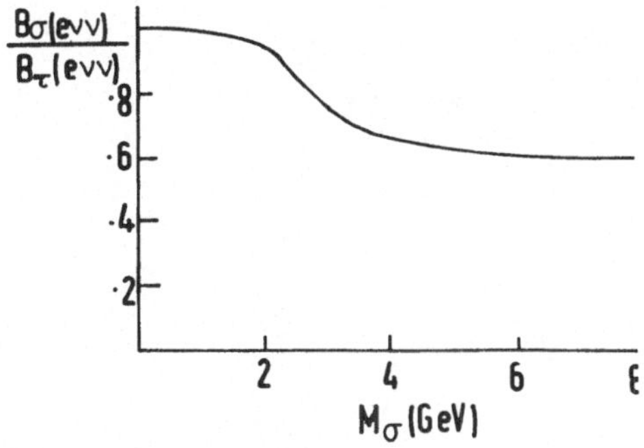

Fig. 3

$$n(\sqrt{Q^2}) \approx 5 + 4\Theta(\sqrt{Q^2} - 1.8 \text{ GeV}) \tag{23}$$

The result is fig. (3)

Since we expect that $M_\sigma > 3.5$ GeV,

$$\frac{\sigma(e^+e^- \to \sigma^+\sigma^- \to e\mu + \ldots)}{\sigma(e^+e^- \to \tau^+\tau^- \to e\mu + \ldots)} \approx \frac{1}{3} \tag{24}$$

provided $M_\sigma < M_b + M_t$

(ii) Telling $\sigma^+\sigma^-$ and $\tau^+\tau^-$ Apart

Both $e^+e^- \to \tau^+\tau^-$ and $e^+e^- \to \sigma^+\sigma^-$ give an $e\mu$ signal. In order to check for a new component in $e^+e^- \to e\mu$ + missing energy, it is simplest to plot data in a way such that it would exactly scale if there were no new $e\mu$ source. The following distributions scale exactly for τ (Ref. 18):

$$E_e(E_\mu) \text{ spectrum: } \frac{1}{\Gamma}\frac{d\Gamma}{dz}\frac{2\beta}{1+\beta} \; ; \quad z = \frac{2\beta}{1+\beta}\frac{E_e}{E_\tau}, \; z > \frac{1-\beta}{1+\beta}$$

$$\Theta_{e\mu} \text{ distribution: } \frac{1}{p}\beta^2\frac{d\Gamma}{dx} \; ; \quad x = \frac{\beta^2\gamma^2}{2}(1+\cos\Theta_{e\mu}) \tag{25}$$

Fig. (4) shows these functions for a V±A τ

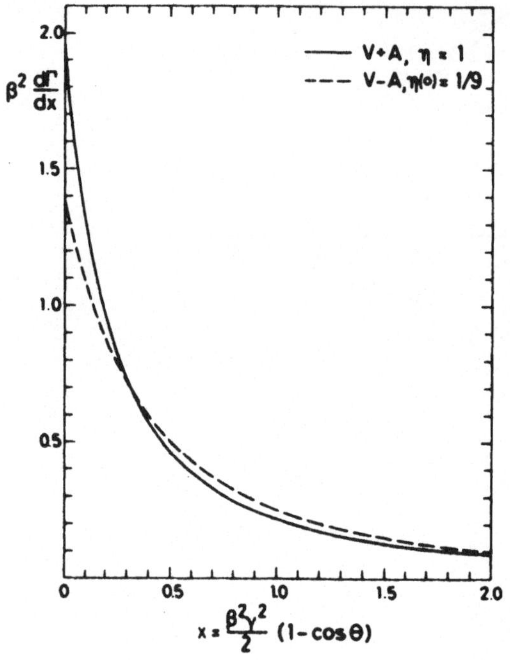
V+A, η = 1
V−A, η(o) = 1/9
β² dΓ/dx
2.0
1.5
1.0
0.5
0
0.5
1.0
1.5
2.0
x = β²γ²/2 (1−cosθ)

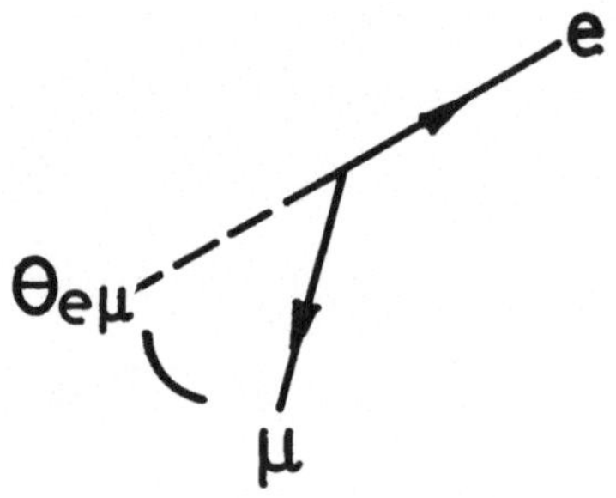
e
θeμ
μ

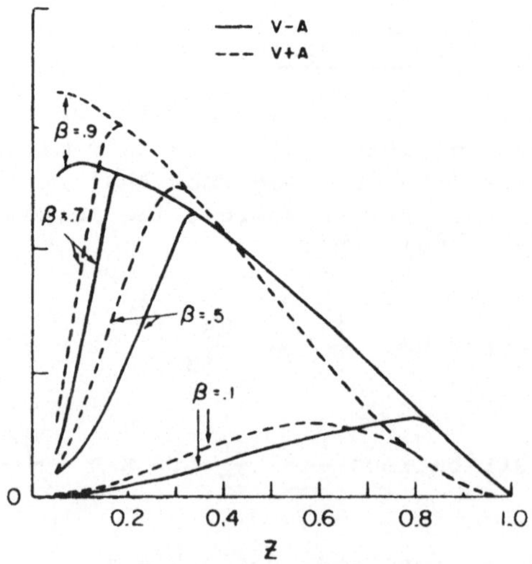
V−A
V+A
β=.9
β=.7
β=.5
β=.1
0
0.2
0.4
0.6
0.8
1.0
z

Fig. 4

On passing a new $\sigma^+\sigma^-$ threshold, dramatic violations of the scaling laws (24) will appear. Qualitatively,

$$p_e, \; p_\mu \sim O(\frac{E_\sigma}{3}) \;, \quad \theta_{e\mu} \sim O(\frac{m_\sigma}{E_\sigma}) \tag{26}$$

both much larger on average than p_e, p_μ or $\theta_{e\mu}$ for the τ .

Another potentially interesting signal for a new heavy lepton occurs when $e^+e^- \to \sigma^+\sigma^-$ followed by $\sigma^+ \to e^+\nu\nu$ or $\mu^+\nu\nu$ and $\sigma^- \to \nu_\sigma \bar{u}d$ or $\nu_\sigma \bar{c}s \to 2$ jets (Ref. 19)

$$e^+e^- \to \sigma^+\sigma^- \to \ell^\pm + 2\text{jets} + \text{missing energy} \tag{27}$$

For a V-A σ, the invariant mass distribution of the 2 jets and its mean are shown on Fig. (5)

Two well defined jets will require $\langle M_{had} \rangle > 7$ GeV or $M_\sigma > 15$ GeV.

(iii) Other backgrounds

Separating $e\mu$ events from $\tau^+\tau^-$ and $\sigma^+\sigma^-$ is clearly not a problem. It may be worth remark that other (quantum electrodynamics) backgrounds generally depend only logarithmically on energy for experimental cuts which scale with the e^+e^- beam energy. Thus if

$$p_e, \; p_\mu \gtrsim \epsilon E_B \;, \quad \theta_{e\mu} > \theta_o \tag{28}$$

Such backgrounds will be not materially worse for $\sigma^+\sigma^-$ and $\tau^+\tau^-$ provided ϵ and θ_o are the same. (e.g. at $E_B \sim 2$-3.5 GeV θ_o is 20° and ϵ typically 0.5 (Ref. 7).) Other backgrounds with a threshold behaviour, e.g. the 2γ process

$$e^+e^- \to e^+e^-\tau^+\tau^- \to e\mu + \ldots$$

have to be dealt with separately.

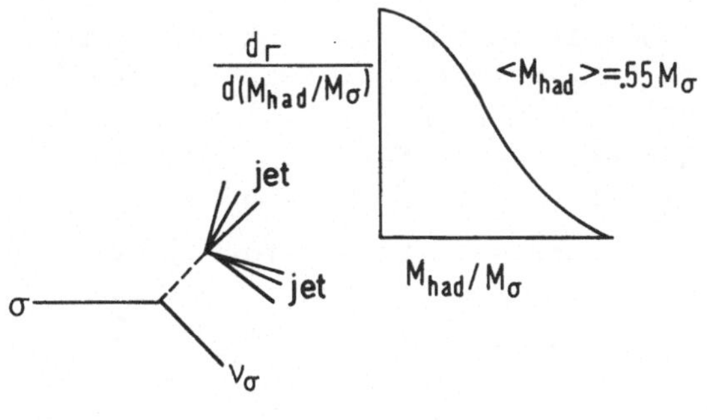

Fig. 5

4. UNBOUNDED FERMION MASS SPECTRUM

It would be very exciting to find a new heavy lepton at PETRA or PEP. Perhaps the spectrum of "elementary fermions continues up to very high mass indeed. But there are limitations. We do not expect that calculable radiative corrections to known processes which are finite for a few leptons become infinite if we allow an unbounded lepton (and quark) mass spectrum (Ref. 20). Take the fermion loop corrections to the muon anomalous moment as an example,

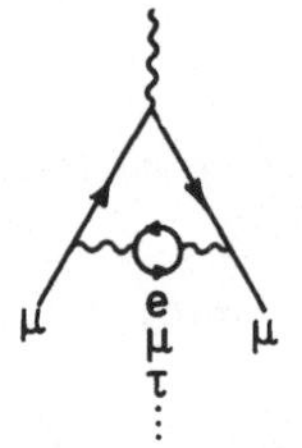

$$a_\mu^{LOOP} = m_\mu^2 \left(\frac{\alpha}{3\pi}\right)^2 \int \frac{ds}{s^2} R(s) \qquad (29)$$

Then we require that for an unbounded spectrum (Ref. 21)

$$a_\mu^{LOOP} \quad \sum_{n=1}^{\infty} \frac{1}{M_n^2} < \infty \qquad (30)$$

where M_n is the mass of the nth pointlike fermion. The simplest (but not the only way to assure $a^{loop} < \infty$ is for (Ref. 21)

$$M_n^2 = \text{const } n^\alpha \qquad \alpha > 1 \qquad (31)$$

and it is easy to show that

$$\frac{M_{n+1}}{M_n} \xrightarrow[n \to \infty]{} 1 \; ; \quad M_{n+1} - M_n \xrightarrow[n \to \infty]{} \infty \text{ if } n > 2 \qquad (32)$$

This suggests that if we want to look for fermion mass regularities we try plotting $\ell n M_n$ versus $\ell n n$. This is done in fig.(6). I plotted $\ell n M_n/M_e$ (n=1 for e, 2 for μ, ...) versus $\ell n n$, $\ell n M_n/M_u$ for charge + 2/3 quarks, and $\ell n\, M_n/M_d$ for charge −1/3 quarks. In the latter case n=1 for u or d, 2 for c or s, ...

Lepton masses are well known; for quark masses I took

$$m_u = 4 \text{ MeV} \qquad m_c = 1.2 \text{ GeV}$$
$$m_d = 6 \text{ MeV} \qquad m_s = 135 \text{ MeV} \qquad m_b = 4.5 \text{ GeV} \qquad (33)$$

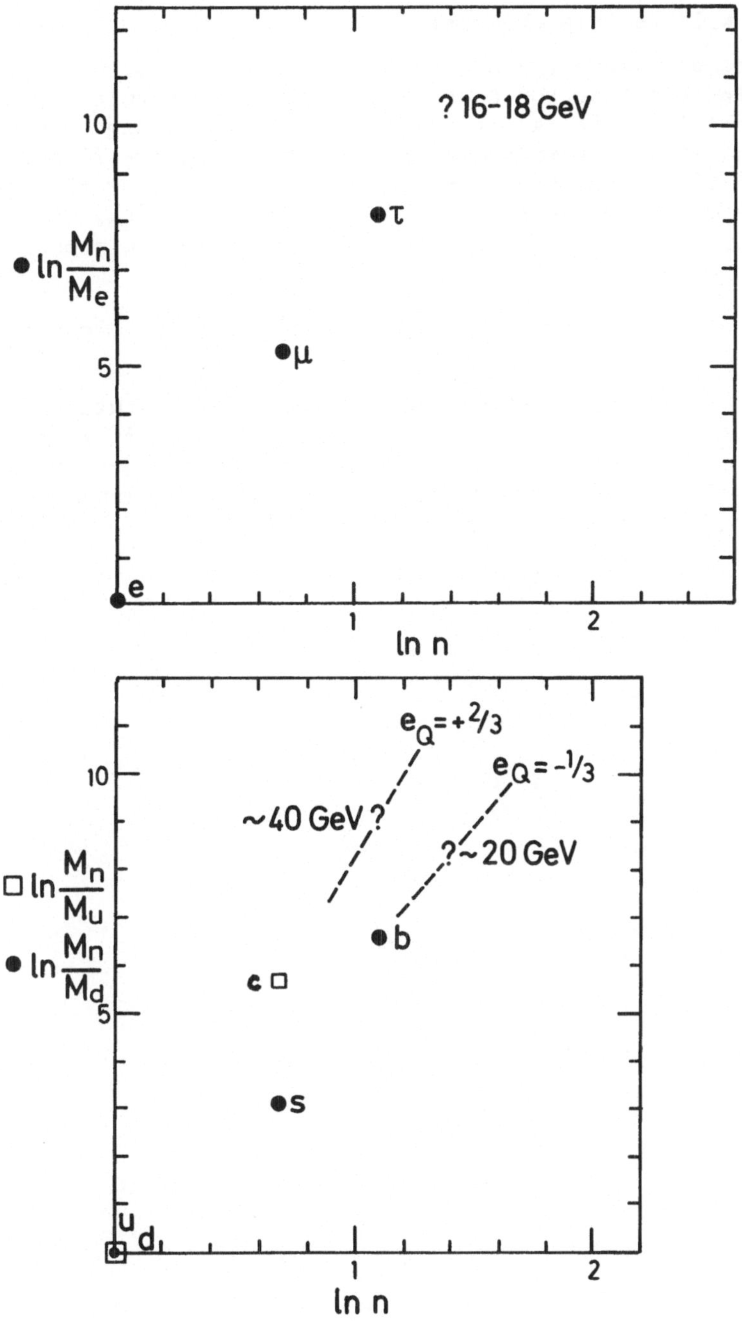
? 16-18 GeV
10
5
τ
μ
e
ln $\frac{M_n}{M_e}$
1
2
ln n
$e_Q = +2/3$
$e_Q = -1/3$
~40 GeV ?
?~20 GeV
ln $\frac{M_n}{M_u}$
ln $\frac{M_n}{M_d}$
b
c
s
u
d

Figure 6

These are, of course, more ambiguous than lepton masses. So any regularity will be less obvious.

There does seem to be a pattern to the masses, which is clearer for leptons than for quarks. This apparent regularity may well be accidental. If it is not, then there may be a new charged lepton around 16-18 GeV mass, and perhaps a new $e_Q = -1/3$ quark too. The indication from the figure is that the next $e_Q = +2/3$ quark is rather more massive than the next $e_Q = -1/3$ quark. (Perhaps it lies around ~40 GeV mass; this differs from one popular guess that

$$2m_t : M_{\Upsilon} : M_{J/\psi} \,/\, M_{\phi} \propto 27 : 9 : 3 : 1 \text{ GeV})$$

It would be surprising if any such naive guess of future fermion masses actually turned out to be correct. However it is already clear that the masses of the "elementary fermions are one of the basic puzzles of elementary particle physics. The more fermions there are, the more fascinating the puzzle.

REFERENCES

1 S. Weinberg, Phys.Rev.Lett. 19 (1967) 1264
A. Salam, Elementary Particle Physics (ed. N. Svartholm, Stockholm 1968).
S. Glashow, Nucl.Phys. 22 (1961) 579
2 See R.E. Marshak, Riazuddin and C.P. Ryan, Theory of Weak Interactions, Wiley, Interscience, New York, 1969
3 For a critical approach see J.J. Sakurai, Topical Conference on Neutrino Physics, Oxford, 3-7 July, 1978
4 L.F. Abbot and R.M. Barnett, SLAC-PUB-2136
5 E. Paschos, Brookhaven preprint.
D.P. Sidhu and P. Langacker, Brookhaven preprint.
6 The plot is from ref. (4), modified slightly. In $SU_2 \times U_1$ gauge models g_A has to change in half-units. Only the "standard" model is allowed by the figure, since $g_A \not\approx - 1/2$.
7 M. Perl, this school.
8 Y.S. Tsai, Phys.Rev. D4 (1971) 2821
H.B. Thacker and J.J. Sakurai, Phys.Lett. 36B (1971) 103.
9 S. Weinberg, Phys.Rev. 112 (1958) 1375
M.A. Bég and J. Bernstein, Phys.Rev. D5, 714 (1972)
10 F. Gilman and D. Miller, Phys.Rev. D17 (1978) 1846
N. Kaeamoto and A. Sanda, Phys.Lett. 76B (1978) 446
11 $\Gamma(A_1 \to \rho\pi)$ follows from ref. (12) if $F_A = 1/2\ F_\rho$
$f_{A\rho\pi} = g_{A\rho\pi}$, and $M_A = 2M_\rho$.
12 D. Geffen and W. Wilson, Minnesota preprint.
13 S. Weinberg, XIV Conference on High Energy Physics, Vienna, 1968
14 A. Halprin, B.W. Lee and P. Sorba, Phys.Rev. D14 (1976) 2343
15 J. Pestieu, Louvain preprint (unpublished)
16 F. Vanucci, this school
17 From J. Ellis et al., Nucl.Phys. B131 (1977) 285
Strictly speaking, θ', $\theta'' < \theta_C$ hold in a 6-quark, 6-lepton model without σ . But the limits are probably more general.
18 K. Fujikawa and N. Kawamoto, Phys.Rev. D13 (1976) 2534, S-Y Pi and A.I. Sanda, Ann.Phys. (NY) 106 (1977) 171
19 See also J. Vermaseren, Purdue preprint
20 S. Glashow, Topical Conference on Neutrino Physics, Oxford, 3-7 July, 1978.
H. Nielsen and C. Froggatt, private communication
21 The argument presented here is a heuristic one. It is, of course, hardly clear that leptons and quarks remain point-like even if they have very large mass. One can only try to motivate a search for mass regularities.
22 J. Gasser and H. Leutwyler, Nucl.Phys. B94 (1975) 269
23 Suppose there are very many fermion flavors. Then we also expect many "neutrinos" $\nu_e, \nu_\mu, \nu_\tau \ldots \nu_n \ldots$. Probably most (if not all) are massive. If their masses follow the pattern of e, μ, τ we would expect a lot of them (of order 10' or so) within the mass range up to a few GeV. They will

mix with one another and decay (as b,s do). For ν_n lifetime $\tau \lesssim 10^{-9}$ the signatures at PETRA are

$$\left.\begin{array}{l} e^+e^- \\ \\ Q\bar{Q} \end{array}\right\} \to Z^0 \to \nu_n \bar{\nu}_n$$

$$\nu_n \to \ell^+\ell^-\nu \qquad \bar{\nu}_n \to \ell^+\ell^-\nu$$

where $\ell^{\pm}$,ν stands for any charged or neutral lepton of mass $< m_{\nu_n}$. Since most of these decay too, many $e^{\pm}$,$\mu^{\pm}$ can result.

HEAVY LEPTON PHENOMENOLOGY

Martin L. Perl

Stanford Linear Accelerator Center

Stanford University, Stanford, California 94305

HEAVY LEPTON PHENOMENOLOGY

These three lectures on heavy lepton phenomenology were presented at the Advanced Summer Institute at the University of Karlsruhe in September, 1978. They are written primarily from an experimenter's point of view. Thus in the first lecture, I present an empirical definition of a charged lepton using the properties of the electron and muon. To do this, I survey the evidence that the electron and muon are elmentary particles without constituents; and I discuss some of the evidence that these particles do not partake of the strong interactions. The discussion of neutrinos is similarly empirical. Only a small amount of theory is presented in the first lecture; just enough to provide a basis for discussion of the tau lepton in the second lecture. Much more complete discussions of lepton theory and of the relations between leptons and quarks have been presented by T. Walsh at this Institute[1] and by other writers previously[2-4].

The second lecture, when presented, contained a full discussion of the history of the work on the tau and the evidence that it is a lepton. However, the history and evidence have been fully reviewed in several recent papers[5-8]. Therefore, in this written version, I only discuss properties of the tau and areas for future tau studies.

If we believe the tau has a unique neutrino and if we believe some current theories which hold that there are only six lepton types and six quark types; then there is no need for further lepton searches. However, from the experimenter's point of view again, as long as we have no evidence against the existence of a new particle, we are free to search for it.

Therefore, in the third lecture, I review limits from past searches on the existence of other types of heavy leptons; and I discuss the extensions of such searches.

LECUTRE I: WHAT IS A LEPTON?

1. Experimental Definition of a Charged Lepton

1.A. Basic Requirements

A charged particle can be classified as a lepton if it satisfies three basic requirements:

a. the particle does not interact through the strong interactions;
b. the particle has no internal structure and no constituents; and
c. its electromagnetic interactions are described by conventional quantum electrodynamic theory. In section 2 we discuss if there are additional basic requirements.

These requirements were of course developed by experimenting with the electron and muon; and hence discovering what properties separate the charged leptons from the hadrons. I will discuss in section 1.C. - 1.E. relevant experiments in three areas: the e^+e^- colliding beams reactions $e^+ + e^- \rightarrow \mu^+ + \mu^-$; the precision measurements of g_e-2 and $g_\mu-2$; and comparisons of $e + p$ and $\mu + p$ inelastic scattering.

1.B. The Spin of the Lepton

Although the spins of the e and the μ are 1/2, there is no need to restrict the empirical definition of a charged lepton to spin 1/2 particles. We can conceive of leptons with spin 0, 1, 3/2, and so forth; they need only obey the three requirements in section 1.A. This raises the question of how to distinguish a spin 1 charged lepton from the proposed charged intermediate bosons - the $W^\pm$. The only way that I can see to do this is to impose the concept of lepton conservation on all leptons. In that case a spin 1 lepton $L^\pm$ could not decay unless there existed a related smaller mass charged or neutral lepton L' such that

$$L \rightarrow L' + \text{other particles} \qquad (1)$$

The W which decays through channels such that

$$W^+ \rightarrow e^+ + \nu_e, \quad \mu^+ + \nu_\mu \tag{2}$$

does not have this lepton conservation property. I will continue this discussion of lepton conservation in section 2.

1.C. The Reactions $e^+ + e^- \rightarrow e^+ + e^-$, $\mu^+ + \mu^-$

The reactions

$$e^+ + e^- \rightarrow e^+ + e^- \tag{3a}$$

$$e^+ + e^- \rightarrow \mu^+ + \mu^- \tag{3b}$$

have been used for many years to test the correctness of the three requirements listed in section 1.A. To carry out these tests we use quantum electrodynamics, the single photon exchange diagrams of Fig. 1a - 1c, and the assumption that the leptons are point particles to calculate the barycentric differential cross sections

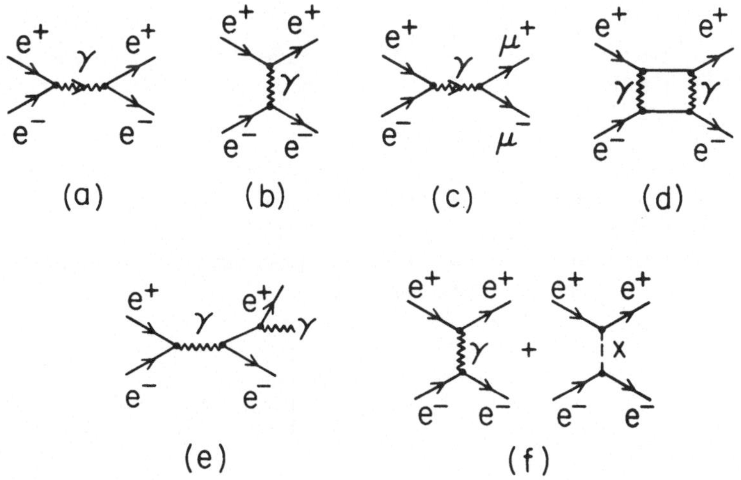

Fig. 1 Feynman diagrams for: (a) and (b) $e^+e^- \rightarrow e^+e^-$ via single photon exchange, (c) $e^+e^- \rightarrow \mu^+\mu^-$ via single photon exchange, (d) $e^+e^- \rightarrow e^+e^-$ via two photon exchange, (e) a typical radiative correction to $e^+e^- \rightarrow e^+e^-$, (f) an anomalous contribution through x exchange to $e^+e^- \rightarrow e^+e^-$.

$$d\sigma(e^+e^- \rightarrow e^+e^-, \text{theor})/d\Omega$$

and

$$d\sigma(e^+e^- \rightarrow \mu^+\mu^-, \text{theor})/d\Omega \quad .$$

We do not calculate the contributions of two photon exchange diagrams, Fig. 1d, or of weak interactions because existing measurements are not sufficiently precise to detect these contributions. We must, however, correct the data for the radiative effects coming from diagrams such as Fig. 1e. We then study the ratios of the experimental results to these theoretical cross sections; namely,

$$\rho_{ee} = \frac{d\sigma(e^+e^- \rightarrow e^+e^-, \text{exp})/d\Omega}{d\sigma(e^+e^- \rightarrow e^+e^-, \text{theor})/d\Omega} \tag{4a}$$

$$\rho_{\mu\mu} \quad \frac{d\sigma(e^+e^- \rightarrow \mu^+\mu^-, \text{exp})/ d\Omega}{d\sigma(e^+e^- \rightarrow \mu^+\mu^-, \text{theor})/d\Omega} \quad . \tag{4b}$$

No statistically significant deviations from

$$\rho_{ee} = 1 \, , \qquad \rho_{\mu\mu} = 1 \tag{5}$$

have been found; the most recent experiments[9,10] having been carried at the SPEAR e^+e^- colliding beams facility in the energy range $3.0 \leq E_{cm} \leq 5.2$ GeV.

There are various ways to interpret the limits placed by these experiments on possible deviations of the e or μ from the basic requirements in section 1.A. If we are looking for deviations from the third requirement - conventional quantum electrodynamics - then we modify the photon propagator

$$\frac{1}{q^2} \rightarrow \frac{1}{q^2} \cdot \left[\frac{1}{1 \pm q^2/\Lambda_\pm^2}\right] \quad . \tag{6}$$

Here q^2 is the square of the four-momentum carried by the virtual photon; and Λ is a parameter equal to infinity for conventional quantum electrodynamics. Hence, the smaller Λ, the greater the deviation from conventional quantum electrodynamics. The effect on the ρ ratios is, for example,

$$\rho_{ee} = \left[1 \mp q^2/\Lambda_\pm^2\right]^2 \simeq 1 \mp 2q^2/\Lambda_\pm^2 \quad . \qquad (7)$$

The approximation on the right hand side is for $\Lambda^2 << |q^2|$.

Alternatively, there could be an additional anomalous interaction between charged leptons or perhaps there is some residue of the strong interactions between charged leptons. A model for an anomalous interaction, Fig. 1f, assumes there is the exchange of a vector particle x of mass M_x and coupling constant g. Then the total amplitude for the diagram in Fig. 1f is

$$\frac{e^2}{q^2} \pm \frac{g^2}{q^2 + M_x^2} \quad . \qquad (8)$$

Then, for example,

$$\rho_{ee} = 1 \pm \left[\frac{(g/e)^2 q^2}{q^2 + M_x^2}\right]^2 \simeq 1 \pm 2\left(\frac{g}{eM_x}\right)^2 q^2 \quad . \qquad (9)$$

The right hand approximation is for $M_x^2 >> |q^2|$.

To test the basic requirement that the charged leptons have no structure, we attach a q^2 dependent form factor

$$F_{e,\mu}(q^2) = \frac{1}{1 \pm q^2/\Lambda_{e,\mu}^2} \simeq 1 \mp q^2/\Lambda_{e,\mu}^2 \qquad (10)$$

to the $ee\gamma$ or $\mu\mu\gamma$ vertices in Figs. 1a-qc. Then, for example,

$$\rho_{\mu\mu} = \left[1 \pm q^2/\Lambda_e^2\right]^{-2} \left[1 \pm q^2/\Lambda_\mu^2\right]^{-2} \simeq 1 \mp 2q^2/\Lambda_e^2 \mp 2q^2/\Lambda_\mu^2 \quad . \qquad (11)$$

We see that all these interpretations can be expressed in the approximate linear form

Table I Lower Limits on Λ at the 95% Confidence Level. (The range for Λ is between Λ_+ and Λ_-.)

	Reaction Used	Lower Limit on Λ (GeV)	Relevant Equation in Text	Reference
	$e^+e^- \to e^+e^-$	Λ > 15 to 19	6	9
	$e^+e^- \to e^+e^-$	Λ > 14 to 23	6	10
	$e^+e^- \to e^+e^-$ and $e^+e^- \to \mu^+\mu^-$	Λ > 35 to 47	6	9
	$e^+e^- \to e^+e^-$ and $e^+e^- \to \mu^+\mu^-$	Λ_e > 19 to 21 Λ_μ > 16 to 21	10	9

$$\rho = 1 + aq^2/\Lambda^2 \tag{12}$$

where a is a coefficient depending upon the type of interpretation and Λ^{-1} gives the size of the deviation. Since all measurements are consistant with $\rho = 1$, we can only deduce upper limits on Λ^{-1} and hence lower limits on Λ. Table I gives some of these limits. The significance of these limits depends upon the model used for the deviation. Thus if we use the anomalous interaction model, Eq. 8 and 9, then $(e/g)M = \Lambda$. Hence, Λ > 20 GeV means M > (g/e) 20 GeV/c^2, and either M is very massive or the coupling constant g is much less than the electromagnetic coupling constant e. If we use the form factor model, Eq. 10, then the uncertainty principle says that the upper limit on the size of possible structure in the e or μ is given by

$$r_{e,\mu,\text{upper limit}} = \hbar c/\Lambda_{\text{lower limit}} \quad . \tag{13}$$

For Λ = 20 GeV,

$$r_{\text{upper limit}} = 10^{-15} \text{ cm} \quad .$$

1.D. Precision Measurements of g_e-2 and $g_\mu-2$

The foregoing discussion concerned high energy, large q^2, moderate precision test of the basic requirements. We now turn to an example of low energy, very high precision tests of these requirements - the very beautiful measurements of the e and μ gyromagnetic ratios g_e and g_μ. We recall that g is defined by

$$\mu/s = g\,\frac{e\hbar}{2mc}$$

where μ, s, e, and m are respectively the magnetic moment, spin, charge, and mass of the particle. For Dirac particles g = 2; however, there are quantum electrodynamic corrections due to diagrams such as those in Fig. 2a - 2c. Therefore, it is conventional to write

$$a_\ell = \frac{g_\ell-2}{2} = \sum_{n=1}^{\infty} A_n^{(\ell)} \left(\frac{\alpha}{\pi}\right)^n + \sum_{n=2}^{\infty} B_n^{(\mu)} \left(\frac{\alpha}{\pi}\right)^n \quad . \qquad (14)$$

Here ℓ means e or μ; α = 1/137 is the five structure constant; and the B_n terms only apply to the μ.

Table II shows that there is excellent agreement between experiment [11-13] and theory. The theoretical values are taken from a recent review [14] of the theory. Incidently the calculations of a_μ include a hadronic contribution a_μ(hadronic) = $66.7 \pm 9.4\times10^{-9}$ from Fig. 2d, and a weak interaction contribution a_μ(weak) = $2.2\pm0.2\times10^{-9}$ from diagrams such as Fig. 2e.

Table II Comparison of Theory and Experiment for a_ℓ

Parameter	Theoretical Value × 10^9	Experimental Value × 10^9	Experimental Reference
a_e	1,159,652.4 ± 0.4	1,159,652.41 ± 0.2	12
a_e	1,159,652.4 ± 0.4	1,159,656.7 ± 3.5	11
a_μ	1,165,920.6 ± 12.9	1,165,922. ± 9.	13

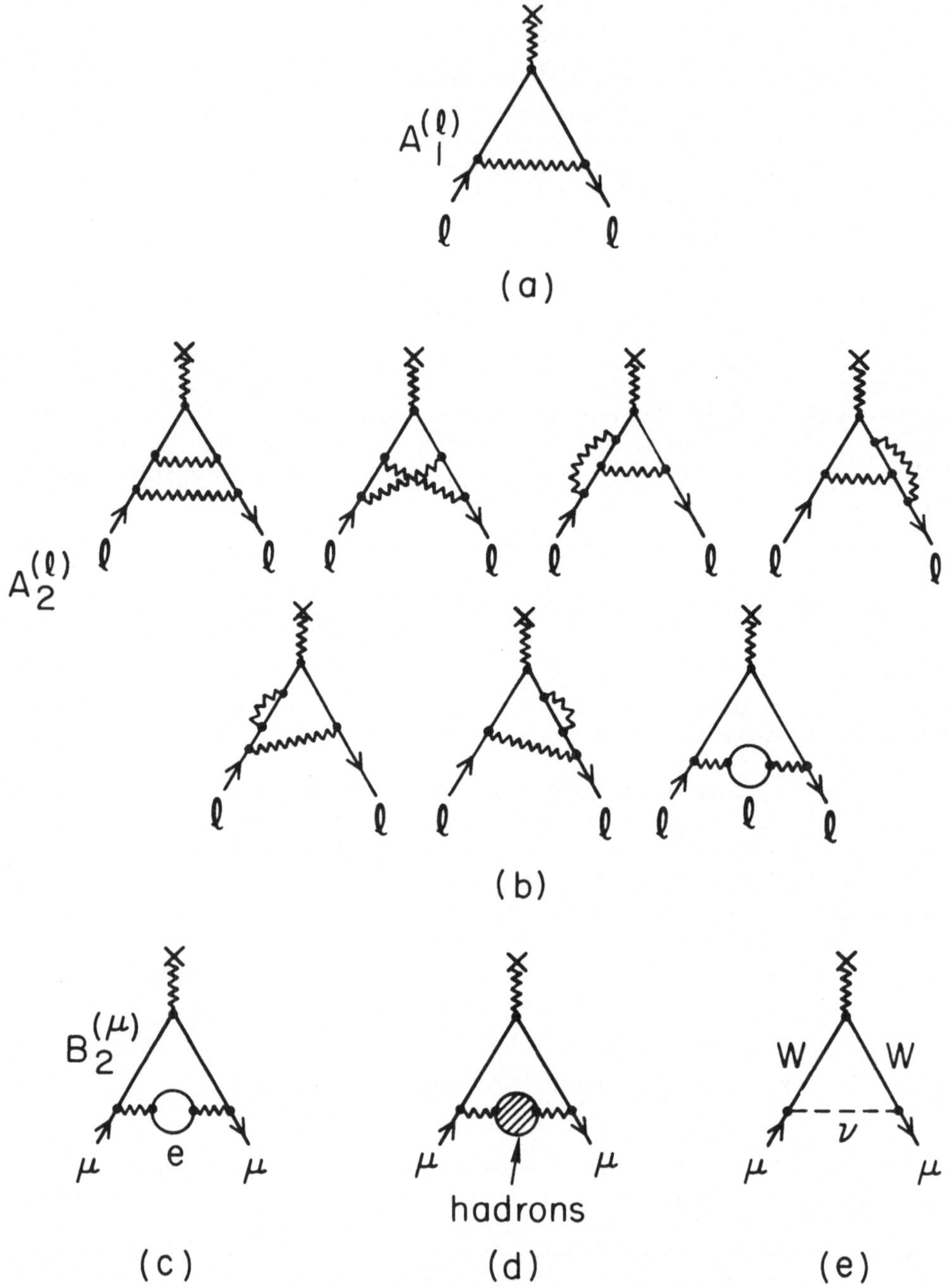

Fig. 2 Feynman diagrams for calculating g-2: (a) through (c) lowest order quantum electrodynamic diagrams for the coefficients in Eq. 14, (d) the hadronic contribution, (e) the lowest order weak interaction contribution.

1.E. Lepton Interactions in the Presence of Hadrons

We have reviewed some examples of how the basic requirements in section 1.A. are satisfied in purely electrodynamic interactions. We might speculate, however, that the basic requirements are not satisfied in the presence of hadrons. Consider three tests of such speculations.

a. Atomic Spectroscopy

The complete explanation of atomic spectra by quantum electrodynamics shows that electrons at atomic distances from nucleons have no anomalous interactions.

b. Mu-Mesic Atoms

Similarly, in mu-mesic atoms [15] there are no proven deviations from conventional quantum electrodynamics.

c. High-Energy Comparison of e+p and μ+p Reactions

Since there are no established low energy deviations, we look at high energy reactions. For example, using the inelastic reactions

$$e + p \rightarrow e + \text{hadrons} \quad , \tag{15a}$$

$$\mu + p \rightarrow \mu + \text{hadrons} \quad ; \tag{15b}$$

we measure the ratio

$$\frac{\sigma(\mu + p \rightarrow \mu + \text{hadrons})}{\sigma(e + p \rightarrow e + \text{hadrons})} = \frac{N^2}{\left[1 + |q^2|/\Lambda^2\right]^2} \tag{16}$$

to look for deviations of either the e or μ from the basic requirements. In Eq. 16, N is a measure of the relative strengths of the two reactions; and $\Lambda \neq \infty$ would mean that the reactions differ in their dependence on q^2 - the square of the four-momentum transferred to the hadrons. Unfortunately, there are no recent tests of the ratio, and older tests suffer from unknown systematic errors caused by the differences between the experimental techniques used to study Eq. 15a and those used to study Eq. 15b. An old analysis [16] found that N = 1 to ±10% and

$$\Lambda_{\text{lower limit}} \simeq 10 \text{ GeV} \quad . \tag{17}$$

Summarizing all of section 1, all experiments show that the e and μ:

a. do not interact through the strong interactions;
b. have no internal structure;
c. obey conventional quantum electrodynamics.

We take these properties as part of the basic experimental definition of a charged lepton.

2. Lepton Conservation and Leptonic Weak Interaction

The e and μ, and their associated neutrinos ν_e and ν_μ, have two other properties: (a) there is lepton conservation in all known reactions; and (b) all these particles interact through the weak interactions; we do not know if these two properties are associated in some direct way with the three basic charged lepton properties discussed in the previous section. The reason for this ignorance is that we obviously can conceive of a point charged particle which interacts only through the electrodynamic force, but not through the strong or weak force. Wouldn't we still want to call this particle a lepton? We also can conceive of a charged point particle which does not interact through the strong force, which obeys conventional quantum electrodynamics, which interacts through the weak force; but which does not have a conserved property in reactions. We might also want to classify such particles as leptons. Although, as mentioned in section 1.B., we then have a problem as to how to classify the proposed $W^\pm$ intermediate boson. There is no way at present to resolve these questions: is lepton conservation or interacting through the weak force an intrinsic property of leptons? It is from one point of view only an issue of nomenclature. This issue can be avoided by simply realizing that in searches for a new, non-hadron-like, non-photon-like, particle one need not impose the requirement that there be a conservation property analogous to lepton conservation, nor need one impose the requirement that the particle interact through the weak force.

I shall not discuss in these talks the weak interaction properties of the e, μ, ν_e, or ν_μ because there have been several recent reviews of this subject [17]; however, I shall discuss a few examples of the tests of lepton conservation, namely: the search for the decay $\mu^\pm \to e^\pm + \gamma$; μ to e conversion in nuclei; and the reaction $e^- + e^- \to \mu^- + \mu^-$.

2.A. Search for the Decay $\mu^\pm \to e^\pm + \mu$

If the decay

$$\mu^\pm \to e^\pm + \gamma \tag{18}$$

Table III Upper limits on $\Gamma(\mu \to e\nu)/\Gamma(\mu \to e\nu\nu)$ at the 90% Confidence Level

Year	Upper Limit	Reference
1964	2.2×10^{-8}	18
1977	3.6×10^{-9}	19
1977	1.1×10^{-9}	20
1978	2.0×10^{-10}	21

existed, it would be the simplest example of the violation of separate e and μ lepton number conservation. Upper limits on this decay rate, -xpressed as the ratio $\Gamma(\mu \to e\gamma)/\Gamma(\mu \to e\nu\nu)$, are given in Table III. Thus the experiments [18-21] are already extraordinarily precise; and still no violation of lepton conservation has been seen. These upper limits can be reduced in future experiments by at least a factor of 100.

2.B. μ to e Conversion in Nuclei

The reaction

$$\mu^- + \text{Nucleus} \to e^- + \text{Nucleus} \tag{19}$$

can only take place if separate e and μ lepton number conservation is violated. A recent experiment using sulfur measured the ratio

$$R_{\nu_e} = \Gamma(\mu^- S \to e^- S)/\Gamma(\mu^- S \to \nu_\mu P^*) \tag{20}$$

They found

$$R_{\mu_e} < 4 \times 10^{-10}, \text{ 90\% confidence limit} \quad . \tag{21}$$

Again, lepton conservation is extraordinarily precisely obeyed. Another type of test involving nuclei is the search for neutrinoless double beta decay of a nucleus

$$_{Z}N^{A} \rightarrow {}_{Z+2}N^{A} + e^{-} + e^{-} \quad . \qquad (22)$$

No evidence for such decays have been found [23].

2.C. The Reaction $e^- + e^- \rightarrow \mu^- + \mu^+$

The foregoing tests of lepton conservation involve relatively small energies - less than the μ-e mass difference. Perphaps violations of lepton conservation only occur at high energies or large momentum transfers? Unfortunately, it is difficult to devise precise high energy tests, because in high energy reactions there is a relatively large background of e's and μ's from the decays of hadrons. For example, it is difficult to put a relatively small upper limit on the reaction

$$\mu + p \rightarrow e + \text{hadrons} \qquad (23)$$

the high energy analogy to the reaction in Eq. 19.

A somewhat high energy test carried out about ten years ago[24] looked for the reaction

$$\sigma(e^-e^- \rightarrow \mu^-\mu^-) < 0.67 \times 10^{-32}\ \text{cm}^2 \qquad (24)$$

at the same energy. Modern techniques could substantially improve the precision of this test; however, I don't know of any plans to repeat this experiment. It could be done using the two rings of the DORIS or DCI colliding beams facilities, but those facilities have so far only been used for e^+e^- collisions.

3. Neutrinos

3.A. General Properties

Just as we used the e and μ to develop an empirical definition for charged leptons; so we can use the ν_μ and ν_e to develop an empirical definition for neutral leptons. There is no need to repeat that elaborate discussion, and I will only summarize the properties of the neutrinos:

a. no interactions through the strong force;
b. no structure or constituents;
c. obey conventional weak interactions;
d. spin 1/2;
e. obey lepton conservation.

Once again properties a and b are essential to the definition of a neutral lepton; whereas, properties c,d, and e are a matter of taste.

3.B. Neutrino Masses

A fascinating property of the neutrinos is that they may have zero mass, Table IV. We only know upper limits.

Table IV Upper Limits on Neutrino Masses at the 90% Confidence Level

Neutrino	Upper Limit On Mass	Method	Reference
ν_e	55 to 60 eV/c^2	measure p_e in ${}_1H^3 \rightarrow {}_2H^3\ e^- + \bar{\nu}_e$	25
ν_μ	0.51 to 0.65 MeV/c^2	measure p_μ in $\pi^+ \rightarrow \mu^+ + \nu_\mu$	26
ν_τ	250 MeV/c^2	measure p_μ in $\tau^- \rightarrow e^- + \bar{\nu}_e + \nu_\mu$	27

(I have broken my rule of not discussing τ physics in this part of the talk to strengthen this point.) It would be useful to substantially lower these upper limit measurements; however, it is very difficult to improve on these measurements [25,26] (except for ν_τ where experiments have just begun [27]).

There are two related questions associated with the neutrino masses for which there are no answers at present:

(1) Are the masses of the known neutrinos different from zero, or is zero mass an intrinsic property of a neutral lepton?
(2) Are there neutrinos with relatively large masses; that is, do heavy neutral lepton exist?

I will return to this subject in section 9.

4. Simple Lepton Models

The last three sections have presented a very general discussion of lepton phenomenology based on the e, μ, ν_e, and ν_μ. I will now discuss three well known models for heavy leptons based on a conservative extension of this phenomenology. I mean by the term conservative that these models retain the following properties:

a. the particles have no strong interactions or anomalous interactions;
b. the particles have no structure;
c. the particles obey conventional quantum electrodynamic theory;
d. the particles obey conventional weak interaction theory;
e. there is some form of lepton conservation;
f. the particles have spin 1/2.

4.A. The Sequential Heavy Lepton Model

In this model [16] one assumes there is a mass sequence of charged leptons, each lepton type having a separately conserved lepton number and a unique associated neutrino of smaller mass. We visualize a sequence:

Charged Lepton	Associated Neutrino	
$e^\pm$	ν_e, $\bar{\nu}_e$	
$\mu^\pm$	ν_μ, $\bar{\nu}_\mu$	(25)
$\ell^\pm$	ν_ℓ, $\bar{\nu}_\ell$	
.	.	
.	.	
.	.	

The radiative decays $\ell^\pm \rightarrow e^\pm + \gamma$ $\ell^\pm \rightarrow \mu^\pm + \gamma$ are then forbidden; and only weak decays such as

$$\ell^- \rightarrow \nu_\ell + e^- + \bar{\nu}_e \tag{26a}$$

$$\ell^- \rightarrow \nu_\ell + \mu^- + \bar{\nu}_\mu \tag{26b}$$

$$\ell^- \rightarrow \nu_\ell + \pi^- \tag{27}$$

$$\ell^- \rightarrow \nu_\ell + K^- \tag{28}$$

$$\ell^- \rightarrow \nu_\ell + \rho^- \tag{29}$$

$$\ell^- \rightarrow \nu_\ell + \text{2-or-more hadrons} \tag{30}$$

occur.

4.B. Ortholepton Model [28]

In this model the charged lepton is assigned the same lepton number as the same sign e or μ; and there is no additional neutrino. For example, we conceive of a particle $e^{*\pm}$ which may be thought of as an excited e.

Then

$$e^{*\pm} \rightarrow e^{\pm} + \gamma \tag{31}$$

will be the dominant decay mode unless we suppress the $e^* - \gamma - e$ coupling. I will continue this discussion in section 8.

4.C. Paralepton Model [28-30]

In this model we conceive of a particle E^+ or M^+ which has the lepton number of the e^- or μ^- respectively. This association of lepton number with opposite electric charge prevents radiative decays such as $E^+ \rightarrow e^+ + \gamma$. Hence, such charged leptons decay through the weak interactions. For example

$$E^- \rightarrow \bar{\nu}_e + e^- + \bar{\nu}_e \tag{32}$$

$$E^- \rightarrow \bar{\nu}_e + \mu^- + \bar{\nu}_\mu \tag{33}$$

$$E^- \rightarrow \bar{\nu}_e + \text{hadrons} \quad . \tag{34}$$

4.D. Decay Rates and Branching Ratios in Charged Sequential Heavy Lepton Decays

There have been several papers [31-34] on the theoretical decay rates (Γ) for decays such as those in Eqs. 26 through 30, and I will only summarize a few results here. In all these results conventional weak interaction theory is assumed with V-A coupling for the ℓ-ν_ℓ vertex and zero mass for the ν_ℓ.

a. $\underline{\ell^- \to \nu_\ell + e^- + \bar{\nu}_e,\ \ell^- \to \nu_\ell + \mu^- + \nu_\mu}$

This decay occurs through the diagram in Fig. 3a. Neglecting the masses of the e and μ

$$\Gamma(\ell^- \to \nu_\ell e^- \bar{\nu}_e) = \Gamma(\ell^- \to \nu_\ell \mu^- \bar{\nu}_\mu) = \frac{G^2 m_\ell^5}{192\ \pi^3} \ . \tag{35}$$

Here G is the Fermi coupling constant and m_ℓ is the heavy lepton mass.

b. $\underline{\ell^- \to \nu_\ell + \pi^-}$

Using Fig. 3b,

$$\Gamma(\ell^- \to \nu_\ell \pi^-) = \frac{G^2 f_\pi^2 \cos^2\theta_c m_\ell^3}{16\pi} \left(1 - \frac{m_\pi^2}{m_\pi^2}\right)^2 \ . \tag{36}$$

Here $f_\pi \approx 0.137\ m_p$, θ_c is the Cabibbo angle, and m_π is π mass.

c. $\underline{\ell^- \to \nu_\ell + \rho^-}$

Using Fig. 3c,

$$\Gamma(\ell^- \to \nu_\ell \rho^-) = \frac{G^2 m_\ell^3 \cos^2\theta_c m_\rho^2}{64\pi^2} \left(1 - \frac{m_\rho^2}{m_\ell^2}\right)^2 \left(1 + \frac{2m_\rho^2}{m_\ell^2}\right) \ . \tag{37}$$

Here m_ρ is the ρ mass.

d. $\underline{\ell^- \to \nu_\ell + \text{hadrons}}$

Fig. 3d illustrates a simple way to estimate the ratio of the decay rate to hadrons compared to the total decay rate. In Fig. 3d we note that given sufficient energy, the virtual W^- has equal probability to decay to an $e^-\bar{\nu}_e$ pair, or to a $\mu^-\bar{\nu}_\mu$ pair;

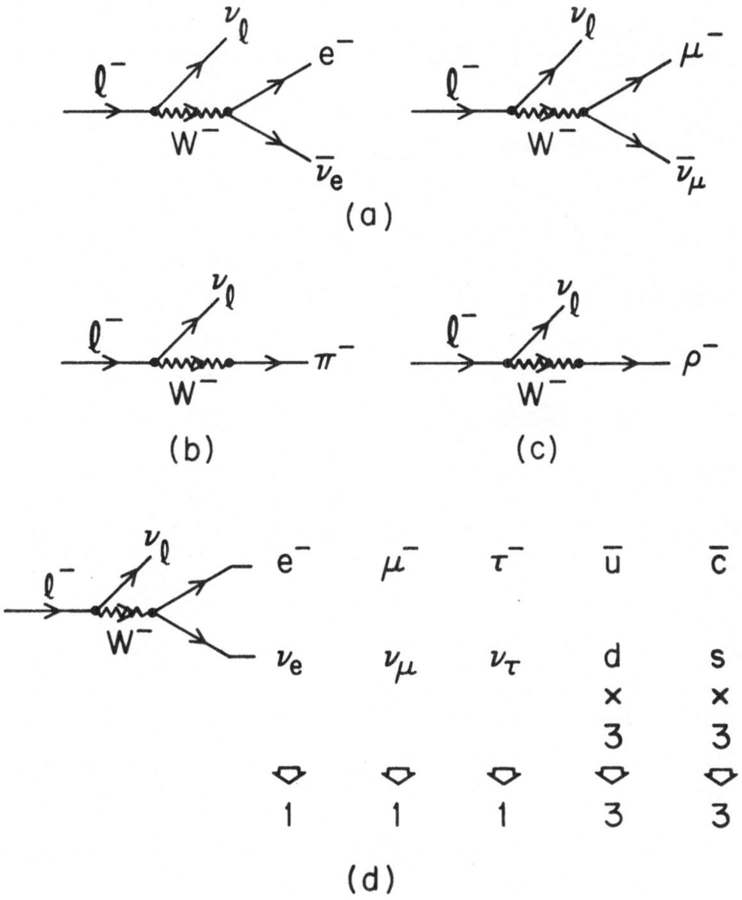

Fig. 3 Feynman diagrams for calculating some decay modes of a sequential heavy lepton ℓ with associated neutrino ν_ℓ.

or if quarks did not have color to a $\bar{u}d$ pair or $\bar{c}s$ pair. However, color makes decay to a $\bar{u}d$ pair or $\bar{c}s$ pair three times as probable. We neglect Cabbibo suppressed decay modes of the ℓ.

Hence

$$\frac{\Gamma(\ell^- \rightarrow \nu_\ell + \text{hadrons})}{\Gamma(\ell^- \rightarrow \text{all})} = \frac{3}{5}, \quad m_\ell < m_c + m_s \quad \text{and} \quad m_\ell < m_\tau$$

$$\frac{\Gamma(\ell^- \rightarrow \nu_\ell + \text{hadrons})}{\Gamma(\ell^- \rightarrow \text{all})} = \frac{6}{9}, \quad m_\ell > m_c + m_s \quad \text{and} \quad m_\ell > m_\tau \qquad (38)$$

Here m_c and m_s are the masses of the c and s quarks. (For very high mass leptons, heavier quarks such as the b quark must be included.)

e. Branching Rations and Lifetime

Figs. 4 and 5 give an overall view of the theoretical predictions for the branching ratios and liftimes as a function of the lepton mass. Fig. 5 assumes that the coupling constant has its conventional value $G = 1.02 \times 10^{-5}/m_p^2$, where m_p is the proton mass.

LECTURE II: THE τ HEAVY LEPTON

This lecture, when presented, contained a full discussion of the history of the work on the τ and the evidence that it is a lepton. Several recent papers [5-8] have reviewed these subjects; therefore, I will not present these subjects here. I only note here that:

a. there is very substantial evidence that the τ is a lepton;

b. all measured properties of the τ are consistent with it being a lepton;

c. no other hypothesis explains the measured properties of the τ.

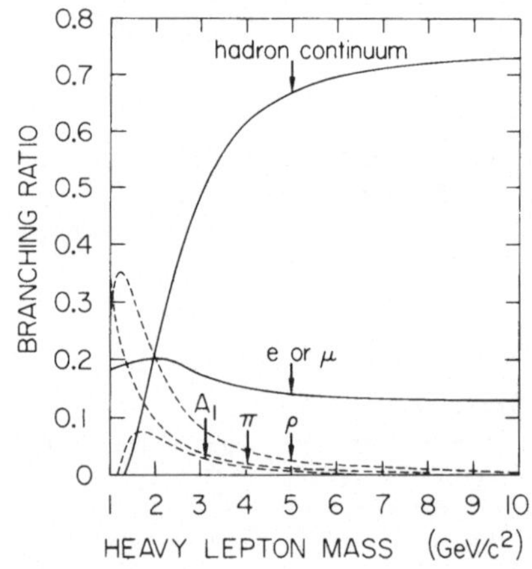

Fig. 4 Branching ratios for a sequential heavy lepton as a function of its mass.

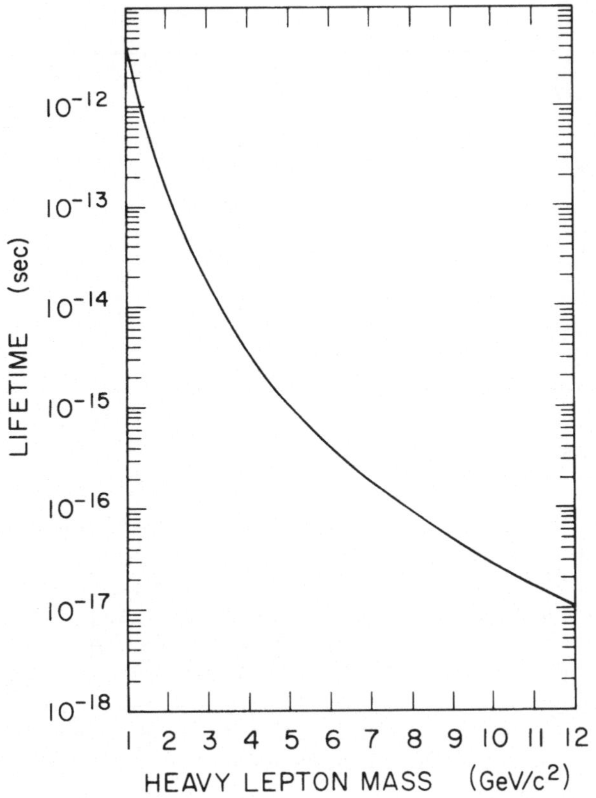

Fig. 5 The lifetime of a sequential heavy lepton as a function of its mass.

5. Properties of the τ

5.A. Mass of the τ

Table V gives three recent mass measurements. The special significance of these mass values is that they prove that the τ mass is less than the mass of the lightest charmed meson - the D. Hence, the τ and D are different particles, and there is an energy region in e^+e^- annihilation physics where one can study τ's without charmed particles being present.

5.B. Mass of the ν_τ

The smallest upper limit [27] on the ν_τ mass is 250 MeV/c^2, determined by the DELCO collaboration.

Table V The τ Mass

Mass in MeV/c^2	Group	Reference
1807 ± 20	DASP	35
1782^{+2}_{-7}	DELCO	36
1787^{+10}_{-18}	DESY-Heidelberg	37

5.C. $\tau - \nu_\tau$ Coupling

The nature of the τ-ν_τ coupling is investigated by studying the e or μ momentum spectrum from the decay mode $\tau \to e\nu\nu$ or $\tau \to \mu\nu\nu$. The basic question is whether the coupling is V-A, as it is for the e-ν_e and μ-ν_μ coupling. In that case, the ν_μ would be left-handed. There was some interest as to whether the coupling could be V+A and hence the ν_τ be right-handed. Measurements by the SLAC-LBL collaboration [38] and by the DELCO collaboration [27,39] have excluded V+A coupling, and the latter measurement finds a preliminary value for the Michel parameter

$$\rho_{exp} = 0.83 \pm 0.19 \qquad (39a)$$

Excluding radiative corrections, the theoretical values are

$$\rho_{theor} = 0.75 \text{ for V-A coupling}$$

$$\rho_{theor} = 0.375 \text{ for pure V or pure A coupling} \qquad (39b)$$

$$\rho_{theor} = 0.0 \text{ for V + A coupling} \quad .$$

5.D. The τ Lifetime

Two 95% confidence upper limits have been measured for the τ lifetime τ_τ. The SLAC-LBL collaboration finds [40]

$$\tau_\tau < 1.0 \times 10^{-11} \text{ sec} \quad . \qquad (40a)$$

The PLUTO collaboration finds [41]

$$\tau_\tau \quad < \quad 3.5 \times 10^{-12} \text{ sec} \quad . \tag{40b}$$

The theory discussed in section 4.D. predicts

$$\tau_\tau \quad \approx \quad 2.5 \times 10^{-13} \text{ sec} \tag{40c}$$

for the mass of about 1800 MeV/c^2; and assuming

$$g_{\tau W \nu_\tau} \quad = \quad g_{eW\nu_e} \quad = \quad g_{\mu W \nu_\mu} \quad = \quad g \tag{41}$$

Here $g_{\tau W \nu_\tau}$ is the weak interaction coupling constant at the τ-W-ν_τ vertex in Fig. 3. From Eqs. 40b and 40c

$$\frac{g^2_{\tau W \nu_\tau}}{g^2} \quad > \quad 0.071, \text{ 95\% confidence limit} \quad . \tag{42}$$

5.E. Type of Lepton

Muon neutrino experiments have shown [42] that the τ is not a μ-related ortholepton or μ-related paralepton. The argument goes as follows. If the τ^+ or τ^- had the lepton number of ν_μ, one of the reactions in Eq. 43 should occur in ν_μ-nucleon reactions

$$\begin{array}{l} \nu_\mu + N \rightarrow \tau^- + \text{hadrons} \\ \qquad\qquad\quad \hookrightarrow e^- + \bar{\nu}_e + \nu_\mu \end{array} \tag{43a}$$

or

$$\begin{array}{l} \nu_\mu + N \rightarrow \tau^+ + \text{hadrons} \\ \qquad\qquad\quad \hookrightarrow e^+ + \nu_e + \nu_\mu \end{array} \quad . \tag{43b}$$

Hence anomalous e^- or e^+ events should appear in ν_μ experiments with a cross section

$$\sigma(\nu_\mu \to e) = \left(\frac{g_{\tau W \nu_\mu}}{g}\right)^2 \sigma(\nu_\mu \to \text{all})\, B(\tau \to e\nu\nu) \;. \qquad (44)$$

Here $g_{\tau W\nu_\mu} = g_{\tau W\nu_\tau}$ in Eq. 41, g is defined in Eq. 41 and $B(\tau \to e\nu\nu)$ is the $\tau \to e\nu\nu$ branching ratio. However, the measured upper limits on $\sigma(\nu_\mu \to e)$ give an upper limit of $(g_{\tau W\nu_\mu}/g)^2$ smaller then the lower limit in Eq. 42. Hence the τ does not have the lepton number of the μ^+ or μ^-.

If the τ were an electron-related paralepton, then the τ^- would have the lepton number of the e^+; and in the decay

$$\tau^- \to \nu_e + e^- + \bar{\nu}_e \qquad (45)$$

there would be two identical particles. This would lead to [43]

$$\frac{\Gamma(\tau^- \to \bar{\nu}_e e^- \bar{\nu}_e)}{\Gamma(\tau^- \to \bar{\nu}_e \mu^- \bar{\nu}_\mu)} \approx 2 \qquad (46)$$

instead of the equality of decay rates predicted by the sequential heavy lepton model. We have shown [44] that the data excludes Eq. 46. Hence, the τ is not an e-related paralepton.

All existing data on the τ is consistant with the τ being a sequential heavy lepton or an e-related ortholepton. The sequential model is certainly more elegant, since the e-related ortholepton model requires that the unobserved decay mode $\tau^- \to e^- + \gamma$ be suppressed by an additional mechanism. Indeed the suppression mechanism must be sufficiently strong to explain why experimentally $\Gamma(\tau \to e\gamma)/\Gamma(\tau \to \text{all}) < 2.6\%$.

5.F. Spin of the τ

The cross section for the reaction

$$e^+ + e^- \to \tau^+ + \tau^- \qquad (47)$$

is

$$\sigma_{\tau\tau,\text{theor}} = \frac{2\pi\alpha^2 \beta(3-\beta^2)}{3E_{cm}^2} \qquad (48)$$

if the τ has spin 1/2. Here E_{cm} is the total energy and $\beta = v/c$ where v is τ velocity. We find

$$\sigma_{\tau\tau,\mathrm{measured}}/\sigma_{\tau\tau,\mathrm{theor}} = 1.00 \pm 0.15 \qquad (49)$$

where the error includes systematic errors. Hence the data is consistant with spin 1/2 for the τ. The cross section for $e^+e^- \to \tau^+\tau^-$ has also been studied carefully near the threshold $E_{cm} = 2m_\tau$ by the DELCO collaboration [27,39] and has been further analyzed by Tsai [45]. As shown in Fig. 6, the threshold behavior strongly favors a spin of 1/2. Therefore, we believe the τ spin is 1/2.

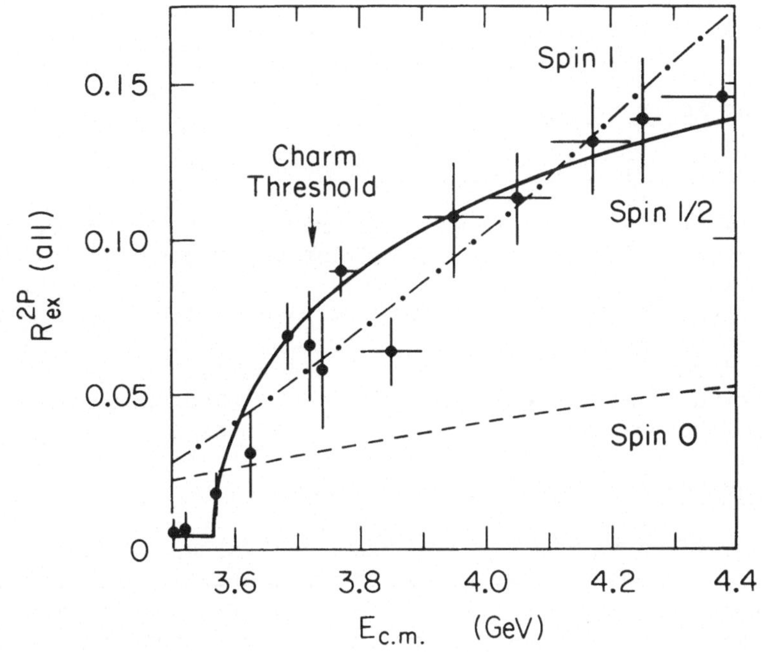

Fig. 6 Comparison of the energy behavior of the ratio $R^{2P}_{ex} = (e^+e^- \to \tau^+\tau^- \to e^\pm x^\mp)$ / $(e^+e^- \to \tau^+\tau^-)$.

Table VI Theoretical Predictions for τ Decay Branching Ratios

Modes (written for τ^-)	Branching Fraction in Percent
$\nu_\tau + e^- + \bar{\nu}_e$	16.4
$\nu_\tau + \mu^- + \bar{\nu}_\mu$	16.0
$\nu_\tau + \pi^-$	9.8
$\nu_\tau + K^-$	0.6
$\nu_\tau + \rho^-$	23.0
$\nu_\tau + K^{*-}$	1.6
$\nu_\tau + A_1^-$	9.3
ν_τ + several hadrons	23.3
Total	100.0%

6. Decay Modes of the τ

6.A. Theoretical Predictions for Branching Ratios

Table VI gives the theoretical predictions [46] for the branching fractions for the various τ decay modes assuming conventional weak interaction theory. A τ mass of 1800 MeV/c^2, a massless ν_τ, and V-A coupling is used. The (ν_τ + several hadrons) decay mode (also called the continuum decay mode) excludes the specific hadronic decay modes listed above in the table. The calculation of this continuum requires the use of $\sigma(e^+e^- \rightarrow$ hadrons) data in the $E_{cm} < 1800$ MeV region and is somewhat uncertain.

6.B. Measurements of Branching Ratios

Table VII presents averaged measured values of the branching ratios for various τ decay modes. These averages were computed by G. Feldman [5] and his paper should be consulted for the individual values, the method of averaging, and the references.

Comparison of Table VII with Table VI shows excellent agreement. This agreement is one of the reasons for accepting the iden-

Table VII Average Measured Values for Branching Ratios in τ Decay

Mode (written for τ^-)	Branching Fraction in Percent
$\nu_\tau + e^- + \bar{\nu}_e$	16.9 ± 1.9
$\nu_\tau + \mu^- + \bar{\nu}_\mu$	18.3 ± 1.9
$\nu_\tau + \pi^-$	8.3 ± 1.4
$\nu_\tau + \rho^-$	24.0 ± 9.0
ν_τ + (1 hadron)$^-$	33.8 ± 4.8
ν_τ + (3 hadrons)$^-$	31.0 ± 3.2

tification of the τ as a lepton. The ($\nu_\tau + A_1$) decay mode is discussed in Section 8.A.

6.C. Upper Limits on Lepton Conservation Violating Decay Modes

The sequential heavy lepton model requires that a neutrino, assumed to be a ν_τ, always be present in a decay mode of the τ. No violations of this requirement have been found [40]; Table VIII lists measured upper limits on branching ratios for decay modes without a neutrino.

Table VIII Upper Limit [40] on Branching Ratios for τ Decay Modes Which Do Not Have a Neutrino

Mode	Upper Limit in Percent	Confidence level	Experiment
τ → 3 charged particles	1.0	95%	PLUTO
τ → 3 charged leptons	0.6	90%	SLAC-LBL
$\tau \to e + \gamma$	2.6	90%	SLAC-LBL
$\tau \to \mu + \gamma$	1.3	90%	SLAC-LBL

7. Future Studies of the τ

There are three areas in which future research on the τ is required. One area concerns further studies of the τ decay modes (Section 7.A.) and the τ-W-ν_τ coupling constant (Section 7.B.). The second area concerns higher energy tests of the leptonic nature of the τ (Section 7.C.). The third area concerns τ production by means other than e^+e^- annihilation (Section 7.D.).

7.A. Further Studies of τ Decay Modes

It would be very useful for checking the conventionality of the weak interactions of the τ, and for other reasons, to obtain more experimental information on various decay modes:

a. $\tau^- \rightarrow \nu_\tau + \pi^-$:

This decay process has a simple Feynman diagram, Fig. 3b, that it is worthwhile to make a precise measurement. Incidently, a study of the endpoints of the π momentum spectrum will provide an independent measurement of m_τ.

b. $\tau^- \rightarrow \nu_\tau + e^- + \bar{\nu}_e,\ \nu_\tau + \mu^- + \nu_\mu$:

A study of the endpoint of the e or μ can provide a smaller upper limit on the mass of the ν_τ.

c. $\tau^- \rightarrow \nu_\tau + \rho^-,\ \nu_\tau$ + several hadrons:

Reduced statistical errors are required for the branching rations for these decay modes.

d. $\tau^- \rightarrow \nu_\nu + A_1^-$:

Two experiments [47,48] have seen invariant mass enhancements at 1100 MeV/c^2 in the $\pi^-\pi^+\pi^-$ state in the decay mode

$$\tau \rightarrow \nu_\tau + \pi^- + \pi^+ + \pi^- \quad . \qquad (50)$$

This enhancement could be the long sought A_1 meson with $J^P = 1^+$. However, much more data is required because there have been many negative searches for the A_1 in hadronic reactions.

e. $\tau^- \to \nu_\tau + K^-,\ \nu^\tau + K^{*-}$:

These two Cabibbo suppressed decay modes have not been found. Conventional weak interaction theory predicts

$$\frac{\Gamma(\tau^- \to \nu_\tau K^-)}{\Gamma(\tau^- \to \text{all})} \approx (\tan\theta_c)^2 \frac{\Gamma(\tau^- \to \nu_\tau \pi^-)}{\Gamma(\tau^- \to \text{all})} \approx \frac{0.1}{16} \approx 0.006 \qquad (51)$$

The measured upper limit [49] on this branching ratio is 0.016.

7.B. The τ-W-ν_τ Coupling Constant $g_{\tau W \nu_\tau}$:

As noted in Section 5.D., we only know that

$$g_{\tau W \nu_\tau} > 0.27\ g \qquad (52)$$

where g is the charged lepton-W-neutrino coupling constant for e's or μ's. We would like to know if $g_{\tau W \nu_\tau} = g$ so that e-μ weak inter-

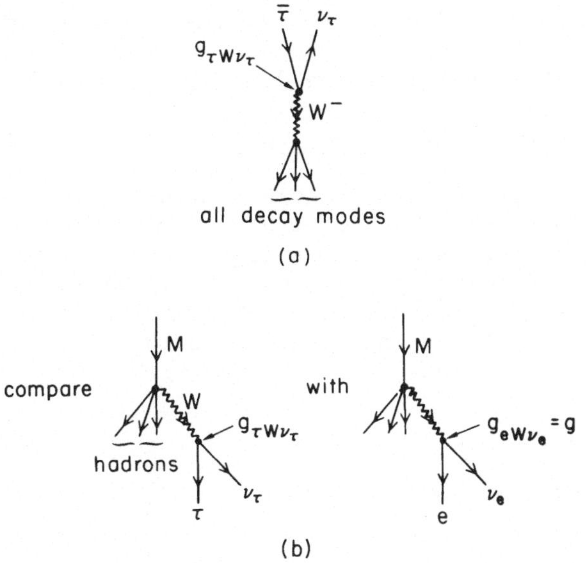

Fig. 7 Two methods for determining the $g_{\tau W \nu_\tau}$ coupling constant:
(a) by measurement of the τ lifetime,
(b) by comparing the $\tau\nu_\tau$ and $e\nu_e$ semileptonic decay rates of a heavy meson M.

action universality extends to the τ. There are two ways to determine $g_{\tau W \nu_\tau}$; both very difficult.

One method, Fig. 7a, requires the direct measurement of the τ lifetime; which would be about 2.5×10^{-13} sec if $g_{\tau W \nu_\tau} = g$.

The other method, Fig. 7b, requires the study of the semileptonic decay modes of a meson M much heavier than the τ, namely

$$M \rightarrow \tau + \nu_\tau + \text{hadrons} \tag{53a}$$

compared to

$$M \rightarrow e + \nu_e + \text{hadrons} \quad . \tag{53b}$$

If the mass is ignored, then for identical hadronic states

$$\frac{\sigma(M \rightarrow \tau\nu_\tau \text{ hadrons})}{\sigma(M \rightarrow e\nu_e \text{ hadrons})} = \left(\frac{g_{\tau W \nu_\tau}}{g}\right)^2 \quad . \tag{54}$$

A candidate for M is the proposed meson that contains a b quark and a non-b antiquark. Since the Υ with a mass of about 10 GeV is presumed to be made up of a $b\bar{b}$ pair; this meson would have a mass of about 5 GeV, allowing the decay in Eq. 53a to occur.

7.C. Higher Energy Tests of the Leptonic Nature of the τ

Whenever higher energies become available, we test once more whether the e and μ obey the basic requirements for a lepton listed in section 1.A.: does the e or μ still behave as a point particle; do they obey conventional quantum electrodynamics, do they exhibit any anomalous interaction? As higher e^+e^- annihilation energies become available at PETRA and PEP, we should maintain a similar inquiring attitude towards the τ. There are several types of tests.

a. $e^+ + e^- \rightarrow \tau^+ + \tau^-$:

One is simply whether the reaction

$$e^+ + e^- \rightarrow \tau^+ + \tau^- \tag{55}$$

occurs through the diagram in Fig. 8a and hence still obeys

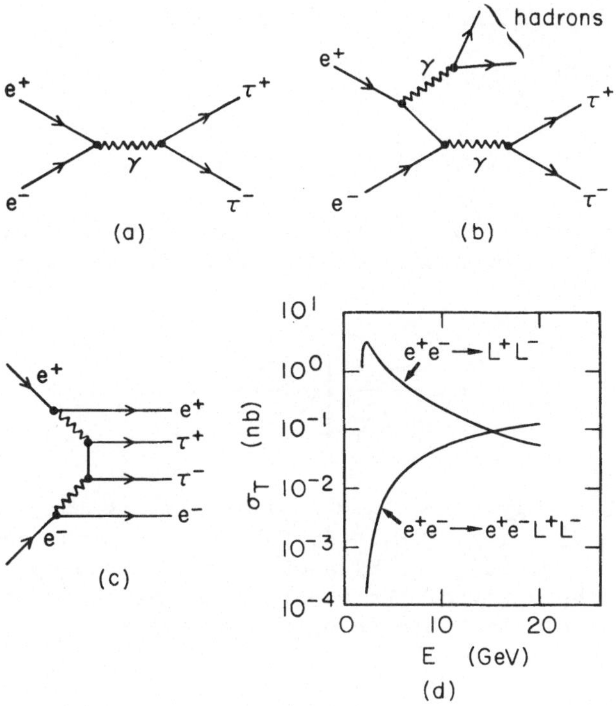

Fig. 8 Production of a $\tau^+\tau^-$ pair
(a) through one photon exchange,
(b) through a higher order reaction,
(c) through a two photon exchange process.
The production cross section for two photon exchange is compared with one photon exchange in (d);
taken from Ref. 51.

$$\sigma_{\tau\tau} = \frac{2\,\pi\,\alpha^2\,\beta(3-\beta^2)}{3E_{cm}^2} \tag{56}$$

at higher energy. A related question is whether the reaction

$$e^+ + e^- \rightarrow \tau^+ + \tau^- + \text{hadrons} \tag{57}$$

occurs more copiously than would be expected from higher order (in α) reactions such as Fig. 8b.

b. Two-Virtual-Photon Production of τ Pairs:

The reaction

$$e^+ + e^- \rightarrow e^+ + e^- + \tau^+ + \tau^- \tag{58}$$

occurs through two-virtual-photon diagrams such as Fig. 8c. Unlike the one-virtual-photon cross section, Eq. 56, the two-virtual-photon cross section

$$\sigma_{ee\tau\tau} \approx \frac{112\alpha^4}{977m_\tau^2}\left[\ell_n\left(\frac{E_b}{m_e}\right)\right]^2\left[\ell_n\left(\frac{E_b}{m_\tau}\right)\right] \tag{59}$$

increases with energy. Here $E_b = E_{cm}/2$ and m_e is the electron mass. Indeed, in the vicinity of E_b = 15 GeV the two cross sections become equal [51] (Fig. 8d), Eq. 59 assumes the τ is a conventional lepton, hence the study of the reaction in Eq. 58 constitutes an additional test of the leptonic nature of the τ.

c. $T \rightarrow \tau^+ + \tau^-$:

Ignoring mass effects the decay rates for

$$\begin{aligned} T &\rightarrow e^+ + e^- \\ T &\rightarrow \mu^+ + \mu^- \\ T &\rightarrow \tau^+ + \tau^- \end{aligned} \tag{60}$$

should be equal if the τ is a lepton. However, since the decay occurs through a virtual photon, this test is not completely independent of a test using the reaction $e^+e^- \rightarrow \tau^+\tau^-$.

7.D. Other Means of Producing τ's

a. The Bethe-Heitler process, Fig. 9a, can yield τ pairs through the reaction [52,53]

$$\gamma + \text{Nucleus} \rightarrow \tau^+ + \tau^- + \text{hadrons} \quad . \tag{61}$$

The total cross section [52,53] for this reaction on a Be nucleus is about 10^{-33} cm^2 at E_γ = 200 GeV.

However, the major experimental problem is how to detect the pairs so produced. The most feasible method so far proposed is to look for $e^{\pm}\mu^{\mp}$ pairs from

$$\gamma + \text{Nucleus} \rightarrow \underset{\substack{\downarrow \\ e^+\nu\nu}}{\tau^+} + \underset{\substack{\downarrow \\ \mu^-\nu\nu}}{\tau^-} + \text{hadrons} \quad . \tag{62}$$

Unfortunately, the photoproduction of pairs of charmed mesons $(C\bar{C})$ and their subsequent decay

$$\gamma + \text{Nucleus} \rightarrow \underset{\substack{\downarrow \\ e^+\nu\nu}}{C} + \underset{\substack{\downarrow \\ \mu^-\nu\nu}}{\bar{C}} + \text{hadrons} \tag{63}$$

is roughly 100 to 1000 times more probable. (We assume the total $C\bar{C}$ photoproduction from Be is 10^{-30} to 10^{-31} cm^2). Hence, the reaction in Eq. 62 is very difficult to detect. Nevertheless, it is important to look for the Bethe-Heitler process for τ pairs as a test of the leptonic nature of the τ.

Fig. 9b shows a related process using almost real virtual photons from a muon beam. The proposal [54] is to use a calorimetric target, and to select events in which there is a relatively small amount of hadronic energy. This criterion would reduce the background from the production of charmed particle pairs.

b. Hadronic Production of ν_τ's:

A similar problem bedevils any so far proposed scheme for making a ν_τ beam by dumping an intense proton beam into a thick target. One might hope to carry out the series of reactions

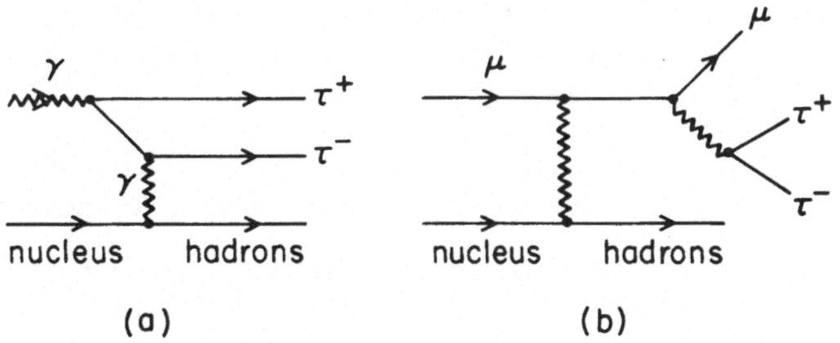

Fig. 9 Production of a $\tau^+\tau^-$ pair through (a) the Bethe-Heitler process, and (b) a virtual photon from a muon.

$$\left.\begin{array}{l} p + \text{Nucleus} \rightarrow \tau + \text{anything} \\ \tau \rightarrow \nu_\tau + \text{anything} \end{array}\right\} \text{occurs in thick target} \tag{64}$$

$$\left.\begin{array}{l} \nu_\tau + \text{Nucleus} \rightarrow \tau + \text{anything} \\ \tau \rightarrow \text{decay mode} \end{array}\right\} \text{occurs in neutrino detector} \tag{65}$$

Unfortunately the ν_e's and ν_μ's from charmed particles produced and decaying in the thick target will overwhelm the $\nu_\tau \rightarrow \tau$ signal.

LECTURE III: PAST AND FUTURE SEARCHES FOR OTHER HEAVY LEPTONS

8. Other Searches For Charged Leptons

8.A. Future Searches for Sequential Heavy Leptons

The only practical method remains e^+e^- annihilation

$$e^+e^- \rightarrow \ell^+ + \ell^- \tag{66}$$

and as with the τ [6] the best signature is $e^\pm\mu^\mp$ events. PEP and PETRA will allow searches for masses up to 15 GeV/c^2. The search will be more difficult than the search for the τ for three reasons: ches

a. the production cross section will be smaller at the higher energy required;
b. the $e^\pm\mu^\mp$ events from the τ will constitute an annoying background.
c. In the case of the τ many of the hadronic decay modes (Table VI) have only one charged particle. These provided the very useful two-prong event signature: electron-hadron and muon-hadron. Larger mass sequential leptons will have much less single hadron decay modes, Fig. 4; hence, these useful signatures will not be available.

However, there is no doubt that a complete search for new charged sequential leptons can be made in the available mass range.

8.B. Electromagnetic Production of Ortholeptons

As discussed in Section 4.B., an ortholepton would be: e^{*-} with the same lepton number as the e^-, μ^{*-} with the same lepton number as the μ^-, or perhaps τ^{*-} with the same lepton number as the τ. To keep the discussion brief, I shall only give examples using the e^*.

Figs. 10a and 10b show some reactions in which an e^* is produced at an electromagnetic vertex e^*-γ-e. Incidently, this same vertex would contribute to g_e-2 (Fig. 10c). Since no ortholeptons have been found, past searches can only put lower limits on the e^* mass. This is done as follows. Current conservation forbids the e^*-γ-e vertex from having the usual form:

$$V^{\mu} = q\bar{\psi}_{e*}\gamma^{\mu}\psi_e \quad . \tag{67a}$$

It has become conventional [16,55] to use the form

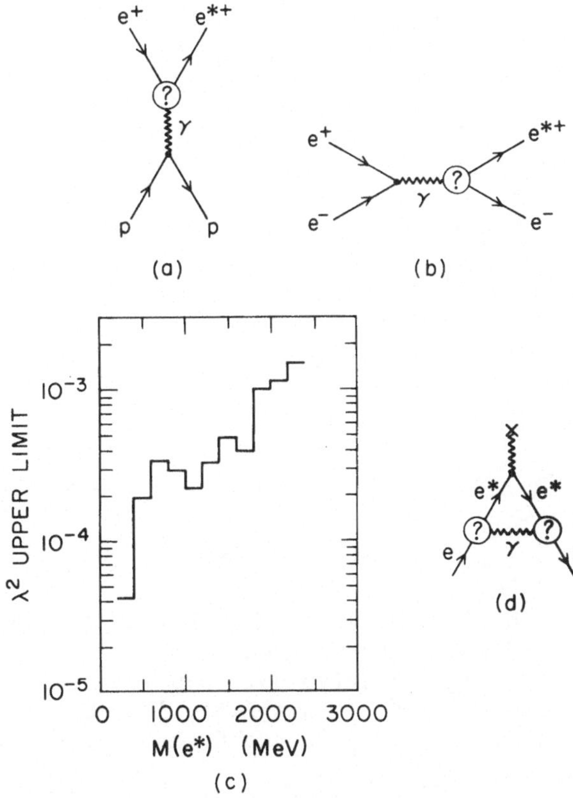

Fig. 10 Feynman diagrams for production of an e^* ortholepton through (a) e-p scattering, (b) e^+e^- annihilation, (c) the upper limit on λ^2 versus $M(e^*)$ in Eq. 67b from Ref. 56, (d) the contribution of an e^* to g-2 of the electron.

$$V^{\mu} = q\left(\frac{\lambda}{M^*}\right)\bar{\psi}_{e^*}\sigma^{\mu\nu}\psi_e p_\nu \qquad (67b)$$

Here q is the electron charge, p is the photon four-momentum, M^* is the e^* mass and λ is a dimensionless coupling constant. Figs. 10c and 11 show current limits on M^* and λ^2 for e^* and μ^*. The use of PEP and PETRA can obviously considerably extend the M^* search

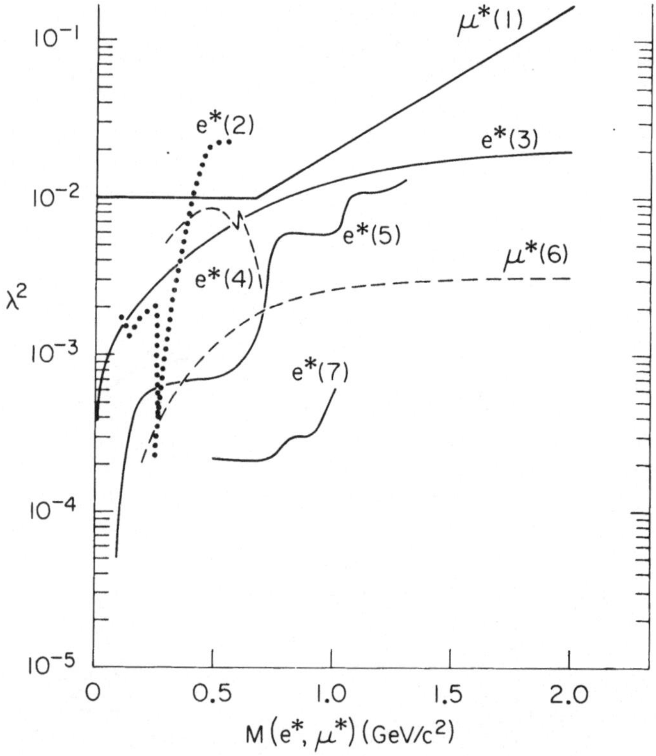

Fig. 11 Comparison of upper limits on the value of λ^2, Eq. 69b, for the e^* and μ^*. References are: 1. H. Gittleson et al., Phys. Rev. D10, 1379 (1974), (μ-p scattering); 2. C. Betorne et al., Phys.Lett. 17, 70 (1965) (e-p scattering); 3. A. De Rujula and B. Latrup, Lett.N.Cim. 3. 49 (1972), (g-2 of electron); 4. R. Budnitz et al., Phys.Rev. 141, 1313 (1966), (e-p scattering); 5. C.D. Boley et al., Phys.Rev. 167, 1275 (1968), (e-p scattering); 6. A. De Rujula and B. Lautrup, Lett.N.Cim. 3, 49 (1972), (g-2 of muon); 7. H.J. Behrend et al., Phys.Rev.Lett. 15, 900 (1965), (e-p scattering).

range. Indeed, no direct use has yet been made of SPEAR or DORIS data. The e^+e^- results in Fig. 10 come from research at ADONE.

8.C. ν_μ Production of Charged Leptons

If a heavy charged lepton couples to the ν_μ, then the reactions in Fig. 12a and 12b would demonstrate its existence through the presence of anomalous $e^\pm$ or μ^+ events [57]. The measured lower limits on the mass depend on the strength of the coupling constant $g_{\nu_\mu W L^\pm}$ in Fig. 12. If we assume $g_{\nu_\mu W L^\pm} = g_{\nu_\mu W \mu^-}$ the lower limits on the $L^\pm$ mass are [57]

$$M_{L^-} \geq 7.5 \text{ GeV}, \quad M_{L^+} \geq 9.0 \text{ GeV; from Fig. 12a}$$

$$M_{L^+} \geq 12.0 \text{ GeV; from Fig. 12b} \quad . \tag{68}$$

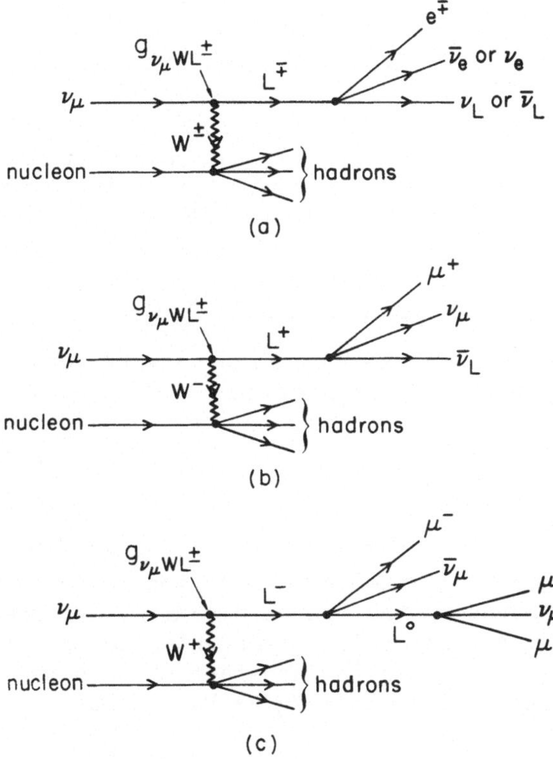

Fig. 12 Possible production processes for heavy leptons using a ν_μ.

Obviously such searches will continue at higher energies.

There have also been extensive searches [58] for heavy charged and neutral leptons using the so-called trimuon events which might be produced through reactions such as Fig. 12c. At present almost all trimuon events are explainable by conventional processes [59]; and the few which cannot be so explained appear not to be connected to heavy lepton productions.

9. Neutral Heavy Lepton Searches

9.A. Past Searches for Heavy Neutral Leptons

Past searches for heavy neutral leptons have usually used hadron reactions [60] or cosmic rays [61]. No definitive evidence for heavy neutral leptons have been found in these searches. However, these searches are not at all restrictive because they usually have depended for their success on either:

a. an unconventional production mechanism, such as an anomalously large cross section for hadronic production of leptons, or

b. an unconventional lepton property, such as large mass combined with a long lifetime.

Searches have also been carried out using muon neutrinos (Ref. 58,62,63). There have been some interesting events found (Ref. 62,63), but again, there is no evidence for a heavy neutral lepton.

9.B. Future Searches Using e^+e^- Annihilation

The only known search method which is very general and yet uses a reasonably probable interaction

$$e^+ + e^- \rightarrow L^0 + \bar{L}^0 \tag{69}$$

through s-channel exchange of a neutral current, Fig. 13a. We can even calculate this production cross section if we assume the model [64] in Fig. 13b, where the proposed neutral intermediate boson, Z^0, has a mass M_Z. A rough calculation which ignores vector coupling, assumes the axial coupling constant at both vertices is g_A, ignores the L^0 mass, and assumes $M_Z >> E_{cm}$ yields

$$\sigma_{L^0\bar{L}^0} \approx \frac{g_A^4}{12\pi E_{cm}^2} \left(\frac{E_{cm}}{M_Z}\right)^4 \tag{70}$$

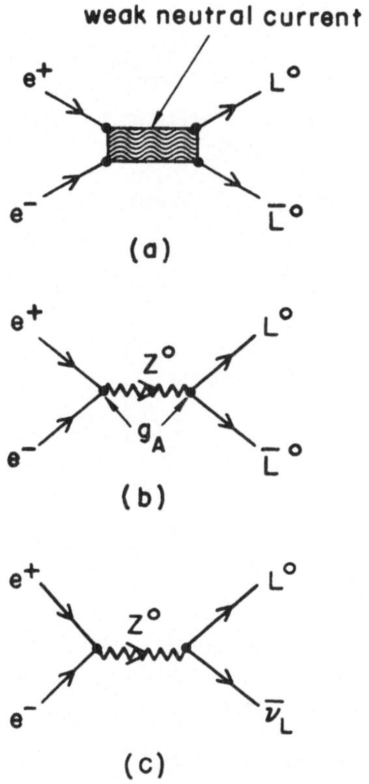

Fig. 13 Production of a pair of heavy neutral leptons through e^+e^- annihilation via the weak interactions.

Recall that the cross section (Eq. 56) for the production of a charged heavy lepton via

$$e^+ + e^- \rightarrow L^+ + L^- \tag{71}$$

is

$$\sigma_{L^+L^-} = \frac{4\pi\alpha^2}{3E_{cm}^2} = \frac{e^4}{12\pi E_{cm}^2} \tag{72}$$

if we ignore the mass of the $L^\pm$.

Hence

$$\frac{\sigma_{L^o\bar{L}^o}}{\sigma_{L^+L^-}} \approx \left(\frac{g_A}{e}\right)^4 \left(\frac{E_{cm}}{M_Z}\right)^4 \quad . \tag{73}$$

Conventional weak interaction theory uses

$$g_A/e = 1/4 \sin \vartheta_W \cos \theta_W = 0.6 \tag{74}$$

$$M_Z = 100 \text{ GeV} \quad .$$

In the next few years PEP and PETRA will have maximum E_{cm} of about 36 GeV. Hence

$$\sigma_{L^o\bar{L}^o} / \sigma_{L^+L^-} \sim 10^{-2}, \quad E_{cm} = 36 \text{ GeV} \tag{75}$$

and $L^o\bar{L}^o$ production will be much smaller than L^+L^- production. The need for higher energy e^+e^- colliding beams facilities in order to make comprehensive L^o searches is clear.

An L^o search at PEP or PETRA is just on the verge of being possible. At E_{cm} = 36 GeV, $\sigma_{L^o\bar{L}^o} \approx 1.4 \times 10^{-37}$ cm^2 assuming an average luminosity of 10^{31} cm^{-2} sec^{-1}; this will yield 0.12 $L^o\bar{L}^o$ pairs per day.

Of course the $L^o\bar{L}^o$ pairs must be detected, and the detection method depends upon how the L^o decays. I will list a few examples:

a. The L^o might be stable. In this case the reaction $e^+e^- \to L^o\bar{L}^o$ is not detectable at the low production rates calculated here with present detector technology.

b. The L^o could have an associated smaller mass charged lepton L^-. Typcial decay modes be:

$$\begin{aligned} L^o &\to L^- + e^+ + \nu_e \\ L^o &\to L^- + (\text{hadrons})^+ \end{aligned} \tag{76}$$

c. The L^o could have the same lepton number as the e^-, μ^-, or τ^-. Typical decay modes would be

$$L^{o} \rightarrow e^{-} + e^{+} + \nu_{e}$$
$$L^{o} \rightarrow e^{-} + (\text{hadrons})^{+} \qquad (77)$$

d. The L^{o} could have an associated smaller mass neutral lepton. For example, a neutrino ν_L. Typcial decay modes would be

$$L^{o} \rightarrow \nu_{L} + e^{+} + e^{-}$$
$$L^{o} \rightarrow \nu_{L} + (\text{hadrons})^{+} \qquad (78)$$

Incidently, in this case the production reaction

$$e^{+} + e^{-} \rightarrow L^{o} + \bar{\nu}_{L} \qquad (79)$$

could also occur [65]. This reaction doubles the mass range of the L^o search if the ν_L is massless. However, the L^o-Z^o-ν_L coupling constant is an unknown quantity.

Returning to the detection of the $e^+e^- \rightarrow L^o\bar{L}^o$ reaction, the decay modes in Eqs. 76-78 are not all detectable. Indeed probably only 10% to 20% of the $L^o\bar{L}^o$ pairs will decay to obviously anomalous looking, and hence useful, events. Therefore per experiment, only 0.01 to 0.02 detectable $L^o\bar{L}^o$ pairs occur in the model we have been developing here. The pooling of anomalous events by the half dozen experiments likely to be simultaneoulsy looking for $L^o\bar{L}^o$ events in their data tapes might help this very low rate problem.

10. How Many Leptons Exist?

The question of how many leptons exist cannot be answered without a theory about the classification of particles and their properties. However, any theory of particles can be destroyed by finding just one particle which is not predicted by the theory. Thus this question is an experimental one, and as long as new experiments looking for new particles can be devised, this question has no answer. This thought was stated most succinctly by my thesis professor I.I. Rabi, "Physics is an experimental science".

Nevertheless, it is very intersting to ask a more restricted question: how many leptons exist within the framework of existing theories, particularly within the framework of the very successful Weinberg-Salam theory of weak and electromagnetic interaction? We shall also assume:

a. the sequential heavy lepton model;
b. massless neutrinos;
c. all $L^{\pm}-\nu_L$ pairs couple with the same $W^{\pm}$ or Z^{o} intermediate bosons with the same strength.

10.A. Can There Be an Infinite Number of Massless Neutrino Types?

With these assumptions, it is easy to see that there cannot be be an infinite number of lepton pairs. Consider the K decay

$$K^{\pm} \rightarrow \pi^{\pm} + \nu_n + \bar{\nu}_n \tag{80}$$

where n = 1,2,3,...,N represents N different types of massless neutrinos. If N were infinite, the total decay rate $\sum_1^N \Gamma(K \rightarrow \pi\nu_n\bar{\nu}_n)$ would be infinite. We know the $K^{\pm}$ has a finite decay rate; hence, N is finite.

Similarly, the fact that the total cross section for reactions such as

$$e^+ + e^- \rightarrow \nu_n + \bar{\nu}_n \tag{81a}$$

$$e^+ + e^- \rightarrow \gamma + \nu_n + \bar{\nu}_n \tag{81b}$$

$$e^+ + e^- \rightarrow \text{hadrons} + \nu_n + \bar{\nu}_n \quad . \tag{81c}$$

is finite, means there can only be a finite number of massless neutrinos. Of course, direct measurements of the total decay rate for Eq. 80, or direct measurements [67] of the total cross section for any of the reactions in Eq. 81, would fix the value of N. Unfortunately, such measurements are exceedingly difficult because the ν's cannot be detected directly, and there are huge backgrounds from much more copious reactions.

10.B. Astrophysics Limits on Neutrinos

If one accepts the "big bang" theory of the creation of the universe, then one can calculate some interesting limits on numbers and properties of neutrinos. The calculation most pertinent to the considerations here is that of Steigman et al. [68]. They find that total number of different types of massless neutrinos is less than or equal to 7! This includes the ν_e, ν_μ and ν_τ. Several astrophysics calculations [69-72] have set limits on the masses and decays of neutrinos and heavy neutral leptons. For example, upper limits of the order of tens of eV have been set on the masses of individual neutrinos, and on the sum over the masses of all types of neutrinos. Thus, further studies of known neutrinos and future searches for new heavy leptons and neutrinos offer us the fascinating possibility of either helping to confirm the "big bang" theory of creation or helping to destroy it.

11. Acknowledgement

I am grateful to D. Fries and the other members of the Organizing Committee of the Advanced Summer Institute for giving me the opportunity to present these lectures.

REFERENCES

1. T. Walsh, in Proceedings of Advanced Summer Institute 1978 (University of Karlsruhe, Karlsruhe 1978)

2. H. Harari, in Proceedings of 1977 Summer Institute on Particle Physics (SLAC, Stanford, 1977), SLAC-Report-204

3. S. Weinberg, in Proceedings of XIX International Conference on High Energy Physics, Tokyo, 1978; to be published

4. T. Walsh, in Proceedings of 1977 International Symposium on Lepton and Photon Interactions at High Energies, Hamburg, 1977

5. G.J. Feldman, in Proceedings of XIX International Conference on High Energy Physics, Tokyo, 1978; to be published

6. M.L. Perl, Nature 275, 273 (1978)

7. B.H. Wiik and G. Wolf, DESY Report-DESY-78/23 (1978)

8. G. Flügge, DESY Report-DESY-78/42 (1978); see also G. Flügge, in Proceedings of Fifth International Conference on Experimental Meson Spectroscopy (Northeastern University, Boston, 1977)

9. J.-E. Augustin et al., Phys.Rev.Lett. 34, 233 (1975)

10. R. Hofstadter, in Proceedings of 1975 International Symposium on Lepton and Photon Interactions at High Energy, Stanford, 1975

11. J.C. Wesley and A. Rich, Rev.Mod.Phys. 44, 250 (1972)

12. R.S. Van Dyke et al., Phys.Rev.Lett. 38, 310 (1977)

13. J. Bailey et al., Phys.Lett. 68B, 191 (1977)

14. J. Calmet et al., Rev.Mod.Phys. 49, 21 (1977)

15. See for example: C.K. Hargrove et al., Phys.Rev.Lett. 39, 307 (1977)

16. M.L. Perl and P. Rapidis, SLAC Preprint SLAC-PUB-1496 (1974), unpublished

17. See for example the reviews by K. Tittel and by C. Baltay in Proceedings of the XIX International Conference on High Energy Physics, Tokyo, 1978; to be published

18. S. Parker et al., Phys.Rev. B133, 768 (1964)

19. P. Depommier et al., Phys.Rev.Lett. 39, 1113 (1977)

20. H.P. Povel et al., Phys.Lett. 72B, 183 (1977)

21. M. Cooper, in Proceedings of 1978 Summer Institute on Particle Physics (SLAC, Stanford, 1978)

22. A. Badertcher et al., Phys.Rev.Lett. 39, 1385 (1977)

23. E. Fiorni, in Proceedings of Neutrino 77 Conference (Institute for Nuclear Research, Moscow, 1978)

24. W.C. Barber et al., Phys.Rev.Lett. 22, 902 (1969)

25. K. Bergkvist, Nucl.Phys. B39, 317 (1972)

26. M. Daum et al., Phys.Lett. 74B, 126 (1972)

27. J. Kirkby, in Proceedings of International Conference on Neutrino Physics and Neutrino Astrophysics (Purdue University, 1978); also issued as SLAC Preprint SLAC-PUB-2127

28. C.H. Llewellyn Smith, in Proceedings of the Royal Society (London) A355, 585 (1977)

29. J.D. Bjorken and C.H. Llewellyn Smith, Phys.Rev. D7, 88 (1973)

30. S.P. Rosen, Phys.Rev.Lett. 40, 1057 (1978)

31. H.B. Thacker and J.J. Sakurai, Phys.Lett. 36B, 103 (1971)

32. Y.S. Tsai, Phys.Rev. D4, 2821 (1971)

33. F.J. Gilman and D.H. Miller, Phys.Rev. D17, 1846 (1978)

34. T.N. Phăm et al., Phys.Rev.Lett. 41, 371 (1978)

35. R. Brandelik et al., Phys.Lett. 73B, 109 (1978)

36. W. Bacino et al., Phys.Rev.Lett. 41, 13 (1978)

37. W. Bartel et al., Phys.Lett. (to be published); also issued as DESY Report-DESY-78/24 (1978)

38. M.L. Perl et al., Phys.Lett. 70B, 487 (1977)

39. S.G. Wojcicki, in Proceedings of 1978 Summer Institute on Particle Physics (SLAC, Stanford, 1978)

40. M.L. Perl, in Proceedings of 1977 International Symposium on Lepton and Photon Interactions at High Energies (Hamburg, 1977)

41. U. Timm, in Proceedings of Third International Conference on New Results in High Energy Physics (Vanderbilt University, 1978); also issued as DESY Report-DESY-78/25

42. A.M. Cnops et al., Phys.Rev.Lett. 40, 144 (1978);
M.J. Murtagh, in Proceedings of 1977 International Symposium on Lepton and Photon Interactions at High Energies (Hamburg, 1977)

43. A. Ali and T.C. Yang, Phys.Rev. D14, 3052 (1976)

44. F.J. Heile et al., Nucl.Phys. B138, 189 (1978)

45. Y.S. Tsai, Phys.Rev. (to be published); also issued as SLAC Preprint-SLAC-PUB-2105 (1978)

46. Y.S. Tsai, private communication

47. G. Alexander et al., Phys.Lett. 73B, 99 (1978)

48. J. Jaros et al., Phys.Rev.Lett. 40, 1120 (1978)

49. S. Yamada, in Proceedings of 1977 International Symposium on Lepton and Photon Interactions at High Energies (Hamburg, 1977)

50. N. Arteaga-Romero et al., Phys.Rev. D3, 1569 (1971); V.N. Baier and V.S. Fadin, Nuovo Cim. Lett. 1, 481 (1971); H. Terazawa, Rev.Mod.Phys. 45, 615 (1973)

51. J. Smith et al., Phys.Rev. D15, 3280 (1977)

52. Y.S. Tsai, Phys.Rev. D4, 282 (1971)

53. J. Smith et al., Phys.Rev. D15, 648 (1977)

54. W. Williams, private communication

55. F.E. Low, Phys.Rev.Lett. 14, 238 (1965)

56. C. Bacci et al., Phys.Lett. 71B, 227 (1977)

57. A recent review has been given by C. Baltay in the Proceedings of the XIX International Conference on High Energy Physics, Tokyo, 1978; to be published

58. D. Reeder, in Proceedings of 1977 International Symposium on Lepton and Photon Interactions at High Energies, Hamburg, 1977

59. K. Tittel, in Proceedings of the XIX International Conference on High Energy Physics, Tokyo, 1978

60. See for example: B.J. Bechis et al., Phys.Rev.Lett. 40, 602, (1978)

61. See for example the Proceedings of the 14th International Cosmic Ray Conference, Munich, 1975, Vol. 7, Sec HE5

62. H. Faissner, in Proceedings of Neutrino 77 Conference, (Institute for Nuclear Research, Moscow, 1978)

63. V.V. Makeev, in Proceedings of Neutrino 77 Conference, (Institute for Nuclear Research, Moscow, 1978)

64. For a full discussion, see J. Ellis and M.K. Gaillard in CERN Report-CERN-76-18 (1976)

65. See for example: K. Yamanashi et al., Phys.Lett. 74B, 101 (1978)

66. S. Weinberg, in Proceedings of the XIX International Conference on High Energy Physics, Tokyo, 1978

67. For a quantitative discussion of Eq. 81b, see: E. Ma and J. Okada, Phys.Rev.Lett. 41, 289 (1978)

68. G. Steigman et al., Phys.Lett. 66B, 202 (1977)

69. D.A. Dicus et al., Phys.Rev.Lett. 39, 168 (1977)

70. B.W. Lee and S. Weinberg, Phys.Rev.Lett. 39, 165 (1977)

71. R. Cowsik, Phys.Rev.Lett. 39, 784 (1977)

72. R. Cowsik and J. McClelland, Phys.Rev.Lett. 29, 669 (1972)

QUARKONIUM

M. Krammer[+] and H. Krasemann[++]

[+]permanent address: Deutsches Elektronen-Synchrotron DESY, Hamburg

[++]present address: II. Institut für Theoretische Physik, Universität Hamburg

1. INTRODUCTION

Since the discovery of the J/Ψ and Ψ' in November 1974 [1] we all witnessed a dramatic revival of the quark model [2]. A new quark flavour, c = charm [3], was added to the hadron spectrocopy, interpreting the J/Ψ and Ψ' as $c\bar{c}$ bound states. This new system promised to be describable as nonrelativistic bound states of c and $\bar{c}$: Charmonium [4]. While the quark model for old mesons suffered from the fact that the quarks move relativistically (mass differences of old mesons are of the order of the masses themselves), in charmonium the relatively heavy (≈ 1.5 GeV) c-quarks should move relatively slowly, $\beta^2 = (v/c)^2 \approx 0.2$. A perturbation expansion in β^2 then converges rapidly and the well known powerful tools of exploring a nonrelativistic bound system could be used. This was the source of real excitement.

Meanwhile we learned about the existence of a still heavier meson family, the Υ, Υ' and Υ" [5], and interpret it as bound states of the b quark and $\bar{b}$, b being the fifth quark flavour [6], much more massive than charm. We further hope to discover the sixth quark flavour, t maybe, and its bound states $t\bar{t}$ in the new e^+e^- machines PETRA and PEP. The larger masses of the b and t quark guarantee that their bound systems $b\bar{b}$ and $t\bar{t}$ are nonrelativistic to a much higher degree than $c\bar{c}$. In this lecture we will discuss the dynamics of a nonrelativistic $Q\bar{Q}$ bound system, Q = c,b,t. As a title for this lecture we chose the generic name for a nonrelativistic $Q\bar{Q}$ system, QUARKONIUM.

On the field theory side, Quantumchromodynamics [7], QCD, turned out to be the most promising key to an understanding of quark dynamics. QCD is a nonabelian gauge field theory of the interactions of quarks and eight massless vector gauge bosons, the gluons. The coupling constant α_s, renormalized at the relevant momentum transfer q^2 or the corresponding distance R, turns out to be a monotonously falling function of q^2 (or rising function of R). It tends logarithmically to zero as $q^2 \to \infty$ or $R \to 0$, this is called asymptotic freedom [8]. α_s becomes large for some large R of the order of one fm, the typical hadron size. Up to today this regime is subject to speculations only, we believe that the rising coupling provides for the permanent confinement of quarks. Perturbation theory is useless in this case, but lattice gauge theories [9] or the string model [10] suggest that the interquark force for large separations might be independent of the distance, thus giving rise to a linearly rising static potential between quarks. At short distances physics is much more pleasant because α_s becomes small. Then perturbation theory is fine and in Born approximation the quark interaction is just one gluon exchange. The nonabelian self-interaction of the colour-charged gluons plays no role in lowest order graphs, and in this approximation gluons are just analogous to photons. The short distance behaviour of QCD is thus very similar to QED, the static potential for short distances being of the Coulomb type.

When QCD is in fact the underlying theory for the Quarkonium systems, we should be able to probe some QCD features by studying these systems. What can we probe? First we should be able to probe the short distance behaviour. The one gluon exchange at short distances leads to a static potential of the form $V_{AF}(R) = -4\alpha_s/3R$. The subscript AF denotes the origin of this potential 'Asymptotic Freedom'. $-4/3$ is a group factor from SU3 (colour) and α_s is the effective coupling. One can take two points of view regarding α_s. Either α_s is really R-dependent [11] but independent of the quark flavour. Or one defines an effective α_s as a constant, different for each quark flavour mass [8]. For simplicity we take the second point of view. Then the α_s in a heavy $Q\bar{Q}$ bound state M_2 is related to that of a lighter one M_1 by the approximate formula

$$\alpha_s(M_2^2) = \alpha_s(M_1^2) \left| 1 - \frac{33-2N}{12\pi} \alpha_s(M_1^2) \log (M_1^2/M_2^2) \right|^{-1} \tag{1.1}$$

(N is the number of 'light' (= lighter than Q) quarks). The potential $V_{AF}(R)$ with α_s given by (1.1) should be correct for very short distances. It further gives rise to the spin-spin and spin-orbit interactions known from positronium, because the quark gluon vertex has the same Dirac structure as the electron photon vertex (γ_μ-coupling).

The second feature of QCD we might be able to probe is the large distance behaviour, $R \to \infty$. The linear potential as suggested by lattice gauge theory or string models should dominate for very large distances R: $V_C(R) = a \cdot R$. The subscript C stands for 'Confinement'. The slope a should be flavour independent and also somehow related to the inverse Regge slope of the low mass mesons (ref. 12). Furthermore this potential should be essentially spin independent [9].

We now have guesses for the static potential at very short distances, $V_{AF}(R) = -4\alpha_s/3R$, and very long distances $V_C(R) = aR$. We have no guess for intermediate distances. The simplest assumption is to write the complete potential as a superposition of these two extremes (E. Eichten et al., ref. 4):

$$V(R) = V_{AF}(R) + V_C(R) \tag{1.2}$$

We further assume that all the spin dependence (except the kinematic Thomas precession) has its origin in $V_{AF}(R)$ and can be calculated via the Fermi-Breit Hamiltonian [13]. Altough these Ansätze have their ciriticism they have worked out to be very useful as a first attempt to the problem. The first part of this lecture will try to show how far these Ansätze reach. In the second part we will discuss decays of Quarkonium and a third test of QCD, namely of gluon helicities and the gluon self coupling. With the experimentally accessible regime of c.m. energies of 10 GeV or more, the gluons which govern annihilations in QCD, might show up as hadron jets [14]. These jets should carry the directed momentum of the initial gluon. In angular distributions of these jets one should then be able to measure gluon helicities [14,15]. One can further speculate on the existence of glueballs [16] to be found in Quarkonium decays and on measuring the nonabelian gluon self coupling by comparing the angular distribution of a 3 gluon decay versus a γ + 2 gluon decay. The latter two things, however, go beyond the Born approximation.

2. The Spectrum

Throughout the discussion we will assume that the Quarkonium ($Q\bar{Q}$) system is essentially nonrelativistic. The perturbative Hamiltonian can then be obtained by solving the Bethe Salpeter equation in nonrelativistic approximation or by expanding the exact relativistic scattering amplitude (Born graph only). One obtains the Schrödinger equation in zeroth order of β^2 and the well known Fermi-Breit Hamiltonian terms up to order β^2. In 0th order

$$H^o = 2\, m_Q + \frac{p^2}{m_Q} + V(R) + \text{const.} \tag{2.1}$$

and all states which only differ in their quark spin configurations are degenerate.

Here we can study the rough structure of the spectrum and try to justify the choice (1.2) for the potential V(R.). In fig. 2.1 it is demonstrated that the center of gravity of the P waves (which is object of 2.1) can be well described if the potential lies between a Coulombic and a linear potential. Also a logarith-

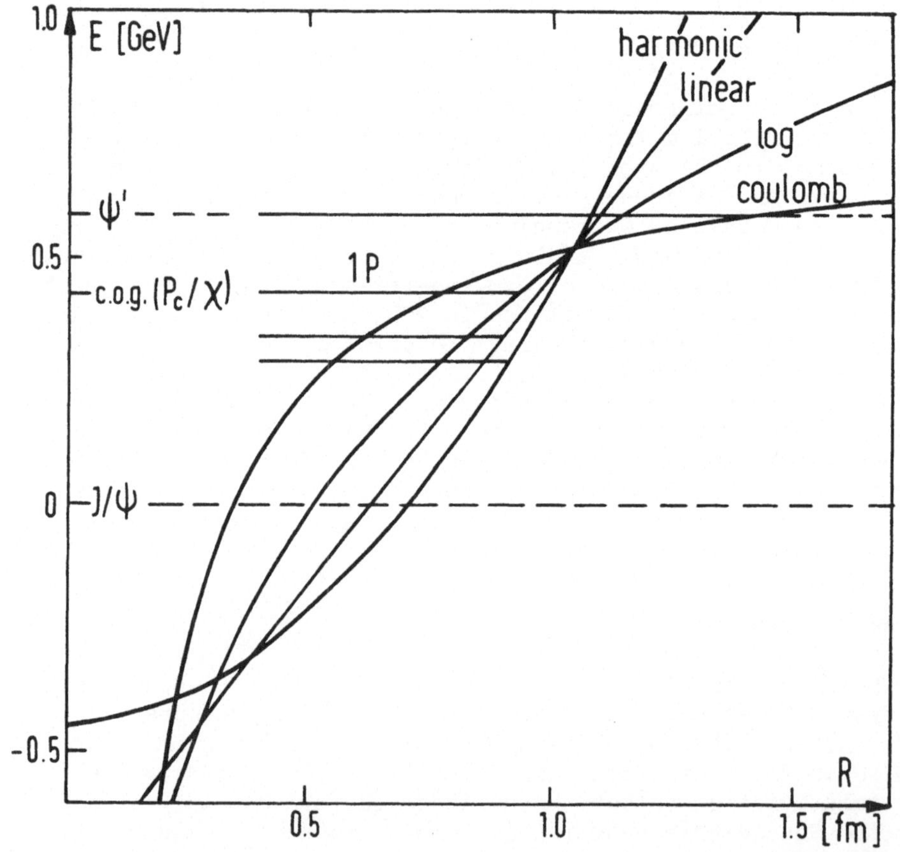

Fig. 2.1 Four different potentials for charmonium, normalized to the J/Ψ and Ψ' binding energies. The solid horizontal lines indicate the P wave of each potential, the experimental c.o.g. (P) is given for comparison.

mic potential is not bad. This may serve to justify the Ansatz (1.2). We note, however, that doing this comparison we assume that splittings due to spin-spin interactions are either small or of the same magnitude in the P and S waves. Calculations of the spectrum of equ. (2.1) have to be done numerically because of the complicated nature of the potential V(R). There are three parameters, m_Q, $\kappa \equiv 4/3\ \alpha_s$ and a. The level splitting of the radial excitation and the ground state ($\Psi'(3.7)$ and $J/\Psi(3.1)$ in Charmonium) determines one of the potential parameters, say a, if the other, say κ, is given. We then can try to determine κ from two independent sources, namely the ratio of the S wave functions at the origin

$$\frac{|\Psi_{\Psi'}(0)|^2}{|\Psi_{J/\Psi}(0)|^2} = \frac{M^2_{\Psi'}\ \Gamma_{e\bar{e}}(\Psi')}{M^2_{J/\Psi}\ \Gamma_{e\bar{e}}(J/\Psi)} = \frac{(3.7)^2 \cdot 2.2\ \text{keV}}{(3.1)^2 \cdot 4.8\ \text{keV}} \tag{2.2}$$

and the relative placement of the center of gravity of the P waves. Both procedures are almost independent of the third parameter, m_Q, and in Charmonium they give

$$\begin{aligned} \kappa &= 0.4 \ \dots\ 0.5 \\ a &= 1 \ \dots\ 0.9\ \text{GeV/fm} \quad . \end{aligned} \tag{2.3}$$

One remark on equ. (2.2) is in order. It is derived from the Van Royen-Weisskopf formula

$$\Gamma_{e\bar{e}}(V) = 16\,\pi\,\alpha^2\,e_Q^2\,\frac{|\Psi_V(0)|^2}{M_V^2} \quad . \tag{2.4}$$

This equation is subject to large corrections in the charmonium system as we will discuss later but in ratios of $\Gamma_{e\bar{e}}$'s these corrections cancel. Therefore (2.2) seems to be quite reliable.

Is the large value of κ reasonable? From the beginning κ is just a free parameter. But with $\kappa = 4/3\ \alpha_s$ we find α_s of the magnitude 0.3 ... 0.4. Is this α_s related to the strong coupling constant in annihilation processes? Or is it related to the strong coupling constant in deep inelastic lepton scattering? From the decay formulae as described in the second lecture one can derive α_s (annihilation at 3 GeV) $\simeq$ 0.2. But this α_s refers to annihila-

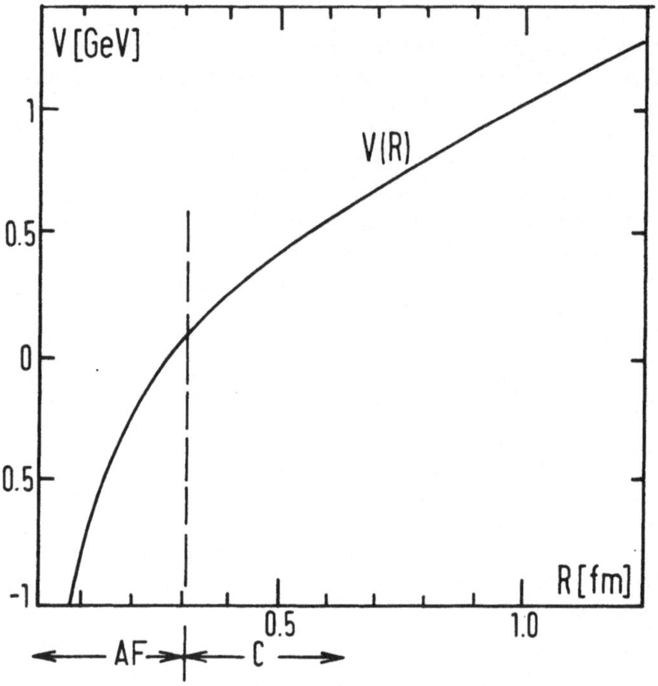

Fig. 2.2 The shape of the standard potential, equ. (1.2). V_{AF} dominates below, V_C above R = 0.3 fm.

tion distances which are shorter than the average interquark distances. From deep inelastic lepton scattering we find $\alpha_s(3\ \mathrm{GeV}) \simeq \alpha_s(0.07\ \mathrm{fm}) \simeq 0.4$ taking the renormalization point $\lambda = 0.5$ GeV, as you have learned in this school [17]. From fig. 2.2 we see that 0.07 fm are just in the middle of the range where the asymptotic freedom potential V_{AF} dominates, between 0 and 0.3 fm. The α_s as determined from the spectrum with the simple Ansatz (1.2) for V(R) agrees roughly with the α_s as measured in scaling violations of deep inelastic lepton scattering. This result encourages us to ask the next question: Is the parameter a in $V_c(R) = aR$ unique for all flavours (quark masses) as QCD suggests? The first estimates of the V'-V splittings in $Q\bar{Q}$ systems heavier than Charmonium predicted a decrease of this splitting with m_Q [18]. At 10 GeV the mass splitting should be 450 MeV only (compared to 590 MeV in Charmonium). As soon as the next Quarkonium system, T and T', was found, this prediction was destroyed. The T'-T mass splitting was around 600 MeV again as in Charmonium. The potential to describe this fact is the logarithmic potential [19]. Here mass splittings are completely independent of the quark mass. But an overall log potential has no justification within QCD. For intermediate distances, on the other hand, it is not worse than the simple superposition (1.2). An interesting - and phenomenologically successful Ansatz was then proposed with the log potential for intermediate distances only [20]:

$$V(R) = \begin{cases} -\kappa/R & R < R_1 \\ b\cdot\log R/R_0 & \text{for} \quad R_1 \leq R \leq R_2 \\ a\cdot R & R > R_2 \end{cases} \qquad (2.5)$$

The ambiguities coming in by 6 parameters, κ, a, R_1, R_2, R_0, b in this potential are removed by demanding V(R) to be continuously differentiable at R_1 and R_2. These are four conditions which remove 4 parameters and for comparison one chooses κ and a to be the only independent potential parameters. The Charmonium system has been solved with this potential and one finds a very good fit to all available data with

$$\frac{3}{4}\kappa = \alpha_s = 0.31 \qquad a = 0.775 \text{ GeV fm}^{-1} \qquad (2.6)$$

Applying the potential (2.5) - with the unique a = 0.775 GeV/fm - to the T system gives the mass difference T' - T to 560 MeV.

Very recently a precise measurement of the T and T' masses at DORIS gave us the experimental value: 560 MeV [5]. This coincidence is of course no prove for the correctness of the potential (2.5) but it shows that - with a more sophisticated potential - the assumption of a flavour independent constant force between quarks at long distances is not in contradiction with what we observe. It is amusing to note that this value of a = 0.775 GeV/fm is even in agreement with what one would expect from the old meson spectroscopy [12].

We want to add a remark on quark masses. Quark masses only slightly influence the two inputs we used, the ratio of wave functions at the origin and the P wave location. What they mainly influence is the wave functions themselves, the dipole matrix elements and the velocity of the quarks. But here is some ambiguity. Fitting $\Psi(0)$ to the naive Van Royen-Weisskopf formula (2.4) gives a rather small value, $m_c \simeq 1.1$ GeV. For the dipole matrix elements on the other hand one would like a large quark mass, $m_c \simeq 2$ GeV. In the best known studies at Cornell [21] the requirement of small quark velocities restricts m_c to be $m_c \simeq 1.6$ GeV. To fix m_c or m_Q resp. is not as easy as to fix α_s and a, because the decay formulae (2.4) and the dipole formula are subject to large corrections as we will discuss in the second lecture. We will use scaling arguments for scale variations of the quark mass. To overcome the ambiguities of determining the quark masses we will set quark mass ratios equal to the corresponding bound states mass ratios. We emphasize that smaller quark masses like $m_c = 1.1$ GeV do not destroy the nonrelativistic approximation. We have calculated $\beta^2 = (v/c)^2$ and find that $\beta^2 < 0.3$ in J/Ψ and $\beta^2 < 0.4$ in Ψ' for $m_c = 1.16$ GeV and $\kappa < 0.55$. We feel that this justifies to leave the quark masses themselves an open question.

3. Spin Interactions

In the physical charmonium spectrum the Schrödinger states are split up due to spin interactions. In this chapter we want to compare the magnitude of these splittings with the simplest Ansatz we can imagine, the Fermi Breit Hamiltonian [13]. These higher order corrections to (2.1) are relativistic kinematic corrections and spin corrections:

$$H = H^{0} + H^{rel} + H^{spin} \quad . \tag{3.1}$$

The spin corrections have three contributions.

$$\text{spin orbit:} \quad H^{LS} = \frac{2}{m_Q^2} \vec{L}\cdot\vec{S} \left[\frac{1}{R} d_R\right] \left(V_{AF}(R) - \frac{1}{4} V(R)\right)$$

$$\text{tensor:} \quad H^{T} = \frac{-1}{12 m_Q^2}\left(3\vec{\sigma}_1\cdot\hat{R}\vec{\sigma}_2\cdot\hat{R} - \vec{\sigma}_1\cdot\vec{\sigma}_2\right)\left[d_R^2 - \frac{1}{R} d_R\right] V_{AF}(R) \tag{3.2}$$

$$\text{spin-spin:} \quad H^{SS} = \frac{1}{6 m_Q^2} \vec{\sigma}_1\cdot\vec{\sigma}_2 \, \Delta V_{AF}(R) \quad .$$

Here $\vec{\sigma}_i/2$ is the quark spin, $S = 1/2(\vec{\sigma}_1+\vec{\sigma}_2)$ the meson spin, L its angular momentum, R the interquark distance. For the potential V(R) we again take the simplest Ansatz (1.2) with only $V_{AF}(R)$ being spin-dependent. As mentioned in the introduction lattice gauge theories suggest that the confinement part $V_c(R)$ of the potential is spin-independent. Nevertheless it contributes to the spin orbit interaction due to the relativistic kinematic effect of the Thomas precession [22], $-1/4V(R)$ in H^{LS}. In Quarkonia the Thomas precession leads to a decrease of the $^3P_2$ - $^3P_1$ splitting relative to the $^3P_1$ - $^3P_0$ splitting. While in Positronium, where $V(R) \sim -1/R \sim V_{AF}(R)$

$$\frac{M(^3P_2) - M(^3P_1)}{M(^3P_1) - M(^3P_0)} = 0.8 \tag{3.3}$$

the additional $V_c(R)$ in the interquark potential (1.2) leads to a decrease of (3.3), which experimentally is found to be 0.5 in Charmonium.

We are confident that the Fermi Breit Hamiltonian (3.2) is not a too bad approximation. As an example let us consider the part of the relativistic corrections due to the kinetic energy of the quarks. This correction is $\langle (p^2)^2/4m_Q^3\rangle \approx E_{kin} \langle 1/4\ \beta^2\rangle$. Up to β^2 of 0.4 the relativistic kinetic energy correction is less than 10%. The β^2 one obtains in the Charmonium calculations are 0.2 to 0.3 for J/Ψ and 0.27 to 0.4 for Ψ' varying m_c from 1.6 to 1.16 GeV.

Let us now compare experiment with the predictions from (3.2). We start considering the experimental states as discussed at this school 23). The three P waves are quite well established, the χ(3.55) as $j^{PC} = 2^{++}$ state, the P_c/χ(3.51) as $j^{PC} = 1^{++}$ state and the χ(3.41) as $j^{PC} = 0^{++}$ state. For the pseudoscalar partners of J/Ψ and Ψ' the experimental situation is not so clear. Candidates for the pseudoscalars are X(2.83), χ(3.45) and χ(3.59 or 3.18).

The P wave splittings can be parametrized as

$$\begin{aligned} \langle H^{LS}\rangle &= A\ \langle \vec{L}\cdot\vec{S}\rangle \\ \langle H^{T}\rangle &= B\ \langle T\rangle \quad , \end{aligned} \tag{3.4}$$

where the tensor operator $T \equiv 3\vec{\sigma}_1\cdot\hat{R}\ \vec{\sigma}_2\cdot\hat{R} - \vec{\sigma}_1\cdot\vec{\sigma}_2$. The expectation values of L•S and T can be found in textbooks on Quantum mechanics (ref. 24). For P wave they are displayed in table 3.1. A Charmonium analysis with the experimental masses of table 3.2 yields

Table 3.1

j	$\langle\vec{L}\cdot\vec{S}\rangle$	$\langle T\rangle$
2	+ 1	-2/5
1	- 1	+2
0	- 2	-4

Table 3.2

State	$^3P_2$	$^3P_1$	$^3P_0$	center of gravity
mass |GeV|	3.552	3.508	3.415	3.522

for A and B

$$A \simeq 34 \text{ MeV} , \qquad B \simeq 10 \text{ MeV} \tag{3.5}$$

On the theoretical side we read off (3.2)

$$\begin{aligned} A &= \frac{2}{m_Q^2} \langle \frac{1}{R} d_R \, (V_{AF}(R) - \frac{1}{4} V(R)) \rangle \\ B &= \frac{-1}{12 m_Q^2} \langle (d_R^2 - \frac{1}{R} d_R) \, V_{AF}(R) \rangle \end{aligned} \tag{3.6}$$

including the Thomas precession. With our standard potential (1.2) this gives

$$\begin{aligned} A &= \frac{2}{m_Q^2} \langle \alpha_s \, R^{-3} - \frac{1}{4} a \, R^{-1} \rangle \\ B &= \frac{1}{3 m_Q^2} \langle \alpha_s \, R^{-3} \rangle \end{aligned} \tag{3.7}$$

We see that the spin dependence from the one gluon exchange (V_{AF}) is governed by $\langle R^{-3} \rangle$ while the Thomas precession is governed by $\langle R^{-1} \rangle$. Taking our $\alpha_s = 0.4$, $m_c^{-1} \langle R^{-3} \rangle \simeq 0.07$ GeV^2 and $\langle R^{-1} \rangle \simeq$ 0.4 GeV from numerical fits yields the values of A and B given in

Table 3.3 A and B from numerical fits. In row A the second number is the contribution from the Thomas precession

m_c |GeV|	1.6	1.1
A |MeV|	35-12	56-32
B |MeV|	6	9

table 3.3 for two different values of m_c +). By comparison of table 3.3 with equ. (3.5) we see that we are in the right ball park. We could not have expected a better agreement from our crude approximation!

Let us now try the spin-spin interaction. According to our philosophy it arises from the short range one gluon exchange (V_{AF}) alone. The relevant term in the Fermi-Breit-Hamiltonian (3.2) was

$$H^{SS} = \frac{1}{6m_Q^2} \vec{\sigma}_1 \cdot \vec{\sigma}_2 \; \Delta V_{AF}(R) \tag{3.8}$$

The eigenvalues of the operator $\vec{\sigma}_1 \cdot \vec{\sigma}_2 = 2\vec{S}^2 - 3$ are +1 in a spin triplet state and -3 in a spin singlet state. Because $\Delta V_{AF}(R) \sim \Delta(-1/R) = 4\pi\delta(R)$ the integral over the wave functions becomes trivial and we have

$$\langle H^{SS} \rangle = \frac{4}{3}\alpha_s \cdot 4\pi \cdot \frac{1}{6m_Q^2} (2\vec{S}^2 - 3) \, |\Psi(0)|^2 \tag{3.9}$$

Taking $|\Psi(0)|^2$ from $\Gamma_{e\bar{e}}$ via equ. (2.4) and α_s from equ. (2.3) gives us for the splittings

$$\begin{aligned} M(1^3S_1) - M(1^1S_0) &\simeq 70 \text{ MeV} \\ M(2^sS_1) - M(2^1S_0) &\simeq 35 \text{ MeV} \end{aligned} \tag{3.10}$$

Trying to identify $\eta_c(1^1S_0) \equiv X(2.83)$ means 70 MeV $\equiv$ 250 MeV, $\eta_c'(2^1S_0) \equiv \chi(3.45)$ means 35 MeV $\equiv$ 230 MeV, or $\eta_c'(2^1S_0) \equiv \chi(3.59)$ means 35 MeV $\equiv$ 80 MeV. Many solutions have been proposed to solve this puzzle, among these are instanton effects [25] and an anomalous colour magnetic moment of the c-quark [26]. The simplest solution might be that the $|\Psi(0)|^2$ in equ. (2.4) and in (3.9) are different objects. The next order correction to $|\Psi(0)|^2$ in (2.4) comes in through a transverse gluon exchange between the two quark lines

---

+) The tensor operator T of equ. (3.4) possesses off diagonal matrix elements, too. They lead to an S-D mixing. Two physical Charmonium states would e.g. be $\Psi'(3.7) = \sqrt{1-\varepsilon^2}\, 2^3S_1 + \varepsilon\, 1^3D_1$ and $\Psi''(3.77) = -\varepsilon\, 2^3S_1 + \sqrt{1-\varepsilon^2}\, 1^3D_1$ with $\varepsilon = (2\sqrt{2}\,\alpha_s/3m_Q^2)(\langle 2^3S_1|R^{-3}|1^3D_1\rangle / M(D)-M(S))$.[32] With $\langle 2^3S_1|R^{-3}|1^3D_1\rangle \approx \langle 1^3P|R^{-3}|1^3P\rangle/7$ we can evaluate $\varepsilon \simeq 0.04$ leading to a $\Gamma_{e\bar{e}}(\Psi'(3.77))$ of 0.16% of that of $\Psi'(3.7)$. Experimentally it is 17%.

before annihilation. It has a large factor in front and the total correction is a factor $(1 - 16\alpha_s/3\pi)$ [27], which in no case is small. But before continuing this discussion let us wait for estimates of some decay rates involving the pseudoscalars. Then we will find that we have much more severe problems which question the identifications above.

4. Scaling the Schrödinger Equation

The radial form of the Schrödinger equation reads

$$\left[-d_R^2 + \frac{l(l+1)}{R^2} + 2m\,(V(R) - E)\right] R(R) = 0 \tag{4.1}$$

For all potentials of the form

$$V(R) = a\cdot R^{\varepsilon}\,, \qquad \varepsilon > -2 \tag{4.2}$$

we can bring it into the dimensionless form

$$\left[-d_\rho^2 + \frac{l(l+1)}{\rho^2} + \rho^{\varepsilon} - \xi\right] R(\rho) = 0 \tag{4.3}$$

with the substitutions

$$\begin{aligned} \xi &= E\cdot(2m)(2ma)^{-2/(2+\varepsilon)} \\ \rho &= R\cdot(2ma)^{+1/(2+\varepsilon)} \end{aligned} \tag{4.4}$$

One can now immediately read off the scaling laws for E and R:

$$\begin{aligned} E &\sim m^{-\varepsilon/(2+\varepsilon)} \\ R &\sim m^{-1/(2+\varepsilon)} \end{aligned} \tag{4.5}$$

(4.5) is also applicable for $\varepsilon = 0$, in which case the potential is $V(R) = a \log R/R_0$. We leave the derivation to the reader.

Let us now consider some aspects of scaling for Quarkonia. We begin with the level spacing. In a potential like $V_{AF}(R) = -4\alpha_s/3R$ alone level spacings scale like $\Delta E \sim \alpha_s^2 m_Q$ in a linear potential like $V_c(R) = aR$ they scale like $\Delta E \sim m_Q^{-1/3}$. To estimate the intermediate scaling behaviour in the standard potential we try a very crude approximation: Let us consider the level spacings given by the linear potential with the Coulombic part $V_{AF}(R)$ as a first order perturbation. Then

$$E_n = E_n(V_c) + \langle n| - \frac{4}{3}\frac{\alpha_s}{R}|n\rangle \qquad (4.6)$$

and ΔE scales like $m_Q^{-1/3}$ with a first order correction ~ $\alpha_s m_Q^{+1/3}$. Because of the mass dependence of α_s, equ. (1.1), this perturbation procedure starts to break down not before $m_Q \gtrsim 100$ GeV. The curve is shown in fig. 4.1 (dashed line). Asymptotically the states fall into the Coulombic potential V_{AF} and the scaling law becomes $\Delta E \sim \alpha_s^2 m_Q$. If α_s would be a universal constant, this would happen much earlier (dotted line in fig. 4.1). From the T'-T mass difference we know that the simple standard potential is not adopted by nature. Using the Ψ'-J/Ψ mass difference as input, the standard model prediction for the T'-T mass difference is much lower than the experimental one (fig. 4.1). The prediction can be raised to the experimental value by fixing α_s to its Charmonium value everywhere, but this seems not appealing theoretically. In Chapter 2 we saw that a reasonable description of the T'-T mass difference was possible by introducing a logarithmic potential for intermedi-

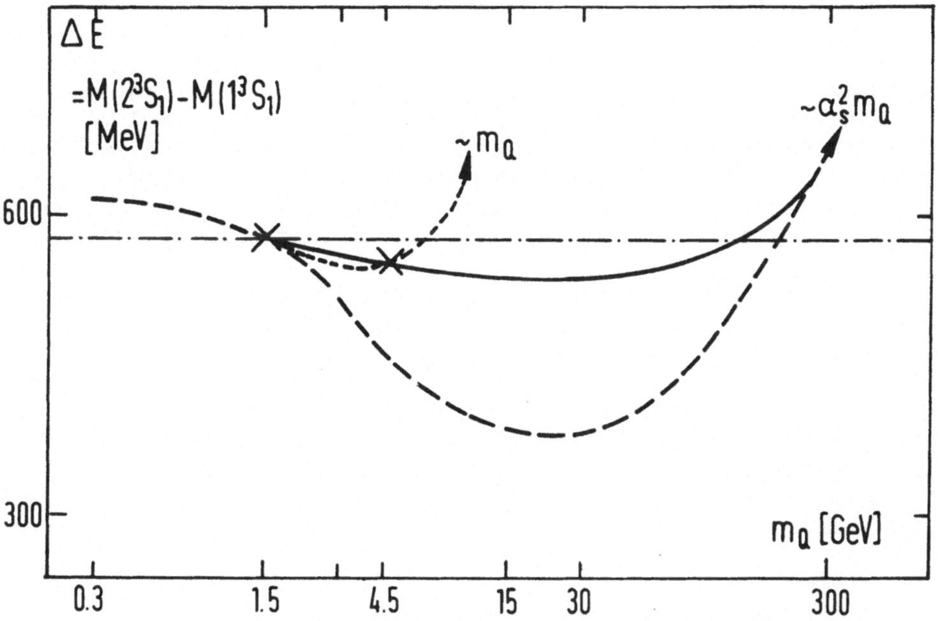

Fig. 4.1 The scaling behaviour of ΔE in different potentials.
— — — standard potential with $\alpha_s(M^2)$ via equ. (1.1)
- - - - - standard potential with fixed α_s
—·—·— logarithmic potential
——— our guess

ate distances. In the log potential ΔE = constant ($\varepsilon = 0$), and an intermediate part in the potential would tend to fill up the valley of the dashed curve in fig. 4.1. We show a guess for the result, the solid line in fig. 4.1. This result means, that we expect no dramatic change of ΔE for the next Quarkonium. Only for quark masses well above 100 GeV the states would sit deeper and deeper in the V_{AF} singularity and ΔE starts to increase. Asymptotically the scaling behaviour of ΔE is $\alpha_s^2 m_Q \sim m_Q/\log^2(m_Q^2)$.

We now turn to level splittings and begin with the P waves. We have shown that the Fermi-Breit Hamiltonian (equ. 3.2) gives a reasonable description. From there we have

$$H^T, H^{LS}_{AF} \sim \frac{1}{m_Q^2} \langle \frac{1}{R} d_R (\frac{1}{R}) \rangle \sim \frac{1}{m_Q^2 R^3} , \qquad (4.7)$$

where H^{LS}_{AF} is the spin orbit term <u>without</u> the Thomas precession. In contrast the Thomas precession term behaves like

$$H^{LS}_C \sim \frac{1}{m_Q^2} \langle \frac{1}{R} d_R (R) \rangle \sim \frac{1}{m_Q^2 R} . \qquad (4.8)$$

The scaling behaviour of R (equ. 4.5) is somewhere between that in a log and in a linear potential, $R \sim m_Q^{-1/2} \ldots m_Q^{-1/3}$, and we can estimate the $^3P_2$ - $^3P_0$ splitting of more massive Quarkonium P waves shown in table 4.1. A comparison of (4.8) with (4.7) shows one more important fact. The ratio of equ. (3.3) which is 0.5 in Charmonium should increase with m_Q and approach 0.8 asymptotically!

The spin spin splittings go essentially as $\alpha_s \cdot \Gamma_{e\bar{e}}$, which can be seen by combining equ. (3.9) with equ. (2.4). Experimentally $\Gamma_{e\bar{e}}$, normalized to the quark charge, is remarkably constant, fig. 4.2. In the frame of nonrelativistic potential models there is no way to explain this for ρ, ω, ϕ. From J/Ψ to Υ, however, we can use the scaling arguments. Table 4.2 shows the scaling behaviour of $|\Psi(0)|^2$ and $\Gamma_{e\bar{e}}$ via equ. (2.4). $|\Psi(0)|^2$ and therefore $\Gamma_{e\bar{e}}$ should feel more of the short distance potential than e.g. the level splittings. Numerical calculations indeed show almost m_Q-independence of

<u>Table 4.1</u> P wave splittings in Quarkonia

Quarkonium:	$c\bar{c}$(3.5 GeV)	$b\bar{b}$(9.8 GeV)	30 GeV
$M(^3P_2) - M(^3P_0)$ |MeV|	150 (input)	50-70	20-40

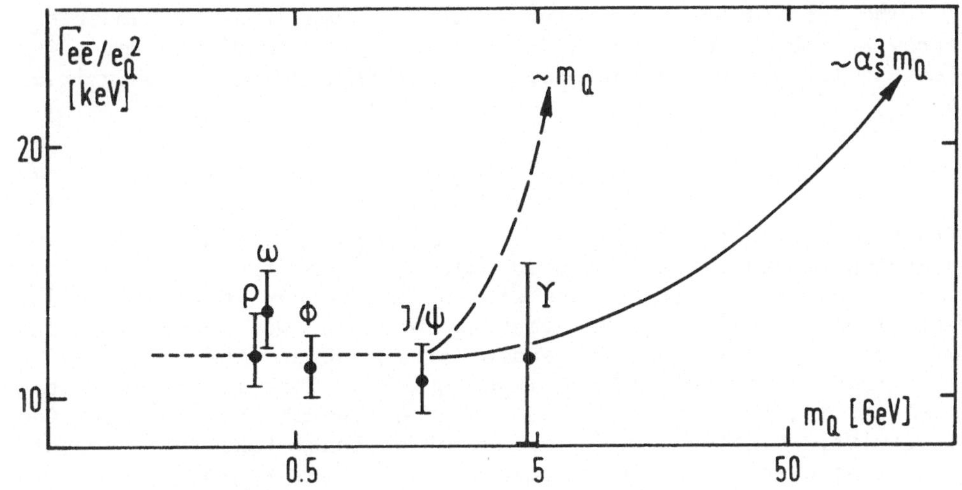

Fig. 4.2 Scaling behaviour of $\Gamma_{e\bar{e}}/e_Q^2$
- - - - Experimental evidence below m_Q = 5 GeV
— — —in a pure Coulomb potential
——— in V_{AF} with $\alpha_s(M^2)$ via equ. (1.1)

$\Gamma_{e\bar{e}}$ in the range from Charmonium Υ [28]. In the asymptotic limit $m_Q \to \infty$, $\Gamma_{e\bar{e}} \sim \alpha_s^3 m_Q \sim m_Q \log^{-3}(m_Q^2)$, which also gives no net m_Q dependence from Charmonium to Υ. We are therefore led to plot this asymptotic m_Q dependence for $\Gamma_{e\bar{e}}$ starting with J/Ψ. This is done in fig. 4.2 also.

The constancy of $\Gamma_{e\bar{e}}/e_Q^2$ below J/Ψ (fig. 4.2), however, cannot be understood with our methods and we want to point out that it is a challenge to explain this fact together with the seemingly constancy of level spacings below 3 GeV, e.g.,
$M(A_2) - M(\rho) \simeq M(\chi 3.55) - M(J/\Psi)$.

Table 4.2 Scaling behaviour of $|\Psi(0)|^2$ and $\Gamma_{e\bar{e}}$ in different potentials

Scaling of	in	$V_{AF}(R) \sim -\frac{\alpha_s}{R}$	$V(R) \sim \log R$	$V_c(R) \sim R$
$\|\Psi(0)\|^2 \sim$	R^{-3}	$\alpha_s^3 m_Q^3$	$m_Q^{3/2}$	m_Q
$\Gamma_{e\bar{e}}$	$\sim R^{-3} m_Q^{-2}$	$\alpha_s^3 m_Q$	$m_Q^{-1/2}$	m_Q^{-1}

The last aspect of scaling we discuss concerns the number of narrow $Q\bar{Q}$ states below the $Q\bar{q}\bar{Q}q$ threshold. The condition for a $Q\bar{Q}$ state to lie below the threshold for strong decays, can be written as

$$E_{Q\bar{Q}} \quad < \quad 2\,m_q \; + \; 2E_{Q\bar{q}} \tag{4.9}$$

with the binding energy $E_{Q\bar{q}} = M_{Q\bar{q}} - m_Q - m_{\bar{q}}$. The binding energy of $Q\bar{Q}$ states depends on the reduced mass of $Q\bar{Q}$ and therefore on the mass of the heavy quark Q. The states fall deeper in the potential well with increasing m_Q. The binding energy of $Q\bar{q}$, however, is in the first approximation independent of m_Q because the system is determined by the mass of the light quark q. This is of course an idealization, to be more sophisticated one would have to treat the relativistic binding problem of $Q\bar{q}$, or at least take into account the slight changes of the reduced mass $\mu = m_Q{\cdot}m_q/m_Q{+}m_q$ with m_Q and effects of the spin-spin interaction which depend stronger on m_Q (but are small). Taking $E_{Q\bar{q}}$ to be constant fixes the threshold for the binding energy $E_{Q\bar{Q}}$. The question one may pose then is: How many $Q\bar{Q}$ S wave states have a binding energy $E_{Q\bar{Q}}$ below this threshold? This question can be answered by semiclassical methods independent of the particular potential. The number n of bound S states below a given energy (r.h.s of equ. (4.9) in this case is given by the Bohr-Sommerfeld condition

$$\int_0^{R_0} dR\,\sqrt{m_Q(E_{thr.} - V(R))} \quad = \quad \pi\,(n - 1/4) \tag{4.10}$$

where R_0 is the classical turning point, $V(R_0) = E_{thr.}$ [29]. For low numbers n (4.10) is only approximately valid (but maybe not worse than our other approximations) and we find

$$n \quad \sim \quad \text{const.} + \sqrt{\frac{m_Q}{m_0}} \tag{4.11}$$

Quigg and Rosner fixed the constant of (4.11) in the Charmonium system ($m_0 = m_c$) and their result is displayed in fig. 4.3. We can read off fig. 4.3 that in the Υ system 3 S waves will be below the threshold of strong decays, the fourth, Υ''', may be even below $Q\bar{q}(\bar{Q}q)^*$ threshold. In any case Υ''' will decay into $B\bar{B}$ or $B\bar{B}^* \to B\bar{B}\gamma$, $B = Q\bar{q}$. The question for the actual threshold energy is not jet answered, to do that we would need calculations of the B masses, e.g. in a potential model. Unfortunately a potential model for the B mesons suffers from the relativistic motion of the light quark q inside the B. However, applying our knowledge about the number of

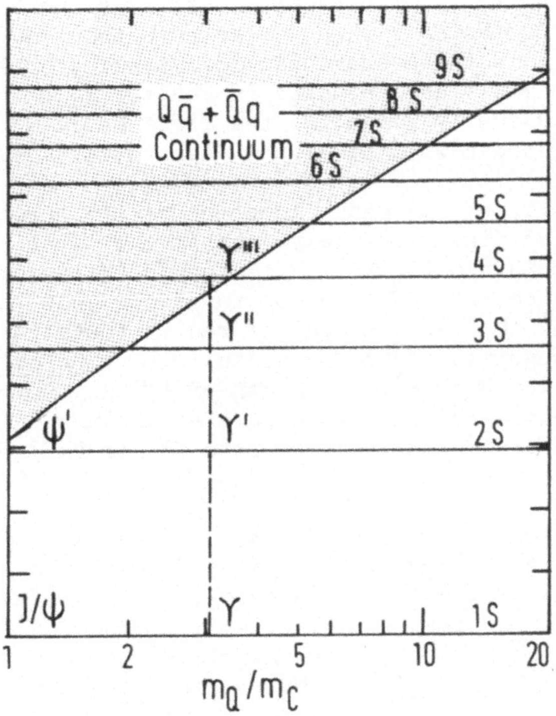

Fig. 4.3 Number of bound states below the strong decay threshold (ref. 29). The T''' will be above the threshold.

bound T S waves, it is sufficient for us to know the masses of T'' and T''', since we already know the threshold relative to these. The latter masses are calculable much more reliably. In table 4.3 the results of two orthogonal approaches are shown.

Table 4.3 Masses and $\Gamma_{e\bar{e}}$ of T radial excitations in the two orthogonal models of a) ref. 30) and b) ref. 20).

		T	T'	T''	T'''
Mass a)	|GeV|	9.46 (input)	10.09	10.45	10.72
Mass b)	|GeV|	9.46 (input)	10.02	10.34	10.60
$\Gamma_{e\bar{e}}$ b)	|keV|	1.1	0.5	0.35	0.3

The model of ref. 30) directly integrates the Bethe Salpeter equation for a $Q\bar{Q}$ system with a distant-dependent $\alpha_s(R)$. The second model, ref. 20), is the phenomenologically successful modification of the standard model as discussed in chapter 2. A look at fig. and tabke 4.3, and slightly rescaling the first model, convinces us that the $B\bar{B}$ threshold will be around 1o.4 to 1o.5 GeV.

Independently of the exact location of the threshold and the exact validity of fig. 4.3 we expect that the first radial T excitation above $B\bar{B}$ threshold is a "B-factory". (We think that this will be T''', of course.) The reason is simply that in the decay of T''' to $B\bar{B}$ or $B\bar{B}^*$ the large number of radial nodes in the T''' wave function will suppress its decay width into two slowly moving ground state S waves like B or B^*. The width of T''' may therefore be well below the resonance machine width in e^+e^- production but, on the other hand, the branching fraction into $B\bar{B}$ (or $B\bar{B}^*$) should be substantial.

One comment on our saying "$B\bar{B}$ or $B\bar{B}^*$" is in order: Either the $B-B^*$ splitting is as large (or larger) as the DD^* splitting, then B^* could decay in πB. But in this case T''' would lie below the $B\bar{B}^*$ threshold, as can be seen from fig. 4.3. Or the $B-B^*$ splitting is less than the $D-D^*$ splitting (in nonrelativistic potential models this splitting goes like $1/m_Q$ - but neither the D nor the B are nonrelativistic), then B^* decays to γB, which experimentally is almost as clean as a pure $B\bar{B}$ decay.

2nd LECTURE

The second lecture covers Quarkonium decays. We will first discuss the radiative photon transitions in E1 and M1 approximation and gluon transitions. These decays have in common that they depend on the medium and long distance behaviour of the wave function. We then (Chapter 6) turn to annihilations which are governed by the short distance behaviour of the wave functions. The annihilation can take place into photons and/or gluons. The gluons may form hadron jets. This is dealt with in Chapter 7.

5. Radiation

a) Electric Dipole Radiation

For photon or gluon wave lengths long against the bound state dimensions of Quarkonium one can try a multipole expansion. The widths of different multipole orders are typically [31)]

$$\Gamma \sim \alpha e_Q^2 \left\{ \begin{matrix} k^3 R^2 \\ \\ k^3 m_Q^{-2} \end{matrix} \right\} \cdot \left(\frac{kR}{2}\right)^{2(n-1)} \text{ for } \left\{ \begin{matrix} E_n \\ \\ M_n \end{matrix} \right\} \text{ transitions.} \tag{5.1}$$

up to numerical factors. k is the photon (gluon) wave number, R the bound state radius in the reduced system (R/2 is the true bound state radius). We see that the expansion parameter in (5.1) is $(k \cdot R/2)^2$ which is roughly 1/4 ... 3/100 in Charmonium and smaller in heavier Quarkonia. This justifies a multipole expansion and we will therefore confine ourselves to the lowest order transitions, E1 and M1.

In hydrogen the formula for an electric dipole transition (E1) is [31)]

$$\Gamma^{E1} (|i\rangle \to \gamma |f\rangle) = \frac{4}{3} \alpha k^3 |\vec{x}_{fi}|^2 , \tag{5.2}$$

where $\vec{x}_{fi}$ is the matrix element of the dipole operator. In Quarkonia we now have three modifications to the case of equ. (5.2). First, both quarks can radiate, not only just one like the electron in hydrogen. Second, the relevant mass is the reduced mass of the quark, $m_Q/2$, not just the particle mass like m_e in hydrogen. Third, the charge of the quark is only $e_Q \cdot e$. The first two modifications cancel each other, so that we are left with

$$\Gamma^{E1} (Q\bar{Q}_i \to \gamma Q\bar{Q}_f) = \frac{4}{3} \alpha e_Q^2 k^3 |\vec{x}_{fi}|^2 \tag{5.3}$$

in Quarkonium.

Of course, there are corrections to this naive formula. The first one are higher multipoles. In Ψ' decays they amount to at most 5% if present (compare equ. (5.1)). The second one is an interference of the finite wave length of the photon field $e^{+ik\cdot R}$ with the bound state wave function. In atomic and nuclear transitions this interference is negligible, $k\cdot R << 1 \Rightarrow e^{ik\cdot R} \simeq 1$. But in Quarkonium transitions higher terms of the expansion of $e^{ik\cdot R/2}$ will partly contribute to dipole transitions and tend to reduce the transition rate. However, Okum and Voloshin [32] have shown that this interference correction amounts to at most 5% in Charmonium. The third but most important corrections are of relativistic nature. They consist of a) recoil corrections, b) relativistic corrections to the wave functions and c) the interaction of the quark magnetic moments with the electric vector of the photon field. The corrections of type c) have been studied by Okun and Voloshin [32]. They find correction factors between essentially 1.0 and 0.6.

The radiative widths of the standard model without corrections of the last type are given in fig. 5.1 and table 5.2. An example for the corrections of this type is shown in table 5.1. The remaining discrepancy between theory (table 5.1) and experiment (fig. 5.1) might be due to relativistic corrections of type a) and b). The recoil corrections have been found to be ≈ +20% in a relativistic model [33]. In any case this indicates that also the model numbers for $\Gamma(P_c/\chi \to \gamma J/\Psi)$ are only good within a factor 2.

b) E1 Sum Rules

A very powerful tool for the discussion of electric dipole transitions has been rediscovered for Charmonium, namely the dipole sum rules [34]. We know two kinds of dipole sum rules, the so called Thomas-Reiche-Kuhn (RTK) sum rule and the Wigner (W) sum rule. Both apply to the dipole matrix element (equ. (5.3)) and any corrections like those discussed have to be done afterwards. The starting point for the dipole sum rules is Heisenberg's uncertainty relation

Table 5.1 Example for the magnitude of relativistic corrections to the naive dipole widths [32].

$\Gamma_{model}(\Psi' \to \gamma^3P_j)$	$^3P_2$	$^3P_1$	$^3P_0$
without corr. |keV|	36	50	58
with corr. type c)	36	40	41

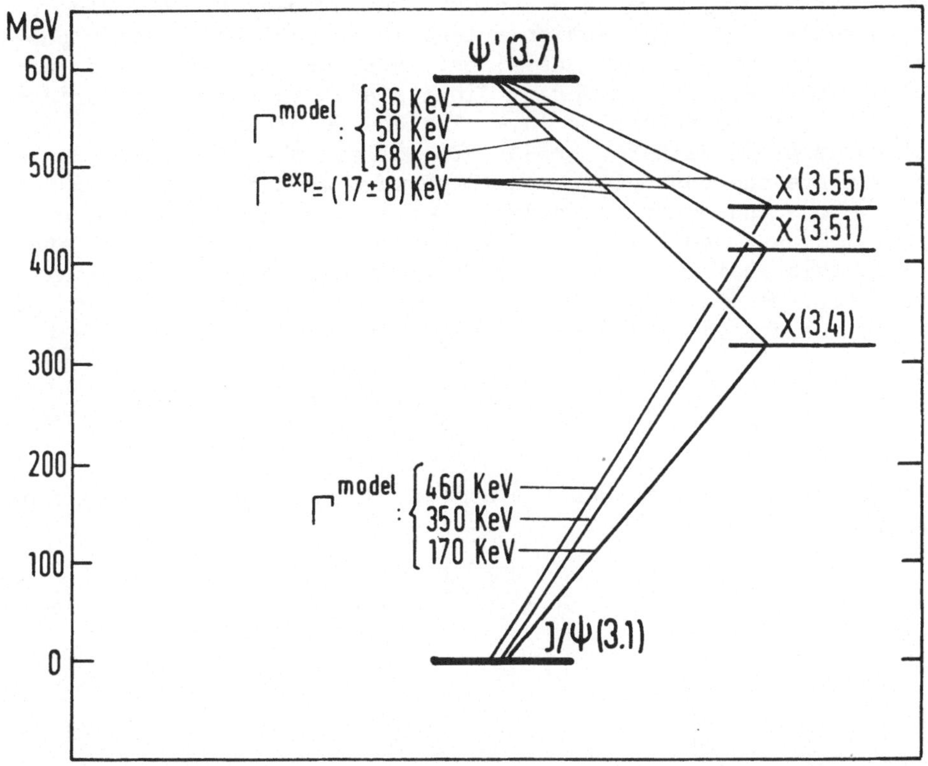

Fig. 5.1 E1 transitions in Charmonium. Modelwidths are calculated via equs. (5.8) and (5.9) and do not include corrections.

$$[\vec{x},\vec{p}] \quad = \quad 3i \tag{5.4}$$

(we set $\hbar = c = 1$). In a static potential for $Q\bar{Q}$ without velocity dependent terms, e.g. no spin-orbit interaction, we can replace p via the equation of motion

$$\vec{p} \quad = \quad i\,\frac{m_Q}{2}\,[H^o,\vec{x}] \tag{5.5}$$

where H^o is the Hamilton operator of the static potential. equ. (2.1). After taking the expectation value in a state $|i\rangle$ and inserting a complete set of states $|f\rangle$ this replacement of $\vec{p}$ leads to

$$\Sigma_f\,(E_f^o - E_i^o)\,[\vec{x}_{fi}]^2 \quad = \quad \frac{3}{m_Q} \tag{5.6}$$

Here E^0 are energy eigenvalues of H^0. The number of final states $|f\rangle$ is restricted by selection rules. In an arbitrary static potential $\Delta l = \pm 1$ for dipole transitions. In a harmonic oscillator potential, however, the number of final states is further restricted by the oscillator selection rule: The change of the number of radial modes Δr is either 0 or $-\Delta l$. It follows that from the S wave ground state one can only reach the P wave ground state, from this 1 P wave one can reach the radially excited S wave, 2 S, the ground state, 1 S, and the D wave 1 D. These are all possible final states. We call this fact the saturation of the sum rule by the harmonic oscillator. To write down the first sum rules it is convenient to express the dipole operator $\vec{x}_{fi}$ through the radial operator R_{fi} 35)

$$\Sigma_{m'} |\langle r', l\pm 1, m' | \vec{x} | r, l, m\rangle|^2 = \{{}^{l+1}_{l}\} \frac{|\langle r', l\pm 1 | R | r, l\rangle|^2}{2l+1} \tag{5.7}$$

where m is the magnetic quantum number. We can now write some rates (5.3) as

$$\Gamma(1P \to \gamma 1S) = \frac{4}{9}\, \alpha\, e_Q^2\, k^3\, |R_{fi}|^2 \tag{5.8}$$

and

$$\Gamma(2^1S_0 \to \gamma 1^1P_1) = \frac{4}{3}\, \alpha\, e_Q^2\, k^3\, |R_{fi}|^2 \tag{5.9a}$$

$$\Gamma(2^3S_1 \to \gamma 1^3P_j) = \frac{4}{3}\, \frac{2j+1}{9}\, \alpha\, e_a^2\, k^3\, |R_{fi}|^2 \tag{5.9b}$$

The TRK sum rule (5.6) gives us a bound

$$(E^o_{1P} - E^o_{1S})\, |R_{1P,1S}|^2 \leq \frac{3}{m_Q} \tag{5.10}$$

which implies an upper bound on $1P \to 1S$

$$\Gamma(1P \to \gamma 1S) \leq \frac{4}{9}\, \alpha\, e_Q^2 \cdot \frac{k^3}{k^{(0)}} \cdot \frac{3}{m_Q} \tag{5.11}$$

We can obtain more bounds with the help of the Wigner sum rule. Recall equ. (5.4). As an expectation value in state $|i\rangle$ it can be written as

$$\Sigma_f \langle i|\vec{x}|f\rangle \langle f|\vec{p}|i\rangle - \langle i|\vec{p}|f\rangle \langle f|\vec{x}|i\rangle = 3i \tag{5.12}$$

The angular selection rule now enables us to project out the final states with $\Delta l = +1$ and those with $\Delta l = -1$. We thus arrive at two sum rules after some elaborate algebra [35]

$$\Sigma_{f,l-1} (E_f^o - E_i^o) \, |\vec{x}_{fi}|^2 = \frac{-l(2l-1)}{2l+1} \cdot \frac{1}{m_Q} \tag{5.13}$$

$$\Sigma_{f,l+1} (E_f^o - E_i^o) \, |\vec{x}_{fi}|^2 = \frac{(l+1)(2l+3)}{2l+1} \, \frac{1}{m_Q} \tag{5.14}$$

which of course add up to (5.6). We have gained two things: first, the number of final states on the l.h.s. of (5.13) and (5.14) is smaller than in the TRK sum rule, and second, (5.13) is negative, which is very helpful. For l=1 in the initial state the first two terms of (5.13) give (using (5.7))

$$(E_{2S}^o - E_{1P}^o)|R_{2S,1P}|^2 + (E_{1S}^o - E_{1P}^o)|R_{1S,1P}|^2 \leq \frac{-1}{m_Q} \tag{5.15}$$

An upper bound for the second term on the l.h.s. is known from (5.10). This leaves us with

$$(E_{2S}^o - E_{1P}^o) \, |R_{2S,1P}|^2 \leq \frac{2}{m_Q} \tag{5.16}$$

and we can deduce an upper bound on transition (5.9):

$$\Gamma(2^3S_1 \to \gamma 1^3P_j) \leq \frac{4}{3} \, \frac{2j+1}{9} \, \alpha \, e_Q^2 \, \frac{k^3}{k^{(0)}} \cdot \frac{2}{m_Q} \tag{5.17}$$

Next we will make use of the negative sign in equ. (5.13) with l=1, the initial state being the 1P wave. The only contribution to (5.13) or (5.15) which is indeed negative is the transition to the 1S ground state. Its magnitude must be larger than the sum of all others! Therefore the knowledge of one of the other transitions, e.g. 2S → γ1P, gives us a lower limit on 1P → γ1S! We write (5.15) as

$$(E_{1P}^o - E_{1S}^o)|R_{1S,1P}|^2 \geq \frac{1}{m_Q} + (E_{2S}^o - E_{1P}^o)|R_{2S,1P}|^2 \tag{5.18}$$

and obtain by "inverting" (5.17)

$$\Gamma(1^3P_j\to\gamma 1^3S_1) \geq \frac{4}{9}\alpha e_Q^2 \frac{k^3}{k^{(0)}}\frac{1}{m_Q} + \frac{3}{2j+1}\frac{k^3_{1P,1S}\,k^{(0)}_{2S,1P}}{k^3_{2S,1P}\,k^{(0)}_{1P,1S}}\,\Gamma^{exp}(2^3S_1\to\gamma 1^3P_j) \tag{5.19}$$

The W sum rule gave us an upper bound on 2S → γ1P and a lower bound on 1P → γ1S. The TRK sum rule gave us an upper limit on the latter transition. We combine all our information in table 5.2. Combining the bounds of table 5.2 and the experimentally measured BRs for $P_c/\chi \to \gamma J/\Psi$ one can deduce bounds for the total widths of the P states in Charmonium. This is shown in table. 5.3

Table 5.2 Upper and lower limits on E1 transitions from the Thomas-Reiche-Kuhn (TRK) and Wigner (W) sum rules (SR). All widths in keV. The second numbers in the lower half of the W SR column arise from the second term r.h.s of (5.19). The quark mass is taken to be m_c = 1.6 GeV.

transition	TRK SR	W SR	model
$2^3S_1 \to \gamma 1^3P_2$		< 40	36
$2^3S_1 \to \gamma 1^3P_1$		< 56	50
$2^3S_1 \to \gamma 1^3P_0$		< 64	58
$1^3P_2 \to \gamma 1^3S_1$	< 490	>160 + 140	460
$1^3P_1 \to \gamma 1^3S_1$	< 370	>125 + 75	350
$1^3P_0 \to \gamma 1^3S_1$	< 180	> 60 + 30	170

Table 5.3 Bounds on $\Gamma_{tot}(P_c/\chi)$ derived from the sum rules, table 5.2, and the experimental BRs of $P_c/\chi \to \gamma J/\Psi$. The sum rules correspond to an uncorrected E1 transition, this gives an additional theoretical uncertainty of a factor 2.

P states	$\chi(3.41)=0^{++}$	$P_c/\chi(3.51)=1^{++}$	$\chi(3.55)=2^{++}$
BR(γJ/Ψ), exp. |%|	3 ± 3	35 ± 7	14 ± 6
$\Gamma_{tot}(P_c/\chi)$, bounds |MeV|	3 ... 6	0.57 ... 1.05	2.15...3.5

The total widths of the P_c/χ states should be calculable as the sum of the radiative widths plus the gluon annihilation widths. A comparison of these total widths with the bounds of table 5.3 will be a comparison of theory with "experiment". We will do that in a forthcoming chapter.

c) Magnetic Dipole Transitions

M1 decays arise from an interaction of the magnetic photon field vector $\vec{m} = \vec{k} \times \vec{\varepsilon}$ and the quark magnetic moment $\mu_Q = e \cdot e_Q / 2m_Q$. The matrix element therefore reads

$$<f| \mu_Q \vec{\sigma} \cdot (\vec{k} \times \vec{\varepsilon}) |i> \qquad (5.20)$$

and acts on the spin part of the states $|i>$ and $|f>$ only. Again we have two graphs for the emission of a photon and therefore 4 times the rate as in atomic M1 transitions [31]

$$\Gamma(V \to \gamma PS) = \frac{16}{3} \mu_Q^2 k^3 \delta^{rr'} = \frac{4}{3} \alpha e_Q^2 \frac{k^3}{m_Q^2} \delta^{rr'} \qquad (5.21)$$

$$\Gamma(PS \to \gamma V) = 3\Gamma (V \to \gamma PS)$$

An M1 transition requires $\Delta l=0$ and the spatial overlap between the two states $|i>$ and $|f>$ with number of radial nodes r and r' is either $1(r = r')$ or $0(r \neq r'$, forbidden M1) in this approximation. Relativistic corrections of course modify the rate (5.21) and lead to small transitions also between orthogonal $(r \neq r')$ states. In allowed M1 transitions $(r = r')$ the spatial overlap of 1 cannot be changed much by relativistic corrections.

d) Scaling of E1 and M1

Before we now discuss the M1 transitions in charmonium, let us look at the scaling behaviour of both kinds of dipole transitions. For E1 transitions the scaling behaviour is most easily obtained from the sum rules.

$$\Gamma \sim \frac{k^3}{k^{(0)}} \frac{1}{m_Q} \approx \frac{k^2}{m_Q} \qquad (5.22)$$

M1 transitions, on the other hand, scale like

$$\Gamma \sim \mu_Q^2 \, k^3 \sim \frac{k^3}{m_Q^2} \tag{5.23}$$

i.e. $\Gamma(M1) \sim \Gamma(E1) \cdot k/m_Q$. Since in the next heavy Quarkonia k does not increase with m_Q the relative magnitude of M1 compared to E1 goes down at least like $1/m_Q$. A comparison of related radiative transitions in different Quarkonia can thus help to distinguish E1 from M1 transitions!

e) Problems with M1 in Charmonium

In fig. 5.2 possible candidates for the pseudoscalars and the corresponding M1 transitions are shown. If the second χ is not at

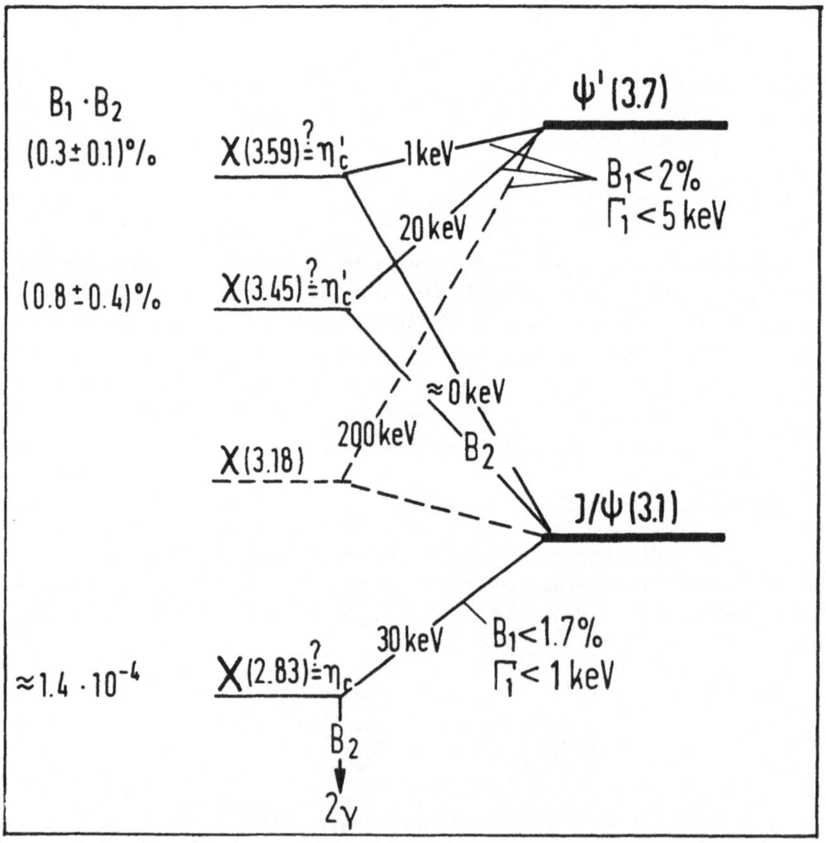

Fig. 5.2 M1 transitions in Charmonium. Theoretical widths, equ. (5.21) are indicated at the transition lines. $B_1(\Gamma_1)$ and $B_1 \cdot B_2$ are from experiment, ref. 1) and 36).

Table 5.4 Experimental upper bounds on B_1 and lower bounds on B_2 via $B_1 \cdot B_2$, and comparison with theory. The kind of transition for B_1, B_2 is indicated in fig. 5.2. The theoretical numbers arise from allowed and "forbidden" M1 transitions and the ratio of 2γ versus 2 gluon annhihilation. For the latter see chapter 6. The forbidden M1 transition should lead to a B2 not bigger than a few 10 keV/a few MeV $\approx 10^{-2}$

State		$\chi(3.59)$	$\chi(3.45)$	$X(2.83)$
$B_1 \cdot B_2$ (Exp.)	$\|\%\|$	0.3±0.1	0.8±0.4	0.014±0.004
B_1 (Exp.)	$\|\%\|$	< 2	< 2	< 1.7
B_1 (Theory)	$\|\%\|$	≃ 0.5	≃ 9	≃ 45
B_2 (Exp.)	$\|\%\|$	> 10	> 20	> 0.7
B_2 (Theory)	$\|\%\|$	< 1	< 1	≈ 0.1

3.59 GeV but at 3.18 GeV (second experimental solution) it can hardly be explained as a pseudoscalar. In fig. 5.2 the calculated M1 widths are shown. They have at first to be contrasted with the experimental bound on these transitions as indicated. Together with the experimental product of branching ratios these bounds allow to derive lower limits on the decay branching fractions of these states. This is shown in table 5.4. There is no way of assigning one of the experimental states to a pseudoscalar state without coming in trouble with a) absolute M1 widths, b) branching fractions for the decay of this state. Considering η_c and η_c' in context leads to even larger discrepancies, e.g. take $\chi(3.59)$ as η_c' and $X(2.83)$ as η_c. Then the M1 transition $J/\Psi \rightarrow \gamma\eta_c$ is down by a factor of 30 compared to the naive theory. The same factor must work in $\Psi' \rightarrow \gamma\eta_c'$ leading to $\eta_1 = 1/30$ keV and consequently to $B_2 >$ 30! For M1 widths only one unpleasant way out seems possible: to give the quarks a vanishing magnetic moment μ_Q in this limit of a static interaction [37].

A much more pleasant way out would be finding the true pseudoscalars much nearer to J/Ψ and Ψ' respectively. Experimentally this is in no way ruled out. Then the X and χ states are either not real or at least no simple $Q\bar{Q}$ states [38]. Remember that QCD is consistent with a possible existence of multiquark or multiquark-gluon states different from $Q\bar{Q}$ [39]. However, their properties are not accessible in our simple Quarkonium model.

f) Gluon Radiation

Radiative gluon transitions can be subject to a similar multipole expansion as electromagnetic radiation. While the expansion in $(k\ R/2)^2$ might converge, the expansion in α_s/π needs not, The distances involved in the process are of the order of the wave function radius and α_s will be large. This is the essential reason why we do not expect to be able to calculate rates for gluon radiation. But we might be able to estimate the scaling behaviour of such radiation. The processes which are described as gluon radiation in QCD are typically

$$2^3S_1 \rightarrow 1^3S_1 + ng \quad \lfloor\!\rightarrow \varepsilon, \eta, \ldots \tag{5.24}$$

The emitted states must be Isosinglets, because gluons carry no Isospin. For the radiation of an $\varepsilon(\pi\pi$ S wave) from 2^3S_1 Gottfried (ref. 40)) estimates an $1/m_Q^2$ behaviour of the matrix element

$$\Gamma(2^3S_1 \rightarrow 1^3S_1 + \varepsilon) \sim \frac{1}{m_Q^2} \cdot \text{phase space} \tag{5.25}$$

If this scaling law is already valid in the Charmonium system, $\Psi' \rightarrow \pi\pi\ J/\Psi \approx 100$ keV implies $\Upsilon' \rightarrow \pi\pi\Upsilon \simeq 10$ keV. In a 30 GeV $Q\bar{Q}$ system this width would be no more than 1 keV. Transitions via gluon radiation will be important for a search for $Q\bar{Q}$ states which are not accessible directly or via photon transitions, like the 1^1P_1 state. In the Υ or higher $Q\bar{Q}$ systems the 3^3S_1 state (Υ'' e.g.) will be narrow and undergo such a transition to the 1^1P_1 state.

$$3^3S_{1--} \rightarrow 1^1P_{1+-} + \eta, \varepsilon \tag{5.26}$$

The finding of a 1^1P_1 state via (5.26) would be very interesting because the knowledge of the 1^1P_1 mass allows to determine whether there are long range spin spin correlations or not. In our Ansatz for the Hamiltonian and potential we only had short range spin spin forces. They do not act on P waves and therefore the 1^1P_1 state is degenerate with the c.o.g. of the 1^3P states. A long range spin-spin force, however, would act on the P waves and would lift this degeneracy.

6. Annihilation

Quarkonium states may annihilate into photons and/or gluons. Since annihilation is a pointlike process (the quarks must come together) not only the annihilation into photons is governed by a

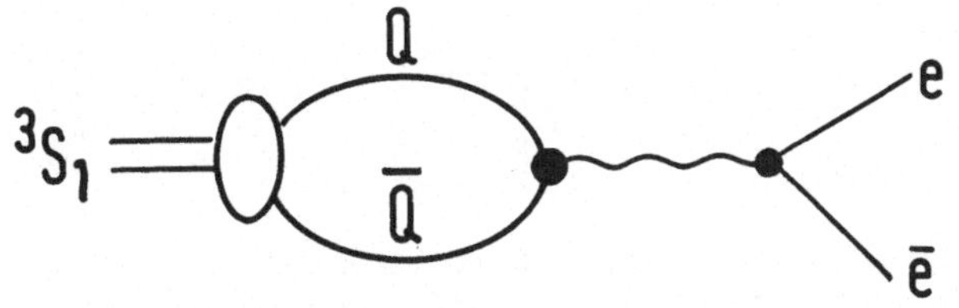

Fig. 6.1 Leptonic decay of $^3S_1(Q\bar{Q})$. The electrons may be replaced by μs, τs or quarks lighter than Q.

a small coupling $\alpha = 1/137$, but hopefully also that into gluons by α_s (small R). We can apply the 'minimal gluon scheme', i.e. approximate the decay by the lowest order (Born-) graph [41]. This will be justified by finding that indeed the α_s (annihilation) is small, even in Charmonium it is much smaller than the effective α_s for the bound state description (see chapter 2). We proceed in the following way. First we collect well known formulae for annihilations in Born approximation. In this approximation there is no gluon selfinteraction yet, so that the conversion from photon annihilations to gluon annihilations is just done by redefining the charge. We will then discuss ratios of these widths as an application in Quarkonia. Our results will also be fundamental for the next chapter on jets.

a) Annihilation Formulae

The vector $^3S_1$ ground state can decay via one photon into lepton or quark pairs (hadrons). The corresponding graph is displayed in fig. 6.1 and the formula is known as Van Royen-Weisskopf formula (ref. 42)) (including colour and for $4m_e^2 << M_V^2$):

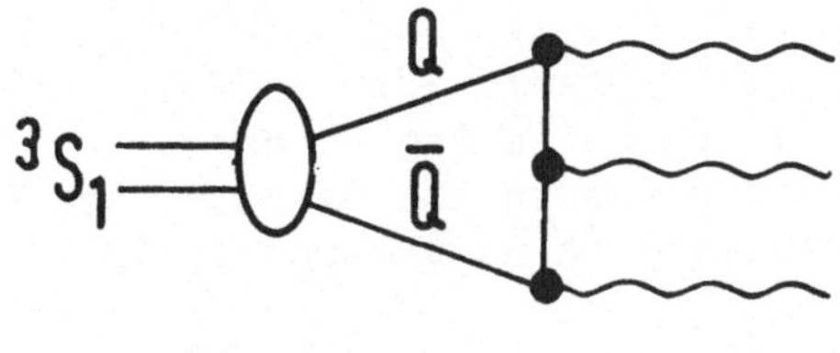

Fig. 6.2 3γ decay of $^3S_1(Q\bar{Q})$. When the photons are replaced by gluons, this denotes the "direct" hadronic decay.

$$\Gamma_{e\bar{e}}(V) = 16\,\pi\,\alpha^2\,e_Q^2\,\frac{|\Psi(0)|^2}{M_V^2} \simeq \alpha^2\,e_Q^2\,\frac{|R(0)|^2}{m_Q^2} \tag{6.1}$$

where M_V is the $V = {}^3S_1$ bound state mass and $m_Q \simeq 1/2\ M_V$ the quark mass. $\Psi(0)$ is the spatial and R(0 the radial wave function at the origin. Quarks couple in the same way to the photon as leptons, so that (6.1) is understood for each lepton or quark flavour separately: $\Gamma_{q\bar{q}} = 3e_q^2\ \Gamma_{e\bar{e}}$.

The decay of ${}^3S_1$ into two photons as well as two gluons is impossible. In the two photon case this is just the photon C parity. Also two gluons, as long as they are in a colour singlet state (which is symmetric), have even C. But the ${}^3S_1$ can decay into three photons as well as three gluons, fig. 6.2. The three photon decay has been calculated by Ore and Powell [43] (here including the statistical colour factor)

$$\Gamma_{3\gamma}(V) = \frac{4}{3}\,\alpha^3\,e_Q^6\,\frac{\pi^2-9}{\pi}\,\frac{R(0)}{m_Q}^2 \tag{6.2}$$

The conversion factor to the three gluon decay is [44]

$$\Gamma_{3g}/\Gamma_{3\gamma} = \frac{\alpha_s^3}{\alpha^3 e_Q^6}\,\frac{1}{9}\,\Sigma_{a,b,c}\left[\mathrm{Tr}\left(\frac{\lambda^a}{2}\,\frac{\lambda^b}{2}\,\frac{\lambda^c}{2}\right)_{\mathrm{sym.}}\right]^2 \tag{6.3}$$

so that we have

$$\Gamma_{3g}(V) = \frac{10}{81}\ {}^{3}_{s}\ \frac{\pi^2-9}{\pi}\,\frac{|R(0)|^2}{m_Q^2} \tag{6.4}$$

The parts of (6.3) have the following origin. $\alpha_s^3/\alpha^3 e_Q^6$ just converts the charges together with $\mathrm{TR}(\lambda^a/2\ \lambda^b/2\ \lambda^c/2)_{\mathrm{sym.}}$. The Σ_{abc} counts the number of coloured graphs in the 3g case, while the 3^{-2} counts the number of coloured graphs in the 3γ case. We do not consider decays of the ${}^3S_1$ into more (≥ 5) photons or (≥ 4) gluons.

The pseudoscalar ${}^1S_0$ ground state can decay into two photons or two gluons, fig. 6.3. The two photon decay was first calculated

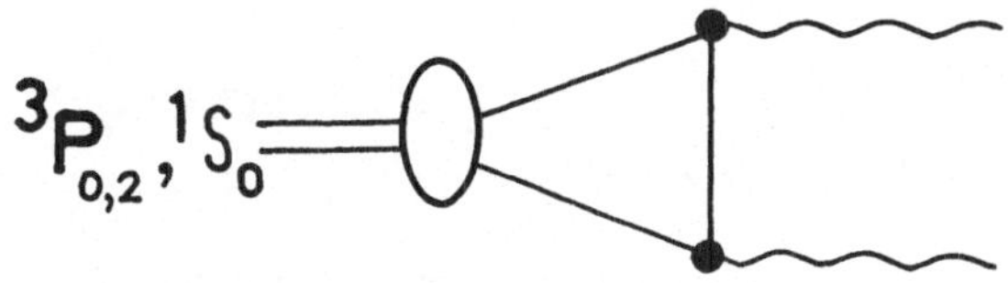

Fig. 6.3 2γ decay of $Q\bar{Q}$. For the hadronic decay the photons are replaced by gluons.

by Pomeranchuk [45] and is (including colour)

$$\Gamma_{2\gamma}(PS) = 3\,\alpha^2\,e_Q^4\,\frac{|R(0)|^2}{m_Q^2} \tag{6.5}$$

With the conversion factor

$$\Gamma_{2g}/\Gamma_{2\gamma} = \frac{\alpha_s^2}{\alpha^2 e_Q^4}\,\frac{1}{9}\,\Sigma_{a,b}\left[\mathrm{Tr}\left(\frac{\lambda^a}{2}\,\frac{\lambda^b}{2}\right)\right]^2 \tag{6.6}$$

whose components are described in the case of equ. (6.3) one obtains

$$\Gamma_{2g}(PS) = \frac{2}{3}\,\alpha_s^2\,\frac{|R(0)|^2}{m_Q^2} \tag{6.7}$$

We do not discuss the decay of $^1S_0$ in more (≥ 4) photons or (≥ 3) gluons. Assuming, that the 2g decay is the basic process for the dominant hadronic decay of the pseuodscalar, allows to derive the branching fraction for the 2γ decay (table 5.4) from equ. (6.6).

We now turn to P wave annihilation, fig. 6.3 and 6.4. Here life is more complicated because the wave function of a P wave at the origin is zero. That means that the quarks do not like to come together to annihilate: The annihilation widths of P waves will be smaller than that of the $^1S_0$ wave! The P waves, however, can annihilate when the two quarks come near each other and simultaneously have a relative velocity $\neq 0$. This is a higher order process in terms of an expansion in $\beta^2 = (v/c)^2$. It is governed by the spatial derivative of the wave function. In this approximation the widths of the spin 0 and spin 2 P waves of Positronium have first been calculated by Alekseev [46]. The same calculation for Charmo-

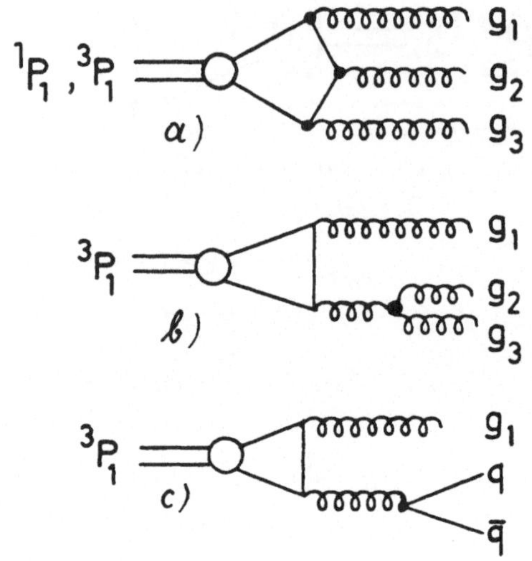

Fig. 6.4 The gluonic decay diagrams of spin 1 P waves

nium has been done by Barbieri, Gatto and Kögerler [47]. They yield

$$\Gamma_{2g}(^3P_o) \quad = \quad 6\,\alpha_s^2\,\frac{|R'(0)|^2}{m_Q^4} \tag{6.8}$$

$$\Gamma_{2g}(^3P_2) \quad = \quad \frac{8}{5}\,\alpha_s^2\,\frac{|R'(0)|^2}{m_Q^4} \tag{6.9}$$

The 2γ widths of $^3P_{0,2}$ can be obtained from (6.8) and (6.9) by the conversion factor given in equ. (6.6).

The decays of the j = 1 P waves are more complicated. A spin 1 state cannot decay into two massless vector bosons, either photons or gluons in a colour singlet [48]. We therefore have to consider the next order (in α_s) diagrams, which for gluon annihilation are shown in fig. 6.4. They bring up another complication. We now have a three body phase space and have to integrate over all possible energies of, say, gluon 1. Gluon 1 is allowed to be soft. It further is allowed to carry away the angular momentum of the P wave. So it has all characteristics of a bremsstrahlungs gluons. The same is true for photon annihilation, except that in this case diagram b) of fig. 6.4 is absent. A bremsstrahlungs gluon or photon in the annihilation of a free $Q\bar{Q}$ pair with l = 1 leads to the typical bremsstrahlungs singularity. The cross section factorizes into the bremsstrahlungs part and the annihilation of an $l = 0$ $Q\bar{Q}$ pair into two photons or gluons. For a bound state, however, the annihilation amplitude cannot be singular, because the quarks are not on shell. Their virtuality is of the order of the bound state dimensions. For a bound state annihilation we therefore may cut the amplitude at momenta of the soft (bremsstrahlungs) photon or gluon which correspond to the bound state radius. In diagram language, the singularity will be cancelled by higher order graphs like vertex corrections. For QED this procedure is well defined [49]. We hope that it will work parallel for QCD. As a cutoff momentum for QCD annihilation we take the typical momentum for a soft "confinement" gluon, 400 MeV, since in a QCD process higher order graphs will involve such "confinement" gluons. We will express the cutoff in terms of a parameter $\Delta = 2M \cdot 400$ MeV [50], M being the Quarkonium bound state mass. Let us first discuss $^1P_1$ decay. This state has $j^{PC} = 1^{+-}$ and therefore only diagram a) of fig. 6.4 can contribute, in either photon or gluon annihilation. Its decay has been calculated by Barbieri, Gatto and Remiddi [49]. They find

$$\Gamma_{3g}(^1P_{1+-}) \simeq \frac{20}{9}\,\alpha_s^3\,\frac{|R'(0)|^2}{m_Q^4}\,\log\frac{M^2}{\Delta} \qquad (6.10)$$

where the log arises from the bremsstrahlungs singularity of the diagram. For the decay of the $^3P_1$ state, $j^{PC} = 1^{++}$, only diagram c) can contribute to the photon annihilation while in principle all three diagrams can contribute to the gluon annihilation. Barbieri, Gatto and Remiddi [49] found that the singular parts of the diagrams a) and b) cancel each other. Okun and Voloshin [32] gave the general argument for this: The amplitudes a) and b) interfere, since they lead to the same final state. Since they can both be factorized into the bremsstrahlungs part times the corresponding annihilation diagram for the 2 gluon annihilation of a coloured $^3S_1$ state, also their sum can be factorized in this way. This sum, however, contains all graphs to this order for $^3S_1$ (coloured) $\to$ 2g, which must be zero [32]. Neglecting the non-singular parts of amplitudes a) and b) against the singular c) means that also for the gluon annihilation the calculation of graph c) is sufficient. It gives [49,50]

$$\Gamma_{g\bar{q}q}(^3P_{1++}) \simeq \frac{N}{3}\,\frac{8\,\alpha_s^3}{3\,\pi}\,\frac{R'(0)^2}{m_Q^4}\,(\log\frac{M^2}{\Delta} - \frac{1}{2}) \qquad ((6.11)$$

where N is the number of light flavours q. The photon versions of (6.10) and (6.11) can be found in ref. 32).

For completeness we note the formula for the decay of the spin 2 D wave into 2 gluons which is given by the second derivative of the wave function, this is the second order in an expansion of $\beta^2 = (v/c)^2$ and therefore even less reliable. Okun and Voloshin 32) calculated

$$\Gamma_{2g}(^1D_2) = \frac{2}{3}\alpha_s^2 \frac{|R''(0)|^2}{m_Q^6} \tag{6.12}$$

b) Ratios and Applications

The ratio of equs. (6.4) and (6.1) gives

$$\frac{\Gamma(^3S_1 \to 3g)}{(^3S_1 \to e\bar{e})} = \frac{10}{81}\frac{\pi^2-g}{\pi}\frac{\alpha_s^3}{\alpha^2 e_Q^2} \simeq 1440 \frac{4}{9e_Q^2}\alpha_s^3 \tag{6.13}$$

If we interpret as usual the 3g annihilation as the total direct hadronic annihilation then this is a measurable quantity and we have e.g. in Charmonium

$$\frac{\Gamma(J/\Psi \to \text{hadr})_{\text{dir.}}}{\Gamma(J/\Psi \to e\bar{e}} \simeq 10 \tag{6.14}$$

from which follows that the α_s at annihilation distances is $\alpha_s \simeq 0.19$. Because of the third power of α_s in (6.13) this value is quite stable even against large corrections on the widths. The kinds of corrections we have discussed to equ. (6.1) at the end of chapter 3 and different ones for Γ_{3g} will not be able to achieve an agreement between $\alpha_s(\text{spectrum}) \simeq 0.4$ and $\alpha_s(\text{annihilation}) \simeq 0.2$ in the Charmonium system. But this discrepancy does not surprise, as we have discussed in chapter 1.

A very interesting ratio is that of equ. (6.8) to equ. (6.10) to equ. (6.9):

$$\Gamma_{2g}(^3P_{0^{++}}) \quad : \quad \Gamma_{gq\bar{q}}(^3P_{1^{++}}) \quad : \quad \Gamma_{2g}(^3P_{2^{++}}) \tag{6.15}$$

$$= \quad 15 \quad : \quad \frac{20\,N\,\alpha_s}{9\pi}\left(\log\frac{M^2}{\Delta}-\frac{1}{2}\right) \quad : \quad 4$$

It leads to ratios of

Table 6.1 Comparison of "experiment" and theory for the P wave total widths, including the radiative transitions. The "experiment" line is taken from tables 5.2 and 5.3. The prospects of the Υ system are also given.

		$\Gamma_{tot}(^3P_0)$ |MeV|	$\Gamma_{tot}(^3P_1)$ |MeV|	$\Gamma_{tot}(^3P_2)$ |MeV|
$c\bar{c}$	theory	4.	0.5	1.5
	"quasiexp."	6 ± 6	1 ± 0.2	3.2 ± 1.6
$b\bar{b}$	$\alpha_s = 0.15$	0.35	0.05	0.15
	$\alpha_s = 0.2$	0.6	0.08	0.2

$$\begin{array}{lll} 15 : 2.1\,\alpha_s : 4 & & J/\Psi \\ 15 : 5.7\,\alpha_s : 4 & \text{in the} & \Upsilon \quad \text{system} \\ 15 : 11\,\alpha_s : 4 & & 30\text{ GeV } Q\bar{Q} \end{array} \tag{6.16}$$

We can of course calculate more than these ratios, namely the total widths of the P waves, assuming that these are given by the gluon annihilation width and radiative transition width essentially. The result is shown in table. 6.1 for charmoniun and the Υ system, and compared to the quasiexperimental bounds of table 5.3. For the calculation of equs. (6.8) - (6.10) we need $|R'(0)|^2$. Numerical calculations give $|R'_{c\bar{c}}(0)|^2 m_c^{-4} \simeq 15$ MeV and $|R'_{b\bar{b}}(0)|^2 m_b^{-4} \simeq 2.5$ MeV. These quantities are relatively quark mass independent. We conclude that although the widths of table 6.1 are very model dependent, the pattern of (6.16) agrees very well with the observed branching ratios of the Charmonium P waves. This is one of the successful predictions of QCD within Charmonium.

We complete our discussion of ratios of widths with a discussion of the $^3S_1$ decays. The decay channels of the vector ground state are: i) into lepton pairs, $e\bar{e}$, $\mu\bar{\mu}$, $\tau\bar{\tau}$, ii) into hadrons, $\Sigma\, q\bar{q}$, the ratio of ii) against i) is essentially given by the famous R, iii) the three gluon annihilation, and iv) the annihilation into one photon and two gluons. The only ratio missing so far is that of iv) to iii). We can estimate it by comparing the electromagnetic and strong coupling for one fermion-boson vertex and by taking into account the different coupling of the colours of two versus three gluons

Table 6.2 Ratios of the ground state decay channels a) in Charmonium, b) in the Υ system, c) in a 30 GeV $t\bar{t}$ system. For Charmonium $\alpha_s = 0.19$ agrees with experiment (lowest order formulae). For Υ decays the value of α_s best compatible with experiment, $B_{\mu\bar{\mu}}$ [52], seems to be 0.18 at present.

$c\bar{c}$ decay channel:	$e\bar{e}+\mu\mu$	:	$\Sigma\, q\bar{q}$	:	3g	:	$\gamma 2g$
$e_Q = 2/3$	2	:	R	:	$\frac{5}{18}\frac{\pi^2-9}{\pi}\frac{\alpha_s^3}{\alpha^2}$	:	$\frac{8}{9}\frac{\pi^2-9}{\pi}\frac{\alpha_s^2}{\alpha}$
a) $\alpha_s = 0.19$	2	:	2.5	:	10	:	1.2

$b\bar{b}$ decay channel:	$e\bar{e}+\mu\bar{\mu}$	:	$\Sigma\, q\bar{q}+\tau\bar{\tau}$	:	3g	:	$\gamma 2g$
$e_Q = -1/3$	2	:	R	:	$\frac{20}{18}\frac{\pi^2-9}{\pi}\frac{\alpha_s^3}{\alpha^2}$	:	$\frac{8}{9}\frac{\pi^2-9}{\pi}\frac{\alpha_s^2}{\alpha}$
b) $\alpha_s = 0.15$	2	:	5	:	20	:	0.8
$\alpha_s = 0.18$	2	:	5	:	34	:	1.1

$t\bar{t}$ (30 GeV) decay channel:	$e\bar{e}+\mu\bar{\mu}$	:	$\Sigma\, q\bar{q}+\tau\bar{\tau}$	:	3g	:	$\gamma 2g$
$e_Q = 2/3$	2	:	R	:	$\frac{5}{18}\frac{\pi^2-9}{\pi}\frac{\alpha_s^3}{\alpha^2}$	:	$\frac{8}{9}\frac{\pi^2-9}{\pi}\frac{\alpha_s^2}{\alpha}$
c) $\alpha_s = 0.12$	2	:	5	:	2.5	:	0.5
$\alpha_s = 0.15$	2	:	5	:	5	:	0.8

$$\frac{\Gamma(^3S_1 \to \gamma gg)}{\Gamma(^3S_1 \to 3g\ )} = \frac{36}{5}\,\frac{\alpha\, e_Q^2}{\alpha_s} \tag{6.17}$$

For Charmonium three exclusive contributions to $^3S_1 \rightarrow \gamma gg$ have been seen so far, namely $J/\Psi \rightarrow \gamma\eta$, $\gamma\eta'$, γf [51]. Together with equ. (6.13) this is all we need to put up table 6.2.

7. Jets

The exploration of QCD suffers from the fact that its constituents, the quark and gluons, cannot exist as free particles because of the confinement. Their properties cannot be investigated directly. But here is a surrogate for the observation of the free constituents, that are the jets. Experimentally jets are observed not only in deep inelastic hadron-hadron and lepton-hadron scattering but especially in e^+e^- annihilation, once the c.m. energy of 5 GeV is exceeded. The angular distribution of these jets is completely consistent with the production of two spin 1/2 (almost) massless particles [53], the quarks, via photon vacuum polarisation. The fragmentation of quarks into hadrons is imagined as a nonperturbative confinement effect, which conserves the original directed momenta.

At present there is no way of calculating this process, but there exists a very suggestive picture: Inside a small space region of ≈ 1/2 fm colour can exist and within this region the $q\bar{q}$ pair (or gluon) production is a short distance effect (see fig. 7.1). When hard coloured quanta (quarks or gluons) with momenta p_i reach the confinement sphere they must fragment into white hadrons since colour fields cannot exist outside this sphere. The coloured quanta break up into hadrons with a finite perpendicular momentum $p_\perp$. This breaking up is energetically much favoured over a further existence as coloured quanta. When the perpendicular momenta are small compared to the longitudinal hadron moment, which

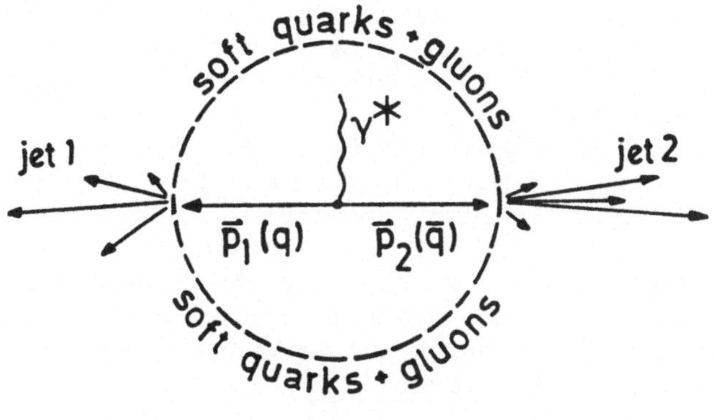

Fig. 7.1 Quark jets

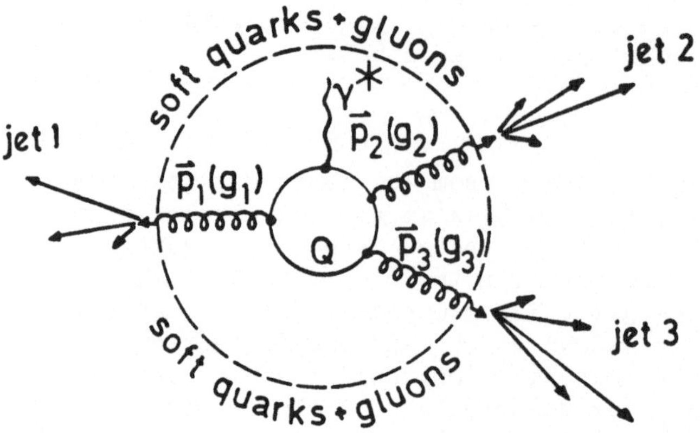

Fig. 7.2
$Q\bar{Q} \rightarrow 3$ gluon jets.

add up to the momentum of the original quantum, we see hadron jets. The confinement effects, however, are assumed to be soft, carried by long wavelength quarks and/or gluons. The wavelength corresponds to the colour bag of 1/2 fm. Therefore the jet momenta equal the original quantum momenta up to the order of 400 MeV. This picture demands the production of the original jet quanta to be a short distance effect (<< 1/2 fm). This is certainly true for the (electromagnetic) quark pair production in e^+e^-. It is also true for a hard gluon bremsstrahlung process [54]. Resonance decays, however, are not pointlike but involve propagators (fig. 7.2 and 6.2). Here it is not so clear, how well the jet picture will work. However, because the propagators are mass dependent the picture will work the better the higher mass of the decaying $Q\bar{Q}$ resonance is. For a Q-mass of 5 GeV the propagator length in fig. 7.2 is probably already short enough to apply the jet picture and for the next new flavour (higher) $Q\bar{Q}$ resonance it will definitely be so.

The quark jets in e^+e^- annihilation became visible above $s = (p_1+p_2)^2 \gtrsim (5\text{ GeV})^2$, i.e. a massless quark needs $\gtrsim 2.5$ GeV of energy against the c.m. to be able to form a jet. For gluons the jet threshold certainly is not lower. But a gluon carries the colour indices of a quark antiquark pair and each index may fragment separately. Then the multiplicity of the jet may be higher and the longitudinal hadron momenta may be lower. In the limit of asymptotic energies the gluon may just fragment like a $q\bar{q}$ pair, each quark carrying half the gluon momentum [55]. From this picture follows that a gluon jet of a certain longitudinal momentum will have a higher multiplicity and a larger opening angle than a quark jet of the same momentum. The threshold for gluon jet production will be higher than that for quark jet production with an upper bound of two times the quark threshold+).

+) Speaking of a jet threshold we refer to the energy of a single quark or gluon versus the center of mass of the colour bag.

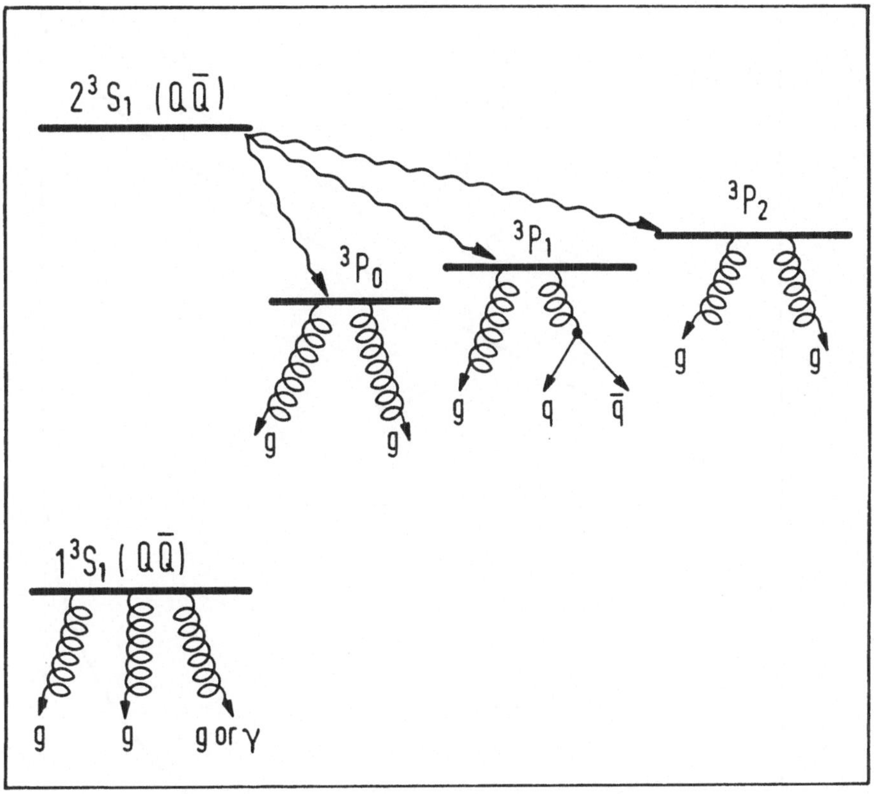

Fig. 7.3 Possible sources of gluon jets in heavy Quarkonia

Some possible sources of gluon jets are shown in fig. 7.3, the pseudoscalars are omitted, they may also form 2 jets out of the 2 decay gluons. We begin with the $^3S_1$ decay into 3 gluons. The three gluons of this decay will form a plane. The angular distribution of the normal $\hat{n}$ of this plane against the beam is

$$\frac{d\Gamma}{d\cos\theta_{\hat{n}e}} \sim 3 - \cos^2\theta_{\hat{n}e} \qquad (7.1)$$

For these decays one defines a variable T "Thrust", which is just the scaled energy of the most energetic gluon, $T = x_1 = 2p_{g1}/M_{Q\bar{Q}}$. The direction of g_1 defines the Thrust axis. The differential rate of the 3 gluon decay together with the angular distribution of this Thrust axis is shown in fig. 7.4. While off resonance the coefficient of the $\cos^2$ term, α, is uniquely 1, it shows a T dependence

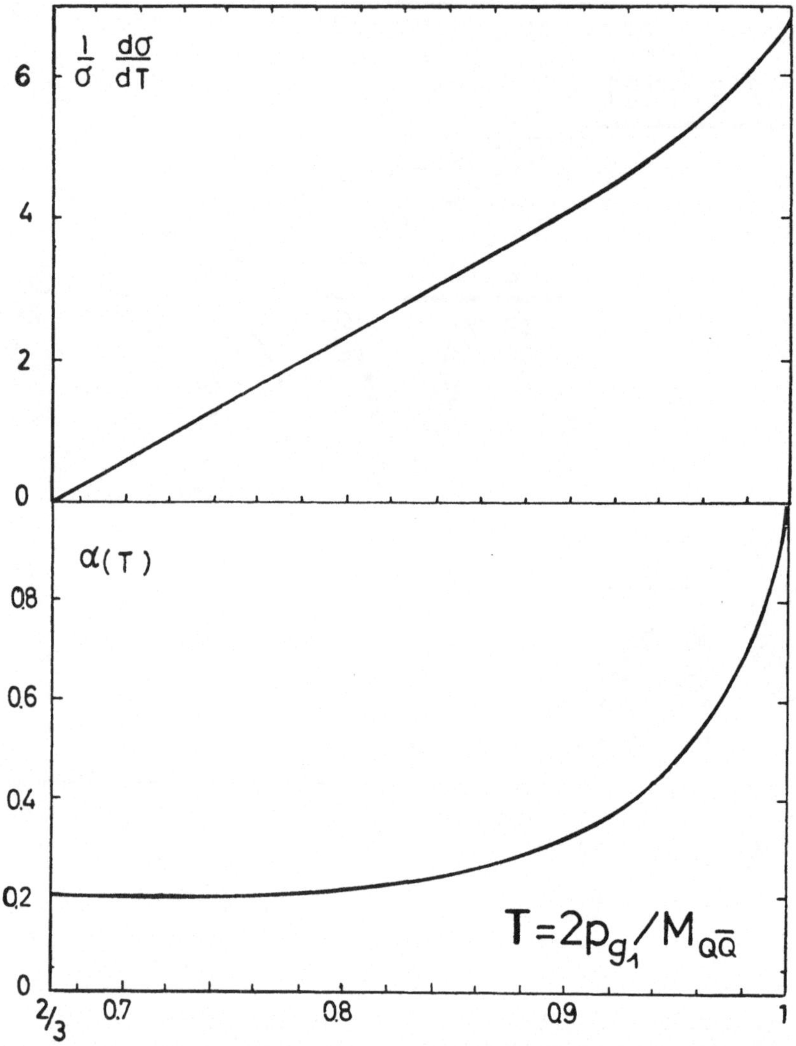

Fig. 7.4 The differential rate of $^3S_1(Q\bar{Q}) \to 3g$ and the thrust angular distribution $W \sim 1+\alpha(T)\cos^2\theta_{Te}$ as functions of T. θ_{Te} is the angle between the thrust axis and the beam.

for $Q\bar{Q}$ decays. The average of $\alpha(T)$ for $Q\bar{Q} \to 3g$ is 0.39. A much more detailed discussion is given in ref. 56).

Once the 3 gluon jet decay and the γ+2 gluon jet decay is found, we can start to compare deviations from the lowest order

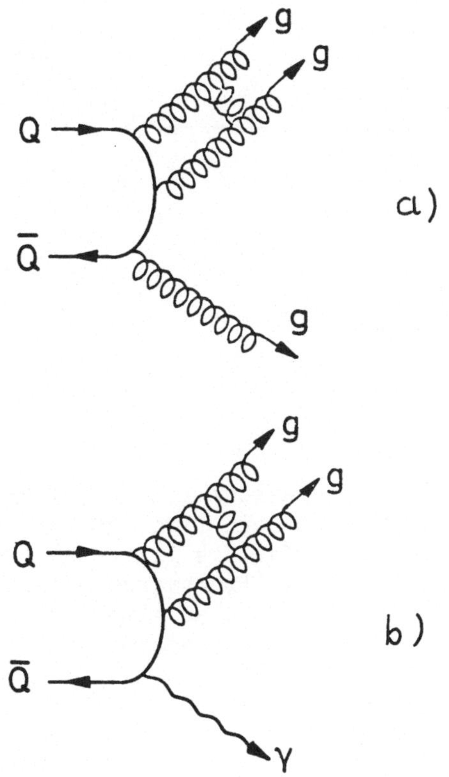

Fig. 7.5 Possible next order (α_s/π) interactions between gluon jets.

angular distributions, which arise through different interactions between two gluon jets, fig. 7.5. The lowest order (Born-approximation) graph gives for the opening angle of the second and third energetic gluon θ_{23} (compare fig. 7.6) the distribution displayed in fig. 7.7. The deviations from this distribution will be different in case a) and b) of fig. 7.5 because in case a) the interaction between the two gluon jets happens in a colour octet (they should repel) whereas in case b) it is in a colour singlet (they should attract).

We will remain at the $\gamma 2g$ decay for another while. The kinematics of this process differ from the 3g decay because the γ can be identified for all photon momenta between 0 and $M_{Q\bar{Q}}/2$. The distributions corresponding to fig. 7.4 are given in fig. 7.8. One notices that the angular correlation drops very fast to a minimum if

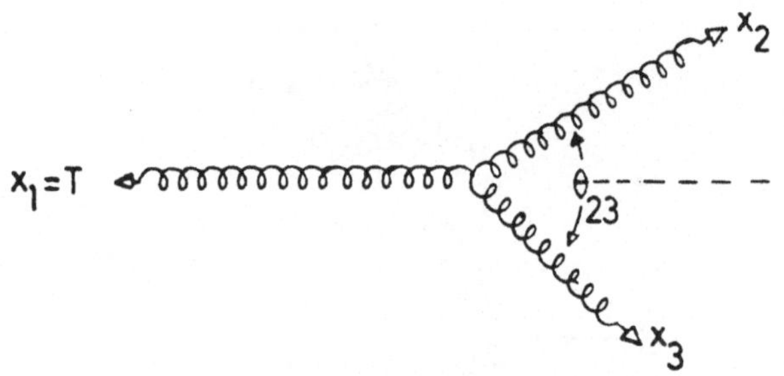

Fig. 7.6 Definition of θ_{23}.

one goes away from the kinematical limit $E_\gamma = M_{Q\bar{Q}}/2$. For the limit $E_\gamma = 1/2\ M_{Q\bar{Q}}$ the coefficient α in front of the $\cos^2\theta$ term is +1. This is easy to understand. In this limit the two gluons have to go parallel. Their helicities (transverse polarisation) have to add up to either 0 or ±2 (for scalar gluons it is 0). Since the photon on the other side is also transverse, the decaying helicity

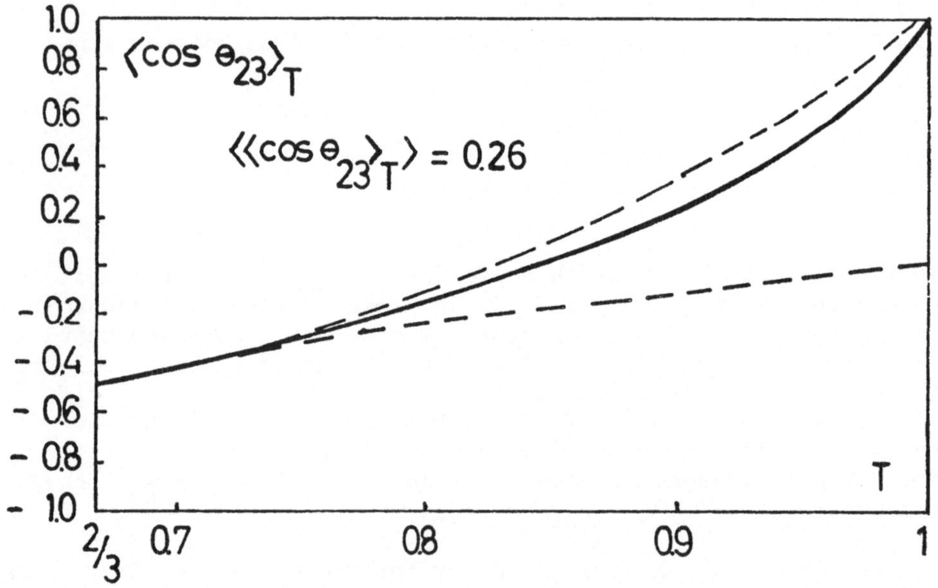

Fig. 7.7 The mean value of θ_{23}, as defined in fig. 7.6 as a function of T. The dashed lines show the kinematic boundaries.

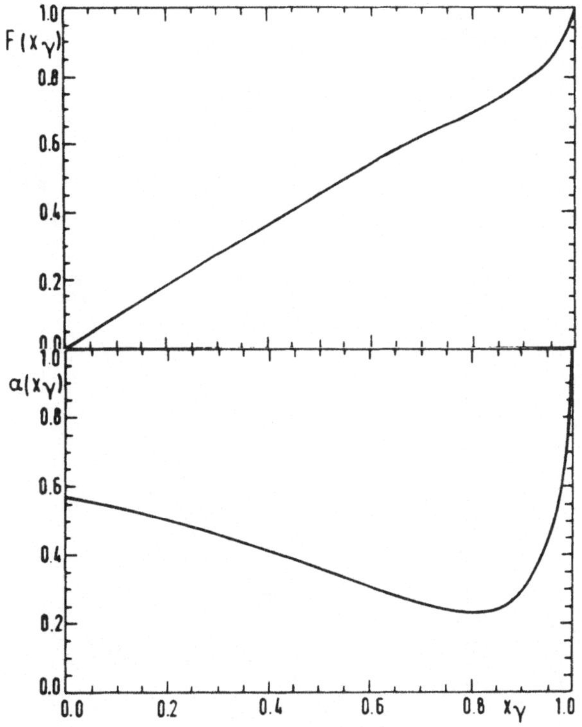

Fig. 7.8 Differential rate $F(x_\gamma) = d\Gamma/dx_\gamma$ and angular distribution $1+\alpha(x_\gamma)\cos^2\theta_{\gamma e}$ as a function of $x_\gamma = 2E_\gamma/M_{Q\bar{Q}}$.

state is the $\lambda = \pm 1$ state. This leads to $1 + \cos^2\theta$. One can show further, that the helicities of the parallel gluons are opposite. If we give the gluon pair a small angle, the net helicity remains zero most of the time. But now we can Lorentz transform to the c.m.s. of the gluon pair and find $\lambda_{cms} = \pm 2$! This means: If low mass hadrons are produced in the process $Q\bar{Q} \to \gamma + 2g \to \gamma$ + hadron, the gluon mechanism favours spin ≥ 2 hadrons over spin 1 or spin 0 hadrons. By this spin argument we can understand the rate for $J/\Psi \to \gamma f$, which is of the same magnitude as $J/\Psi \to \gamma\eta$ and $\gamma\eta'$ although the η and η' should couple to two gluons much stronger because of their large violation of the Zweig rule. The whole argument can of course be made quantitative. The ratios of the helicity amplitudes for $1^{--}(^3S_1) \to \gamma + gg \to \gamma + 2^{++}(^3P_2)$ will depend on the two gluon (or hadron) mass. This is shown in fig. 7.9 [57]. At the point, $J/\Psi \to \gamma f$, these ratios have been measured and agree with this QCD estimate, see fig. 7.10. They also agree with the tensor meson dominance (TMD) model studied here at Karlsruhe [58].

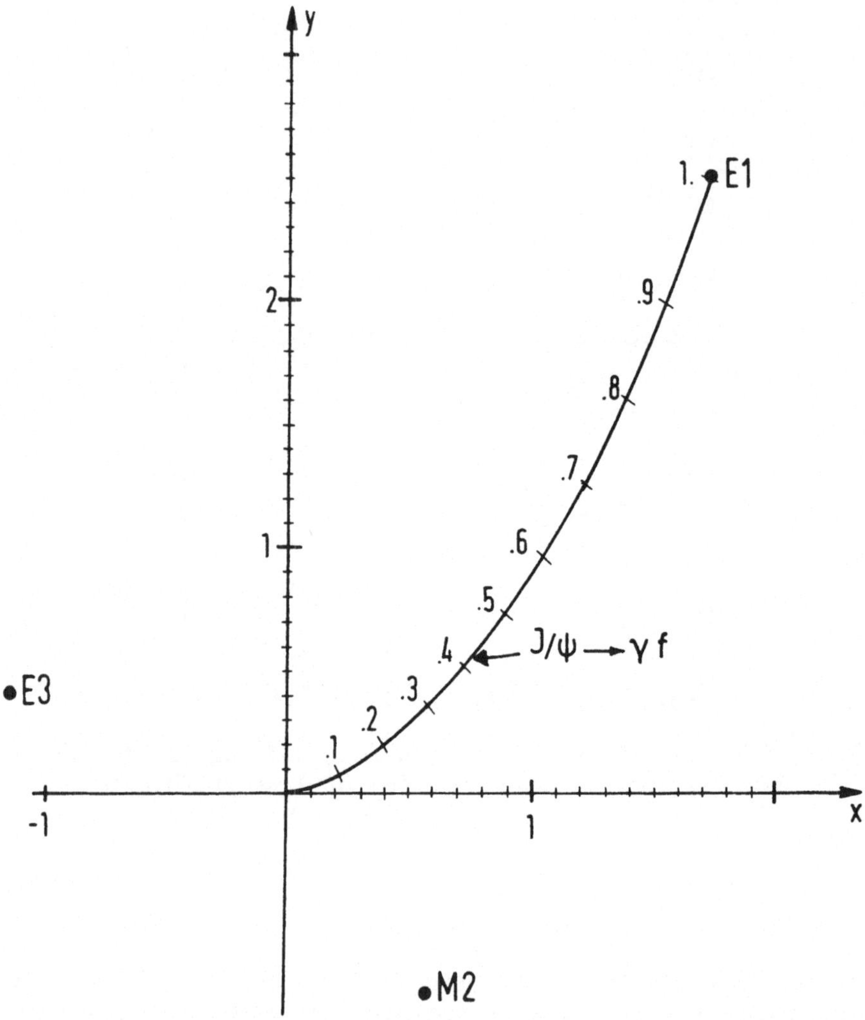

Fig. 7.9 Helicity amplitudes in $^3S_1(Q\bar{Q}) \to \gamma + {}^3P_2(q\bar{q})$ as a function of $M(^3P_2)/M(Q\bar{Q})$. The $A_{|i|}$ are helicity amolitudes. E1, M2, E3 denote the familiar multipole transitions, the (x,y) pair for $J/\Psi \to \gamma f$ is indicated. $x = A_1/A_0$, $y = A_2/A_0$.

From our short excursion we now return to two jets from P wave decays. The first P wave of Quarkonium can be reached from the first radially excited S wave, e.g. ϒ', via an E1 transition. Experimentally it will be necessary to trigger on this monochromatic photon to identify the P wave. The P state then can decay into

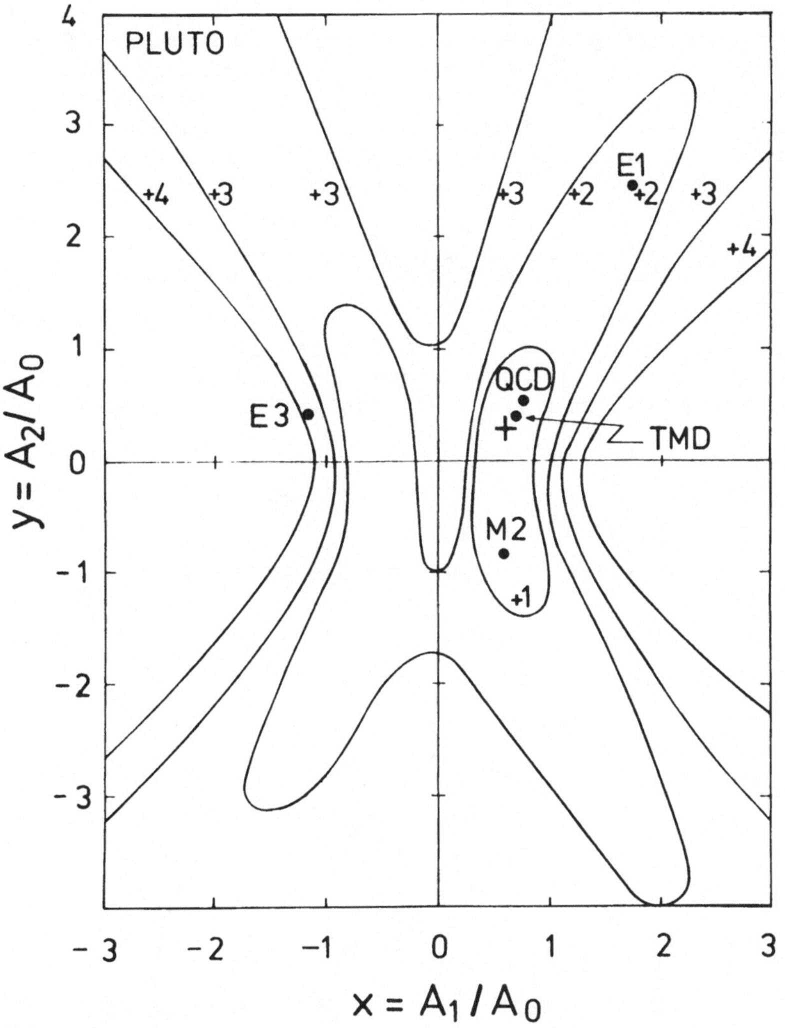

Fig. 7.10 A measurement of the predicted (x,y) pair of fig. 7.9 for $J/\Psi \rightarrow \gamma f$ by the PLUTO collaboration. The cross is the central value of experiment, the lines indicate standard deviations.

2 gluons in case of the $^3P_0$ and $^3P_2$ states. We will discuss the jet decay of the $^3P_1$ state later. These two gluons have a distinct energy of half the P state mass. This is the essential difference to the 3 jet decay of Quarkonium. Here we have monochromatic jets. In Υ' the jet energy is almost 5 GeV, this should be sufficient to

determine the original gluon direction via the jet direction. A measurement of the gluon angular distributions becomes feasible! For the decay of the $^3P_0$ state this angular distribution is trivial: no matter, what the dynamics are, there is only one helicity amplitude which can contribute. But in the $^3P_2$ decays there are two two independent helicity amplitudes for massless gluons. The QCD matrix element for the $^3P_2 \to gg$ decay reads with $q \equiv k_1 - k_2$

$$\varepsilon_{\mu\nu}(\lambda)\,|4k_1\cdot k_2\varepsilon_1^{*\mu}\varepsilon_2^{*\nu}-\varepsilon_1^*\cdot\varepsilon_2^* q^\mu q^\nu+2k_1\cdot\varepsilon_2^*\varepsilon_1^{*\mu}q^\nu-2k_2\cdot\varepsilon_1^*\varepsilon_2^{*\mu}q^\nu| \qquad (7.2)$$

and it turns out that the decay is in the helicity $\lambda = \pm 2$ state. Equ. (7.2) with $\varepsilon_{\mu\nu}(0)$ just vanishes for transverse $\varepsilon_1, \varepsilon_2$. The formula for the kinematics gives us, integrated, the distribution

$$W_{2g}^{2^{++}}(\theta_{\gamma j}) \sim 1 + \cos^2\theta_{\gamma j} \qquad (7.3)$$

where $\theta_{\gamma j}$ is the angle between the trigger photon and one of the jets, measured in the c.m.s. of the jets (fig. 7.11). If the $^3P_2$

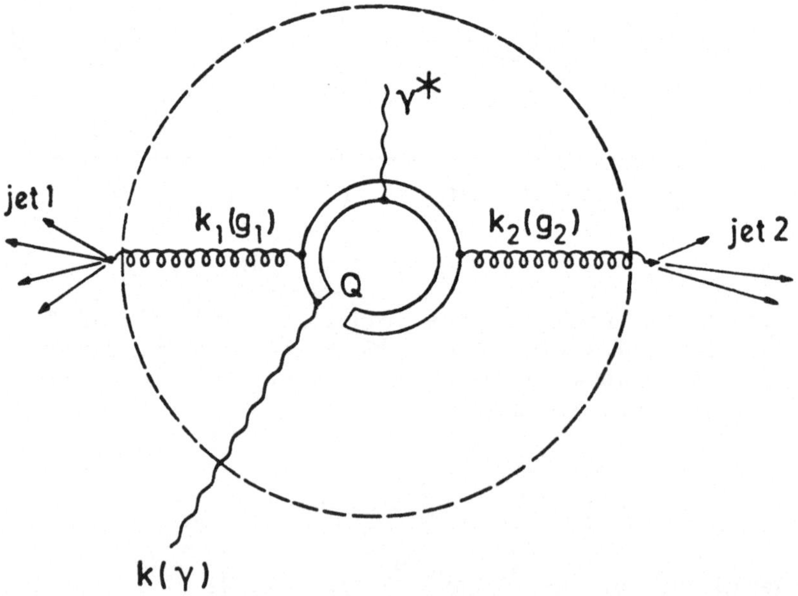

Fig. 7.11 $2^3S_1(Q\bar{Q}) \to \gamma + 1^3P_{0,2}(Q\bar{Q}) \to \gamma + 2g$ jets, as imagined within the colour bag.

would decay into two quark jets by some arbitrary mechanism, the helicity of the two quarks can at most add up to $\lambda = \pm 1$. The kinematic formula then gives

$$W^{2^{++}}_{q\bar{q}}(\theta_{\gamma j}) \quad \sim \quad 1 - \frac{6 + 3A^2}{10 + 9A^2} \cos^2 \theta_{\gamma j} \quad , \tag{7.4}$$

where A gives the weight of helicities $\lambda = \pm 1$ over helicity 0. The sign difference between (7.4) and (7.3) allows a clear test of the QCD mechanism. The rate for this process will be around 5% of all Υ' decays [15].

As we have discussed in chapter 6) the $^3P_1$ decay proceeds via the complicated graph c) in fig. 6.4. The decay is displayed again in fig. 7.12. We will see two quark jets and a hadron cloud from the soft gluon from this decay. The quark jets should be easy to detect. Their angular distribution is given by

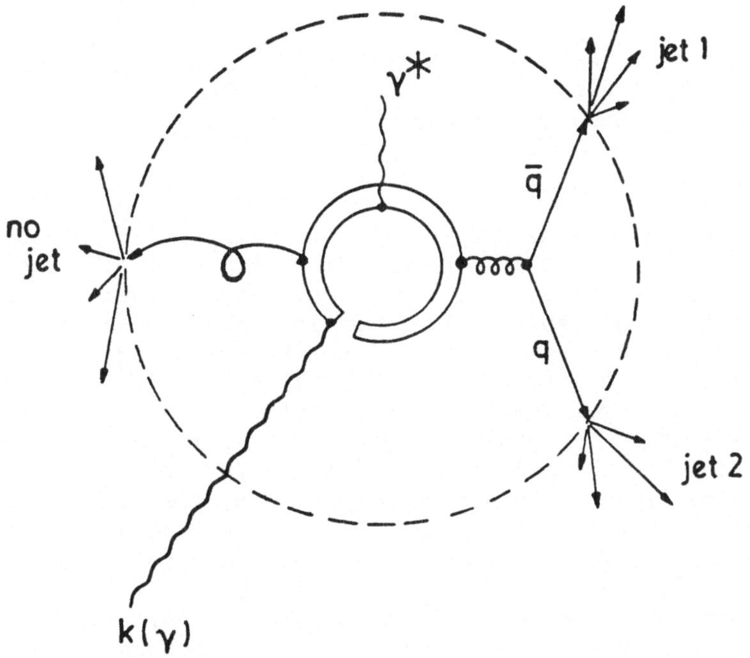

Fig. 7.12 $2^3S_1(Q\bar{Q}) \to \gamma + 1^3P_1(Q\bar{Q}) \to \gamma + 2$ quark jets. Here a soft gluon recoils against the two quark jets.

$$W^{1^{++}}_{gq\bar{q}} \sim 2 - \cos^2\theta_{\gamma e} + \cos\theta_{\gamma e}\cos\theta_{\gamma j}\cos\theta_{je} , \qquad (7.5)$$

already smeared over the important kinematic regime of small gluon momenta [50,59]. $\theta_{\gamma j}$ is the same angle as before, $\theta_{\gamma e}$ is the angle between the trigger photon and the beam, say e^-, and θ_{je} is the angle between the same jet arm as for $\theta_{\gamma j}$ and e^-. $\theta_{\gamma j}$ is measured in the lab frame, but θ_{je} as well as $\theta_{\gamma j}$ are in the c.m.s. of the jets. As an alternative process, the decay into two massless quarks would give [59]

$$W^{1^{++}}_{q\bar{q}} \sim 1 - \cos\theta_{\gamma e}\cos\theta_{\gamma j}\cos\theta_{je} . \qquad (7.6)$$

Here the $\cos^2\theta_{\gamma e}$ term is missing and the term linear in the cosines has a different sign. But the most important difference between the two alternatives (7.5) and (7.6) is the recoil of the soft gluon in the $gq\bar{q}$ decay of the $^3P_1$ state. This recoil will be much larger than the recoil of the trigger photon alone which of course is always present. But with the recoil of the trigger photon alone, the 2 jets would be collinear up to a 10° deviation at most. From the recoil of the soft gluon in the decay of the $^3P_1$ state, however, the angle between the two quark jets may be as small as 110°. This is true for the T system. For a heavier Quarkonium the "soft" gluon may even form a third jet in a small subset of all events.

8. Conclusion

The simplest ansatz for the $Q\bar{Q}$ potential, which is possible using the hints from QCD, works astonishingly well. The short distance spin dependent part of the potential describes the spin orbit splittings reasonably well! If it is correct, heavier Quarkonia should show i) a decrease of the LS splittings $\sim 1/m_Q$ roughly, ii) a tendency of $R = (M(2^{++})-M(1^{++}))/(M(1^{++})-M(0^{++})) \to 0.8$ from 0.5 in $c\bar{c}$. The confinement part of the potential may be spin-independent, as suggested by lattice gauge theories. Its strength depends on the ansatz for the potential at intermediate distances, it is, however, very close to the value suggested by the higher orbital excitations of light mesons (Regge slope). Many details depend on the proper choice for the intermediate distance potential. QCD gives us no hint here. To speculate a little, level spacings might remain almost the same for the next Quarkonium while $\Gamma_{e\bar{e}}/e_Q^2$ might increase very slightly. The number of narrow (= bound) states below the new flavour threshold will increase for the next Quarkonia. T''' might turn out as a perfect B meson factory ($M(T''') \simeq 10.6$ GeV).

We seem to understand parity changing photon transitions in terms of E1 radiation. This means that we understand the "size" of Charmonium. We also seem to understand the branching fractions of P wave decays via the special QCD annihilation mechanism into gluons. This is a short distance phenomenon. We further seem to understand the relative magnitude of $J/\Psi \rightarrow \gamma f$ and $\gamma\eta$ via a simple gluon spin argument.

Up to now we do not know <u>any</u> Quarkonium pseudoscalar state definetely. The experimental candidates X(2.83), $\chi(3.45)$, $\chi(^{3.59}_{3.18})$ cannot be understood in terms of QCD. Especially their M1 transitions und gluon annihilation properties should be much different from what is observed for these states.

Our hopes for the future are that gluon jets show up. Then we can measure the gluon spin and verify certain QCD processes like $^3P_1 \rightarrow gq\bar{q}$. In the $^3S_1$ decay we can study the gluon selfinteraction by comparing γgg vs. $3g$ decays. Finding the gluons is most interesting and important, since they are the gauge bosons of the supposed nonabelian gauge theory of strong interactions, QCD, as the $W^{\pm}$, Z, and γ in weak and electromagnetic interactions.

REFERENCES

1. J.J. Aubert et al., Phys. Rev. Lett. 33 (1974) 1404;
J.E. Augustin et al., Phys. Rev. Lett. 33 (1974) 1406;
G.S. Abrams et al., Phys. Rev. Lett. 33 (1974) 1453;
For reviews see:
B.H. Wiik and G. Wolf, DESY 78/23 (May 1978);
G.J. Feldman and M.L. Perl, Phys. Rep. 33C (1977) 285

2. M. Gell-Mann, Phys. Lett. 8 (1964) 214;
G. Zweig, CERN Preprints TH 401, 412 (1964)

3. J.D. Bjorken and S.L. Glashow, Phys. Lett. 11 (1964) 255;
S.L. Glashow, J. Iliopoulos and L. Maiani, Phys. Rev. D2 (1970) 1285;
M.K. Gaillard, B.W. Lee and J.L. Rosner, Rev. Mod. Phys. 47 (1975) 277

4. T. Appelquist and H.D. Politzer, Phys. Rev. Lett. 34 (1975) 43;
A. De Rújula and S.L. Glashow, Phys. Rev. Lett. 34 (1975) 46;
T. Appelquist et al., Phys. Rev. Lett. 34 (1975) 365;
E. Eichten et al., Phys. Rev. Lett. 34 (1975) 369

5. S.W. Herb et al., Phys. Rev. Lett. 39 (1977) 252;
W.R. Innes et al., Phys. Rev. Lett. 39 (1977) 1240;
Ch. Berger et al., Phys. Lett. 76B (1978) 243
C.W. Darden et al., Phys. Lett. 76B (1978) 246;
J.K. Bienlein et al., Phys. Lett. 78B (1978) 360;
C.W. Darden et al., Phys. Lett. 78B (1978) 364

6. M. Kobayashi and K. Maskawa, Progr. Theor. Phys. 49 (1973) 652;
Y. Achiman, K. Koller and T.F. Walsh, Phys. Lett. 59B (1975) 261;
For a review see:
J. Ellis et al., Nucl. Phys. B131 (1977) 285

7. H. Fritzsch, M. Gell-Mann and H. Leutwyler, Phys. Lett. B47 (1973) 365;
D.J. Gross and F. Wilczek, Phys. Rev. D8 (1973) 3497;
S. Weinberg, Phys. Rev. Lett. 31 (1973) 494

8. G. 't Hooft, unpublished;
D.J. Gross and F. Wilczek, Phys. Rev. Lett. 30 (1973) 1343;
H.D. Politzer, Phys. Rev. Lett. 30 (1973) 1346

9. K.G. Wilson, Phys. Rev. D10 (1974) 2445;
J. Kogut and L. Susskind, Phys. Rev. D11 (1975) 395;
For a review see:
H. Joos, in "Proc. of the International Summer Institute on Theoretical Particle Physics in Hamburg 1975" (J.G. Körner et al., eds., Springer, Berlin, 1976) and DESY 76/36 (July 1976)

10. H. Nielsen and P. Olesen, Nucl. Phys. B61 (1973) 45;
J. Kogut and L. Susskind, Phys. Rev. D9 (1973) 2273;
Y. Nambu, Phys. Rev. D10 (1974) 4262;
B. Brout, F. Englert and W. Fischler, Phys. Rev. Lett. 36 (1975) 649;
For a review see:
M. Böhm and H. Joos, DESY 78/27 (May 1978)

11. V.B. Berestetski, Sov. Phys. Uspekhi 19 (1976) 934

12. W. Celmaster, H. Georgi and M. Machacek, Phys. Rev. D17 (1978) 879

13. L.D. Landau and E.M. Lifshitz, Relativistic Quantum Theory (Pergamon Press, 1971);
A.I. Achieser and W.B. Berestezki, Quantenelektrodynamik (Teubner, Leipzig, 1962);
J. Pumplin, W. Repko and A. Sato, Phys. Rev. Lett. 35 (1975) 1538 and references therein;
D. Gromes, Nucl. Phys. B131 (1977) 80

14. J. Ellis, M.K. Gaillard and G.G. Ross, Nucl. Phys. B111 (1976) 253;
T.A. DeGrand, Y.J. Ng and S.H.H. Tye, Phys. Rev. D16 (1977) 3251;
S.J. Brodsky, D.G. Coyne, T.A. DeGrand and R.R. Horgan, Phys. Lett. 73B (1978) 203;
H. Fritzsch and K.H. Streng, Phys. Lett. 74B (1978) 90;
K. Hagiwara, Nucl. Phys. B137 (1978) 164;
A. De Rújula et al., Nucl. Phys. B138 (1978) 387;
K. Koller and T.F. Walsh, Nucl. Phys. B140 (1978) 449;
K. Koller, H. Krasemann and T.F. Walsh, DESY 78/37 (1978)

15. M. Krammer and H. Krasemann, Phys. Lett. 73B (1978) 58;
H. Krasemann, DESY 78/46 (Sept. 1978), to appear in "Zeitschrift für Physik C"

16. D. Robson, Nucl. Phys. B130 (1977) 328 and references therein;
P. Roy and T.F. Walsh, Phys. Lett. 78B (1978) 62

17. See the lecture by
F. Dydak and W. Williams, this school

18. E. Eichten and K. Gottfried, Phys. Lett. 66B (1977) 286

19. C. Quigg and J.L. Rosner, Phys. Lett. 71B (1977) 153

20. G. Bhanot and S. Rudaz, Phys. Lett. 78B (1978) 119

21. E. Eichten et al., Phys. Rev. D17 (1978) 3090

22. For a derivation on the classical level see the text books, e.g. A. Sommerfeld, Atombau und Spektrallinien (Vieweg, Braunschweig, 1950) or J.D. Jackson, Classical Electrodynamics (J. Wiley and Sons, New York 1962). In the framework of the Bethe Salpeter equation for Charmonium see A.B. Henriques, B.H. Kellett and R.G. Moorhouse, Phys. Lett. 64B (1976) 85

23. G. Buschhorn, this school

24. See e.g.: A.I. Achieser and W.B. Berestezki, ref. 13, §39

25. F. Wilczek and A. Zee, Phys. Rev. Lett. 40 (1978) 83

26. H.J. Schnitzer, Phys. Lett. 65B (1976) 239

27. R. Barbieri, R. Gatto, R. Kögerler and Z. Kunszt, Phys. Lett. 57B (1975) 455

28. J.L. Rosner, C. Quigg and H.B. Thacker, Phys. Lett. 74B (1978) 350

29. C. Quigg and J.L. Rosner, Phys. Lett. 72B (1978) 462

30. P. Ditsas, N.A. McDougall and R.G. Moorhouse, Glasgow preprint (1978)

31. See the text books, e.g. W. Heitler, The Quantum Theory of Radiation (Oxford Univ. Press, London, 1947), J.M. Blatt and V.F. Weisskopf, Theoretical Nuclear Physics (J. Wiley and Sons, New York, 1952) or S.A. Moszkowski in Beta and Gamma Ray Spectroscopy, K. Siegbahn ed. (North Holland, Amsterdam, 1955)

32. L.B. Okun and M.B. Voloshin, preprint ITEP 152 (Moscow 1976); V.A. Novikov et al., Phys. Rep. 41C (1978) 1

33. H. Krasemann, Thesis, Hamburg 1978

34. J.D. Jackson, Phys. Rev. Lett. 37 (1976) 1107; H.A. Bethe and E.E. Salpeter, Quantum Mechanics of One- and Two-Electron Atoms, Springer Verlag (Berlin, Göttingen, Heidelberg, 1957)

35. See e.g.: Bethe and Salpeter, ref. 34

36. W. Bartel et al., DESY 78/49 (September 1978); see also the lecture of Prof. Buschhorn at this school

37. A. De Rújula, in "Current Induced Reactions", Int. Summer Institute on Theoretical Particle Physics in Hamburg 1975 (Springer Verlag, Eds. J.G. Körner, G. Kramer and D. Schildknecht)

38. K. Gottfried, in "Proc. 1977 International Symposium on Lepton and Photon Interactions at High Energies" (DESY, F. Gutbrod ed.)

39. R.L. Jaffe, in "Quark Spectroscopy and Hadron Dynamics", Proc. of Summer Institute on Particle Physics, 1977 (SLAC, M.C. Zipf ed.)

40. K. Gottfried, Phys. Rev. Lett. 40 (1978) 598 and ref. 38

41. T. Appelquist and H.D. Politzer, Phys. Rev. D12 (1975) 1404

42. H. Pietschmann and W. Thirring, Phys. Lett. 21 (1966) 713; R. Van Royen and V.F. Weisskopf, Nuovo Cim. 50 (1967) 617; ibid 51 (1967) 583

43. A. Ore and J.L. Powell, Phys. Rev. 75 (1949) 1696

44. These conversion factors are reviewed in ref. 32

45. I.Ya. Pomeranchuk, Doklady Akademii Nauk SSSR 60 (1948) 263

46. A.I. Alekseev, Sov. Phys. JETP 34 (1958) 826

47. R. Barbieri, R. Gatto and R. Kögerler, Phys. Lett. 60B (1976) 183

48. C.N. Yang, Phys. Rev. 77 (1950) 242; compare also L.D. Landau and E.M. Lifshitz, ref. 13

49. R. Barbieri, R. Gatto and E. Remiddi, Phys. Lett. 61B (1976) 465

50. H. Krasemann, ref. 15

51. W. Braunschweig et al., Phys. Lett. 67B (1977) 243; W. Bartel et al., Phys. Lett. 64B (1976) 483; G. Alexander et al., Phys. Lett. 72B (1978) 493; R. Brandelik et al., Phys. Lett. 76B (1978) 361

52. DASP-II-Collaboration, Internal Report DESY F15-78/01 (August 1978)

53. R.F. Schwitters et al., Phys. Rev. Lett. 35 (1975) 1320; G. Hanson et al., Phys. Rev. Lett. 35 (1975) 1609; G. Hanson, SLAC PUB 2118 (1978); PLUTO-Collaboration, Phys. Lett. 78B (1978) 176

54. J. Ellis, M.K. Gaillard and G.G. Ross, ref. 14

55. K. Koller and T.F. Walsh, ref. 14

56. K. Koller, H. Krasemann and T.F. Walsh, ref. 14

57. M. Krammer, Phys. Lett. 74B (1978) 361

58. W. Gampp and H. Genz, Phys. Lett. 76B (1978) 319

59. A. De Rújula et al., ref. 14

CHROMODYNAMICS AND DEEP INELASTIC PROCESSES

G. Altarelli

Istituto di Fisica dell'Università - Roma

Istituto Nazionale di Fisica Nucleare - Sezione di Roma

1. INTRODUCTION

The present three lectures are devoted to a summary on the phenomenological predictions of Quantum Chromo-Dynamics (Ref. 1) (QCD) for deep inealstic processes and their comparison with the data.

The first lecture contains a physically transparent description of the QCD predictions on scaling violations in terms of the familiar and very useful parton concepts. This part deals with the best established predictions of QCD for deep inelastic processes, originally derived (Ref. 2) by rigorous field theoretic methods based on the operator expansion near the light cone and renormalization group techniques for coefficient functions. In most of the current applications the above formalism is considered in the leading logarithmic approximation which is asymptotically correct for sufficiently large Q^2. It is precisely in this limit that we can directly reproduce the rigorous results by a simple refinement (Refs. 3,4) of the parton model (Ref. 5).

Recently a renewed theoretical effort (Refs. 6,7) has been dedicated to a diagrammatic translation and rederivation of the known results with the aim of enlarging the set of serious QCD predictions to situations not amenable to a solution by the operator expansion approach. In the second lecture I shall deal with some of the results obtained in this direction. Precisely I shall consider the transverse momentum of jets in leptoproduction and more briefly in e^+e^- and the progress made with respect to Drell-Yan processes concerning the effects of gluon corrections and the study of the transverse momentum distribution of muon pairs.

In the last lecture I shall describe some recent work (Refs. 8,9,10,11) on the problem of going beyond the leading logarithmic approximation in leptoproduction and Drell-Yan processes. The results are of special interest for the latter case where the corrections are found to be particularly important for practical purposes.

2. THE NAIVE PARTON MODEL AND QCD

Consider deep inelastic leptoproduction (Ref. 12) on a nuclear target. The current four momentum q is spacelike and we are free to go to a Lorentz frame where q carries no energy:

$$q \equiv (0;\ Q,\ \vec{0}) \tag{1}$$

For all four vectors, in our notation, the first entry is the energy, the second the z component of three momentum, and the last is the bidimensional vector of transverse momentum. The nucleon four momentum P, neglecting the mass M, is then given by:

$$P \equiv \left(\frac{Q}{2x} ;\ \frac{Q}{2x} ,\ \vec{0}\right) \tag{2}$$

where

$$x = \frac{Q^2}{2(P\cdot q)} = \frac{Q^2}{2M\nu} \tag{3}$$

is the familiar Bjorken variable.

In the naive parton model (Ref. 5), when terms down by powers of $1/Q^2$ are neglected, a given structure function, e.g. F_1, is obtained by convoluting the density of a quark parton in the nucleon (with fraction y of the nucleon longitudinal momentum) with the appropriate <u>pointlike</u> cross section for quark-current scattering:

$$2F_1(x) \sim \int \frac{dy}{y}\, q(y)\sigma_{point}(yP+q) + O(1/Q^2) \tag{4}$$

This is true provided the nucleon wave function produces enough damping for the virtual mass ($\sqrt{K^2}$) and the transverse momentum (P_t) distributions for the parton so that all terms of order K^2/Q^2 or P_t^2/Q^2 can be neglected (in that after integration over phase space they remain of order $1/Q^2$). If this is true then the parton is quasi real and collinear and the amplitude squared factorizes. The parton four momentum k reduces in this approximation to a fraction of the nucleon momentum:

$$k = \left(\frac{yQ}{2x} ; \frac{yQ}{2x} , \vec{0} \right) \tag{5}$$

Finally the pointlike cross section is proportional to a δ function expressing the fact that the total incoming invariant mass of the quark-current system must equal the final parton rest mass. Neglecting this rest mass and taking all factors into account one ends up with:

$$2F_1(x) = \int \frac{dy}{y} q(y) e^2 \delta\left(\frac{x}{y} - 1\right) = e^2 q(x) \tag{6}$$

where e^2 is the charge squared of the quark (or the appropriate coupling squared) and the sum over quark flavors is omitted for simplicity.

In QCD the factorization of a parton density times a parton-current cross section is also true if the same assumptions on the nucleon wave function are kept, but one cannot restrict to the pointlike cross section. In fact, the parton-current cross section takes important contributions from all orders in the strong interaction coupling constant g_s, i.e. the quark gluon coupling. In the following we shall use

$$\alpha_s \equiv \frac{g_s^2}{4\pi} \tag{7}$$

in analogy with QED. Note that the true expansion parameter is actually α_s/π.

To illustrate this point we consider the addition of $\sigma_{\text{order } \alpha_s}$ to the expression of $2F_1$ in eq. 6:

$$2F_1(x) = \int_x^1 \frac{dy}{y} q(y) \left(e^2 \delta\left(\frac{x}{y} - 1\right) + \sigma_{\text{order } \alpha_s} \left(\frac{x}{y}, Q^2\right) + \dots \right) \tag{8}$$

Note that our "cross sections" are adimensional in that a common large scale factor was removed.

The cross section at order α_s is computed from the diagrams in Fig. 1 including virtual corrections to the basic process $q+\gamma^* \to q$ and real diagrams for $q+\gamma^* \to q+G$ with G being a gluon. It is a function of x/y and Q^2 (as is clear from the parton momentum in eq.(5)) and $y \geq x$ because when a gluon is also present in the final state the energy of the initial state must be larger than in the elastic case.

Diagrams contributing to $\sigma_{\text{order }\alpha_s}$

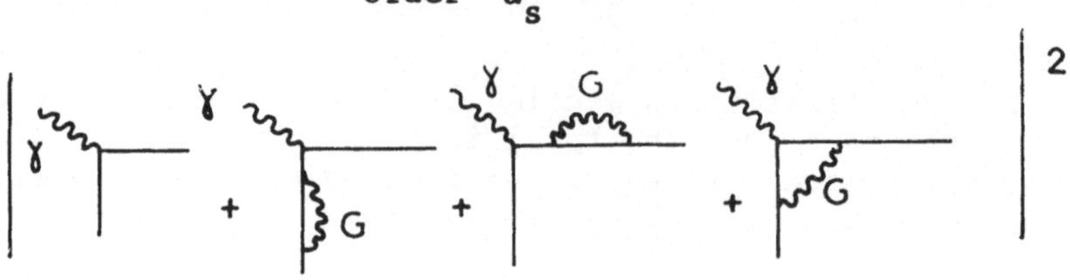

(a)

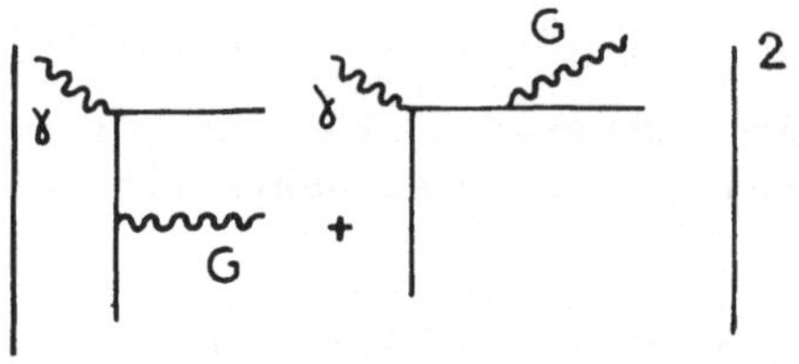

(b)

Fig. 1

It is clear that the structure of $\sigma_{\text{order}\alpha_s}$ is as follows:

$$\sigma_{\text{order}\alpha_s}(z,Q^2) = \frac{\alpha_s}{2\pi} e^2 \left(t\ P(z)+f(z)+O(1/Q^2)\right) \tag{9}$$

with

$$t = \ell n \frac{Q^2}{\mu^2} \tag{10}$$

where μ^2 is a convenient but arbitrary reference mass.
The crucial feature is the presence of the log Q^2 term. This term implies that scaling is violated and the naive parton model is not strictly valid in QCD. Its origin is quite analogous to that of the log E/m term in the total cross section of Compton scattering in QED (in that case E and m are the energy and mass of the electron). Consider in fact the diagram in Fig. 2. This is not the only diagram that contributes but it serves best as an example. Neglecting masses we parametrize the initial quark and the gluon momenta as:

$$k = (P;\ P,\ \vec{0}) \quad ; \quad P = \frac{yQ}{2x} \tag{11}$$

$$p' = \left((1-z)P + \frac{p_t^2}{2(1-z)P}\ ;\ (1-z)P,\ -\vec{p}_t\right) \tag{12}$$

so that the gluon carries a fraction (1-z) of the initial quark momentum and a given $\vec{p}_t$ (opposite, by momentum conservation, to that of the intermediate quark r and of the final quark).

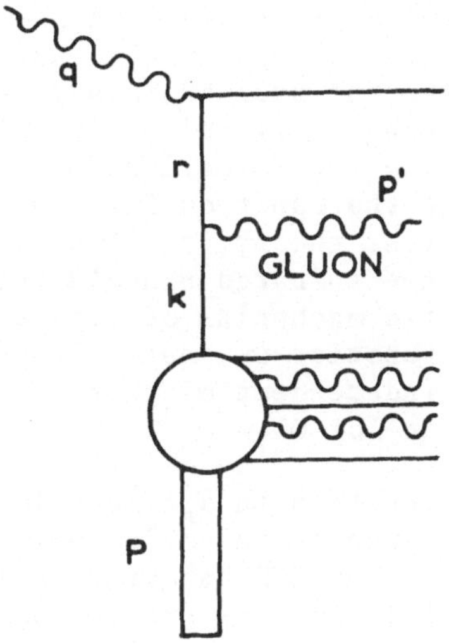

Figure 2

A simple calculation shows that

$$r^2 = (k-p')^2 = -p_t^2/(1-z) \qquad (13)$$

The virtual mass and the p_t^2 distribution are thus related for the quark r. Now the differential cross section for $q+\gamma^* \to q+G$ behaves as $1/p_t^2$. This arises from a factor $1/(r^2)^2 \sim 1/p_t^4$ from the r propagator and a factor p_t^2 in the numerator due to the vanishing of the residue at the lower vertex because of helicity conservation that forbids the emission of a collinear gluon. On the other hand the upper limit of the integration over p_t^2 is of order Q^2 and one obtains:

$$\sigma_{\text{order } \alpha_s} \sim \alpha_s \int^{\sim Q^2} \frac{dp_t^2}{p_t^2} \sim \alpha_s \ln Q^2 + \ldots \qquad (14)$$

The log Q^2 thus arises from the hard component of the p_t distribution of the emitted gluon. We see that even if the p distribution of the initial quark is well damped by the wave function this is not the case, due to gluon emission, for the virtual mass and p_t

distributions of the quark r that actually interacts with the current.

Note that the divergence at the lower limit of integration in eq. (14) has to be cured by restoring the physical mass of the quark, or by working with off mass shell quarks of zero rest mass (indeed the quark inside the nucleon is a virtual particle) or in some other way. All these complicacies can be lumped inside f(z) in eq. (9) by introducing the normalization mass μ^2 that appears in eq. (10). Thus while the function P(z) in eq. (9) is well defined even in the massless theory, the function f(z) depends on the regularization of the infrared singularities in the massless theory. Since the precise mechanism of regularization has to do with the nucleon wave function we shall always limit to properties of f(z) that are independent of that. For the moment we concentrate on the terms in log Q^2.

The problem with the term in $\alpha_s t$ (and in higher orders for those in $\alpha_s^n t^n$) is that even if we could replace α_s with $\alpha_s(t)$, the running coupling constant of QCD behaving as 1/t at large t, they would still be non negligible. The problem is circumverted by reabsorbing the dangerous terms in a modified parton density. We rewrite eq. (8) in a more explicit form:

$$2F_1(x,t) = \int_x^1 \frac{dy}{y} q(y) e^2 \left(\delta\left(\frac{x}{y}-1\right) + \frac{\alpha_s}{2\pi} t\, P\left(\frac{x}{y}\right) + \ldots\right) \tag{15}$$

and we cast this equation in the form

$$2F_1(x,t) = \int_x^1 \frac{dy}{y} \left(q(y)+\Delta q(y,t)\right)\, e^2 \delta\left(\frac{x}{y}-1\right) + \ldots \tag{16}$$

This is certainly possible provided that:

$$\Delta q(x,t) = \frac{\alpha_s}{2\pi} t \int_x^1 \frac{dy}{y} q(y) P\left(\frac{x}{y}\right) + \ldots \tag{17}$$

where in first order the dots stand for the terms with no factor of t in eq. (9). We see that lowest order perturbation theory suggests to replace the density of partons inside the proton by an effective density as seen by the photon which depends on t:

$$q(x) \rightarrow q(x,t)$$

We understand the t dependence as due to the fact that a current with larger Q^2 explores a wider range of p_t^2 inside the proton. Furthermore the variation of $q(x,t)$ for an increase dt, which is due to probing a new infinitesimal interval of p_t^2, is given by

$$\frac{dq(x,t)}{dt} = \frac{\alpha_s}{2\pi}\int_x^1 \frac{dy}{y}\, q(y,t)P(\frac{x}{y}) \tag{18}$$

Note that this equation is insensitive to $f(\frac{x}{y})$ in Eq. (9) and is exact up to terms of order α_s.

The interest of the above procedure stems from the fact that the effect of higher orders in the parton-current cross section is simply to change α_s into $\alpha_s(t)$ (and to add corrections of higher order in $\alpha_s(t)$):

$$2F_1(x,t) \simeq e^2 q(x,t) + O(\alpha_s(t)) \tag{19}$$

$$\frac{dq(x,t)}{dt} = \frac{\alpha_s(t)}{2\pi}\int_x^1 \frac{dy}{y}\, q(y,t)P(\frac{x}{y}) + O(\alpha_s(t)) \tag{20}$$

$\alpha_s(t)$ is the running coupling constant of QCD (Ref. 1) which in the leading logarithmic approximation is given by

$$\alpha_s(t) \simeq \frac{\alpha_s}{1+b\alpha_s t} = \frac{1}{b\ell n Q^2/\Lambda^2} \tag{21}$$

with

$$b = \frac{33-2f}{12\pi} \tag{22}$$

where f is the number of quark flavors (with mass much smaller than Q^2). Note that α_s is the value of $\alpha_s(t)$ at t=0 (i.e. $Q^2=\mu^2$) and

$$\Lambda^2 = \mu^2 \exp(-\frac{1}{b\alpha_s}) \tag{23}$$

$\alpha_s(t)$ decreases logarithmically with Q^2 so that at sufficiently large Q^2 eqs. (19), (20) become exact asymptotic statements. Empirically it turns out that

$$\Lambda \simeq 0.5 \text{ GeV} \tag{24}$$

Thus the asymptotic region, i.e. the region where $\alpha_s(t)/\pi$ is small, is reached for Q^2 in the few GeV range. The derivatives of quark densities become small and we have precocious approximate scaling.

The structure of eq. (20) for the derivatives simplifies considerably if we go to moments defined by:

$$M_n(t) = \int_x^1 dx\, x^{n-1} q(x,t) \qquad (n \geq 1) \tag{25}$$

It directly follows from eq. (20) that the moments satisfy the simple differential equations:

$$\frac{dM_n(t)}{dt} = \frac{\alpha_s(t)}{2\pi} A_n M_n(t) + O(\alpha_s^2(t)) \tag{26}$$

where the set of constants A_n is given by the moments of the function $P(z)$ in eq. (20):

$$A_n = \int_0^1 dz\, z^{n-1}\, P(z) \tag{27}$$

This is the reason why results for scaling violations in QCD are often formulated in terms of moments of structure functions.

Note that eq.(26) for moments is readily solved:

$$M_n(t) = M_n(0) \left[\frac{\alpha_s}{\alpha_s(t)} \right]^{\frac{A_n}{2\pi b}} \tag{28}$$

By eqs. (21), (10) the ratio $\alpha_s/\alpha_s(t)$ can be written as:

$$\frac{\alpha_s}{\alpha_s(t)} = \frac{\ell n Q^2/\Lambda^2}{\ell n \mu^2/\Lambda^2} \tag{29}$$

Thus the moments behave as powers of logarithms of Q^2 and the exponents are all generated by $P(z)$ which is determined by lowest order perturbation theory. The initial value $M_n(0)$ must however be fixed empirically. It corresponds to our ignorance of the parton densities in the nucleon at a given Q^2 that are determined by the wave function.

Note also that the moments are not perturbative in $\alpha_s(t)$ (there is a branch point at $\alpha_s(t)=0$) but their derivatives are instead perturbative. This is because the moments contain the series of leading logs $\alpha_s^n t^n$ in ordinary pertubation theory.

3. SOME RIGOROUS RESULTS

The previous results were rigorously derived in Ref. (2) by renormalization group equations applied to the coefficient functions in the operator light cone expansion for the product of currents. Actually more general results are obtained by the latter method (Ref. 13), which we now briefly describe in this section.

We denote by $J_a(x,t)$ ($a=1,2,3$) the structure functions $2F_1$, F_2/x and F_3 respectively and by $J_a^n(t)$ their moments:

$$J_a^n(t) = \int_0^1 dx\, x^{n-1}\, J_a(x,t) \tag{30}$$

The following result holds when all terms in $1/Q^2$ are neglected (but to all orders in $\alpha_s(t)$):

$$J_a^n(t) = c_a^n(\alpha_s(t))\, q_0^n \exp \int_{\alpha_s}^{\alpha_s(t)} \frac{\gamma^n(\alpha)}{\beta(\alpha)}\, d\alpha \tag{31}$$

with the definition of $\alpha_s(t)$ given by

$$\frac{d\alpha_s(t)}{dt} = \beta(\alpha_s(t)) \quad ; \quad \alpha_s(0) \equiv \alpha_s \tag{32}$$

In the above formulae $c_a^n(\alpha)$, $\gamma^n(\alpha)$ and $\beta(\alpha)$ are functions that can be computed in perturbation theory. Note that all the dependence on the index a is through the function c_a^n. On the other hand q_0^n is related to the matrix elements between nucleons of the local operators in the light cone expansion and is beyond reach of perturbation theory.

Expanding in α one obtains:

$$\beta(\alpha) = -b\alpha^2 \; (1+b'\alpha+ \ldots) \tag{33}$$

$$\gamma^n(\alpha) = \frac{A_n}{2\pi}\,\alpha(1+g'^n\alpha + \ldots) \tag{34}$$

$$c_a^n(\alpha) = 1 + c_a'^n \;\alpha + \ldots \tag{35}$$

Note that $\beta(\alpha)$ is negative for sufficiently small α (the coefficient b appearing in eq. (33) is given by eq. (22)).This is the property of QCD of being asymptotically free: starting from a value $Q^2=\mu^2$ such that $\beta(\alpha_s) < 0$, increasing Q^2 makes the running coupling constant smaller as is clear from eq. (32). In particular eq. (21) for $\alpha_s(t)$ is immediately reproduced when the lowest order approximation for $\beta(\alpha)$ is inserted in eq. (32).

Similarly when the lowest order approximations for $\beta(\alpha)$, $\gamma^n(\alpha)$ and $c_a^n(\alpha)$ are inserted in eq. (31) a result of the form of eq. (28) is directly obtained for J_a^n.

It is clear that the knowledge of $\sigma_{\text{order }\alpha_s}$ is sufficient for deriving A_n (and $c_a'^n$) and therefore to reproduce the asymptotic form of $J_a^n(t)$ for sufficiently small $\alpha_s(t)$.

We shall come back later in sect. 12 to the above general results. For the time being we resume our heuristic approach.

4. The GENERAL EVOLUTION EQUATIONS

Up to this point we implicitly restricted to non singlet (under the flavor group) structure functions. We can now readily take into account the presence of gluons and of many flavors. In general when computing $\sigma_{\text{order }\alpha_s}$ we must also consider the diagrams in Fig. 3 where a gluon in the nucleon makes a quark antiquark pair that interacts with the current.

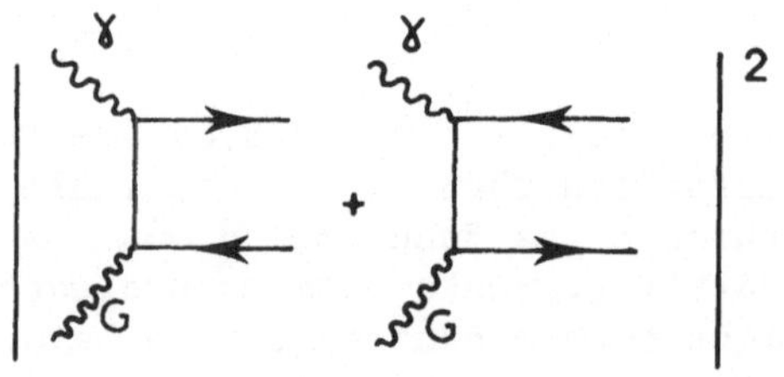

Figure 3

The contribution of this new parton process $G+\gamma^{*}\to q+\bar{q}$ to σ order α_s also contains a log Q^2 term plus a finite term. The complete form of eq. (15) is therefore given by :

$$2F_1(x,t) = \int_x^1 \frac{dy}{y}\Big\{\sum_i q_i(y)e_i^2\Big(\delta(\frac{x}{y}-1)+\frac{\alpha_s}{2\pi}(t\,P_{qq}(\frac{x}{y})+f_1^q(\frac{x}{y}))\Big)+ \\ +\sum_i e_i^2\,G(y)\,\frac{\alpha_s}{2\pi}\Big(t\,P_{qG}(\frac{x}{y})+f_1^G(\frac{x}{y})\Big)\Big\} \tag{36}$$

where $i=1,\ldots$ 2f runs over both quarks and antiquarks of all flavors and G is the gluon density in the nucleon. Correspondingly for each q_i we have (instead of eq. (17)):

$$\Delta q_i(x,t) = \frac{\alpha_s}{2\pi}\,t\int_x^1 \frac{dy}{y}\Big(q_i(y)P_{qq}(\frac{x}{y}) + G(y)P_{qG}(\frac{x}{y})\Big) + \ldots \tag{37}$$

As a consequence the final results in eqs. (19), (20) are modified into

$$2F_1(x,t) = \sum_i e_i^2 q_i(x,t) + O(\alpha_s(t)) \tag{38}$$

$$\frac{dq_i(x,t)}{dt} = \frac{\alpha_s(t)}{2\pi}\int_x^1 \frac{dy}{y}\Big(q_i(y,t)P_{qq}(\frac{x}{y})+G(y,t)P_{pG}(\frac{x}{y})\Big) \\ + O(\alpha_s^2(t)) \tag{39}$$

The above set of equations does not close and we need an equation for the evolution in t of the gluon density. This can be obtained by suitably extending the same line of reasoning and reads:

$$\frac{dG(x,t)}{dt} = \frac{\alpha_s(t)}{2\pi}\int_x^1 \frac{dy}{y}\Big(\sum_i q_i(y,t)P_{Gq}(\frac{x}{y})+G(y,t)P_{GG}(\frac{x}{y})\Big) + O(\alpha_s^2(t)) \tag{40}$$

Note that when all masses are neglected the functions P(z) do not depend on the index i. Equivalent equations for the moments of quark and gluon densities can be written in the form of eq. (26) where M_n is now a vector and A_n a matrix. For each n, the entries of the A_n matrix are the moments of the P functions as in eq. (27). By taking the difference of the equations for q_i and q_j the gluons drop out and the simpler equation of the previous sections is reobtained:

$$\Delta_{ij}(x,t) = q_i(x,t) - q_j(x,t) \tag{41}$$

$$\frac{d\Delta_{ij}(x,t)}{dt} = \frac{\alpha_s(t)}{2\pi} \int_x^1 \frac{dy}{y} \Delta_{ij}(y,t) P_{qq}\left(\frac{x}{y}\right) + O(\alpha_s^2(t)) \tag{42}$$

As anticipated this equation is thus appropriate for non singlet densities (in the sense of eq. (41)). All non singlet densities have their Q^2 behaviour determined by the single function $P_{qq}(z)$. The content of the remaining equations for singlet densities is best analyzed by summing up eq. (39) on the quark index i. Defining:

$$\Sigma(x,t) = \sum_i q_i(x,t) = \sum_{\text{flavors}} (q(x,t)+\bar{q}(x,t)) \tag{43}$$

We have

$$\begin{aligned} \frac{d\Sigma(x,t)}{dt} &= \frac{\alpha_s(t)}{2\pi} \int_x^1 \frac{dy}{y} \left(\Sigma(y,t) P_{qq}\left(\frac{x}{y}\right) + 2fG(y,t) P_{qG}\left(\frac{x}{y}\right)\right) + O(\alpha_s^2(t)) \\ \frac{dG(x,t)}{dt} &= \frac{\alpha_s(t)}{2\pi} \int_y^1 \frac{dy}{y} \left(\Sigma(y,t) P_{Gq}\left(\frac{x}{y}\right) + G(y,t) P_{GG}\left(\frac{x}{y}\right)\right) + O(\alpha_s^2(t)) \end{aligned} \tag{44}$$

Eqs. (42), (44) contain all the information on scaling violations to the leading order in $\alpha_s(t)$. They were originally obtained for moments in Ref. (2) while they were cast in the present form in Refs. (3,4).

5. THE PHYSICAL MEANING OF THE P FUNCTIONS

We have seen in the previous sections that the structure functions are asymptotically given by their naive parton model formulae in terms of Q^2 dependent effective parton densities. In turn the derivatives of the effective parton densities are governed by the functions $P_{BA}(z)$. These functions can be computed from the various parton-current cross sections at order α_s as discussed above.

However this derivation is not completely satisfactory. In fact parton densities are useful for relating different structure functions for the same process or the same structure function for different processes, say electroproduction and ν or $\bar{\nu}$ scattering. It is thus important to show that the P_{BA} functions and consequently the effective Q^2 dependent parton densities only depend on the target and for each fixed Q^2 are insensitive to whether the probe is a photon with one or another polarization or a weak current with any mixture of vector and axial vector couplings. The universality of the P_{BA} functions is apparent in the renormalization group approach where it amounts to the independence of the functions $\gamma^n(\alpha)$ from the index a in eq. (31) and arises because $\gamma^n(\alpha)$ is the anomalous exponent of a given local operator. Following the more transparent language of ref. (4) a direct proof of the universality of the $P_{BA}(z)$ functions, i.e. their independence from the probe, can be obtained by first giving a physical interpretation for P_{BA} and then using the latter for a very simple calculation of the whole set of P_{BA} (actually also those relevant for polarized deep inelastic scattering) with no reference to the current.

In order to briefly discuss this point we go back to eq. (17) for the non singlet quark density in first order perturbation theory and rewrite it in the following equivalent form:

$$q(x,t)+\Delta q(x,t) = \int_0^1 dy \int_0^1 dz\delta(zy-x)q(y,t)\left(\delta(1-z)+\frac{\alpha_s}{2\pi}tP_{qq}(z)+\ldots\right) \quad (45)$$

The quantity

$$P(z,t) = \delta(1-z) + \frac{\alpha_s}{2\pi} t\, P_{qq}(z)+ \ldots \quad (46)$$

can be naturally interpreted as an expansion of the probability density of finding a quark inside a quark with fraction z of the parent quark momentum. Conseqeuntly $P_{qq}(z)$ appears as the lowest order variation of this probability density for unit t. More in general:

$$P_{BA}(z,t) = \delta_{BA}\delta(1-z)+ \frac{\alpha_s}{2\pi} t\, P_{BA}(z)+ \ldots \quad (47)$$

is the probability for finding the parton B inside the parton A with fraction z of the parent momentum. The diagonal probabilities start with a $\delta(1-z)$ term corresponding to nothing happening. Similarly at order α_s they also contain other terms proportional to $\delta(1-z)$. The coefficient of $\delta(1-z)$ in eq. (47) is in fact modified by terms of order α_s because the probability for a quark to remain a quark with the same energy is lowered by the interaction. In our definitions <u>all</u> terms of order α_s including these terms are present in P(z). For example in our gedanken derivation of $P_{qq}(z)$ from $\sigma_{\text{order}\,\alpha_s}$, the $\delta(1-z)$ contributions arise from the interference of the pointlike diagram with the radiatively corrected (to order α_s) amplitude for a final state with a single quark and no gluons (see Fig. 1a where gluons are marked by a G). It therefore follows that $P_{qq}(z)$ is directly proportional to a probability (in particular is positive definite) only at $z < 1$; while at $z=1$, $\delta(z-1)$ contributions are present which spoil the positive definiteness. At all z the real probability density is $P_{qq}(z,t)$ and not $P_{qq}(z)$. The same is true for $P_{GG}(z)$.

This interpretation of $P_{BA}(z)$ can serve as a basis for their direct computation starting from the lowest order vertices of QCD. The method is a generalization in QCD of the well known Weizacker-Williams calculation of the equivalent number of photons in an electron in QED. The functions P_{qq} and P_{Gq} are obtained from the vertex $q \to q+G$, $P_{qG} = P_{\bar{q}G}$ from the vertex $G \to q+\bar{q}$ and P_{GG} from the three gluon vertex $G \to G+G$ typical of non abelian gauge theories.

The strategy is to compute the regular part of $P_{BA}(z)$ at $z < 1$. The $\delta(z-1)$ singularities that are present for the diagonal cases (A=B) are then fixed by the constraints imposed by charge and momentum conservation. The basic vertices of QCD in fact conserve fermion number and flavor. Therefore for example for the u and d quarks in the proton we have:

$$\begin{aligned} \int_0^1 dx(u(x,t)-\bar{u}(x,t)) &= 2 \\ \int_0^1 dx(d(x,t)-\bar{d}(x,t)) &= 1 \end{aligned} \tag{48}$$

This implies that the moment with n=1 of non singlet densities must be Q^2 independent. Through eqs. (27), (42) this in turn implies

$$\int_0^1 dz\, P_{qq}(z) = 0 \tag{49}$$

Thus if we know $P_{qq}(z)$ at $z < 1$ we can fix the coefficient of the $\delta(1-z)$ term by imposing eq. (49). Similarly the momentum conservation constraint:

$$\int_0^1 dx\ x\ \left(\Sigma(x,t) + G(x,t)\right) = 1 \tag{50}$$

where $\Sigma(x,t)$ is defined by eq. (43), allows to fix the $\delta(1-z)$ term in $P_{GG}(z)$. This is a useful (but not necessary (Refs. 1o,14))short cut that leaves for us to compute diagrams with real parton emission only.Thus for $P_{qq}(z)$ we can restrict to the diagrams in Fig.1b. The diagram with the gluon emitted by the final quark seems superficially to contradict our interpretation for the P_{qq} function as a correction factor to the distribution of quarks inside a nucleon due to the probability of emitting a gluon. This interpretation is in fact justified only if the diagram with the gluon emitted by the initial quark leg is dominant. This diagram is itself of parton type, in that the emitted gluon could be considered as part of the nucleon blob. Actually both diagrams are necessary in a general gauge to ensure gauge invariance of the amplitude.But looking more closely one discovers that the second diagram is only necessary to cancel the contribution of the unphysical longitudinal and scalar gluons in the parton like diagram which are present in a general gauge. In other words if we only sum over transverse gluons the contribution of the parton diagram is the only relevant one for the terms which contain t i.e. are logarithmically dominant. Thus in a physical gauge only one diagram is left. We already observed in sect. 2 (see eq. (14)) that this diagram leads to a p_t distribution like p_t^{-2} and that it is precisely this distribution that leads to a factor of t in $\sigma_{\text{order}\,\alpha_s}$. Therefore in order to pick up P_{qq} which is related to the coefficient of t we need the residue of the p_t^{-2} singularity. This residue factorizes into a contribution from the current vertex (the pointlike cross section) and one from the gluon vertex and the latter is the one relevant for $P_{qq}(z)$. In the following we sketch the computation of $P_{BA}(z)$.

In order to identify $P_{BA}(z)$ we consider the two processes in fig. 4a,b . C is the third particle in the QCD vertex ABC. D is a virtual unspecified particle of mass $-Q^2$ and f a given final state.

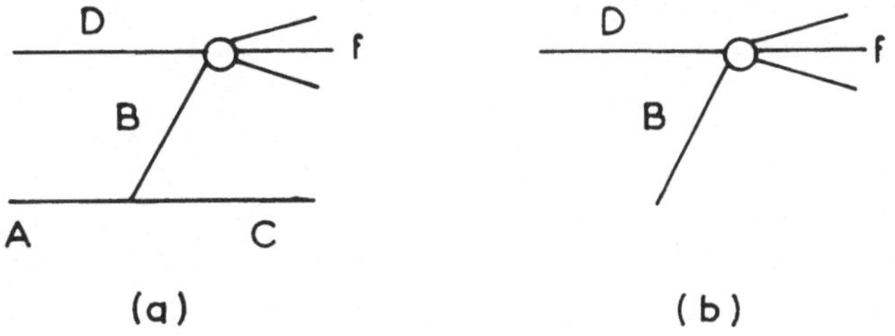

Figure 4

Once again we reason in parton terms: the cross section for process (a) is the probability of finding a parton B in the parton A times the cross section for process (b):

$$d\sigma_a = \frac{\alpha_s}{2\pi} t\, P_{BA}(z) dz\, d\sigma_b \tag{51}$$

It is convenient to phrase the calculation in terms of time dependent (old) perturbation theory. The diagram in fig. 4a refers to the intermediate state CBD. In covariant perturbation theory it would also apply to the intermediate state ABf (see fig. 5) which however is non leading because it does not lead to the $1/p_t^2$ singularity. This can be shown by inspection of the energy denominators $(E_{initial} - E_{interm.})$ which are $(E_A - E_B - E_C)$ and $(E_D - E_B - E_f)$ in the case of the intermediate states CBD and ABf respectively. In old perturbation theory energy is not conserved at the vertices but all particles are on mass shell. We therefore write down the following parametrization for the momentum of particles A, B and C:

$$\begin{aligned} k_A &= (P;\ P,\ \vec{0}) \\ k_B &= (zP + \frac{p_t^2}{2zP}\ ;\ zP,\ \vec{p}_t) \\ k_C &= ((1-z)P + \frac{p_t^2}{2(1-z)P};\ \ (1-z)P,\ -\vec{p}_t) \end{aligned} \tag{52}$$

It is a pure matter of computation to evaluate $d\sigma_a$ and $d\sigma_b$ and find by eq. (51) the general form of $P_{BA}(z)$. We refer the interested reader to ref. (4) for all the details. The general re-

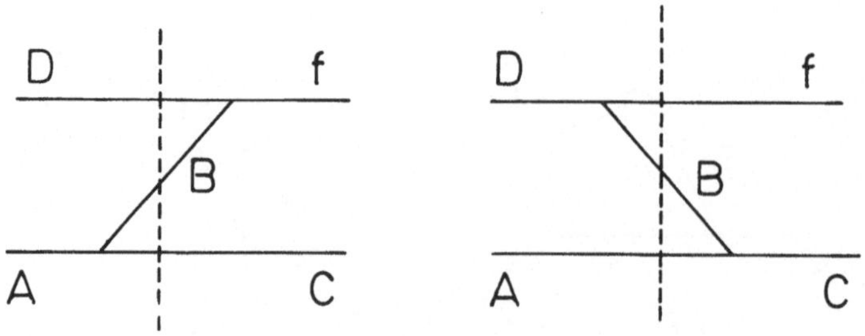

Figure 5

sult for $P_{BA}(z)$ is :

$$P_{BA}(z) = \frac{z(1-z)}{2} \overline{\sum_{spins}} \frac{|V_{A\to BC}|^2}{p_t^2} \qquad (z < 1) \qquad (53)$$

where $V_{A\to BC}$ is the invariant vertex for $A \to BC$ with the coupling constant and the factors $(2E)^{-1/2}$ for each particle being extracted. The sum refers to the spin sum and average. The factor p_t^{-2} is canceled by the vanishing of the numerator at $p_t^2=0$ in all cases of interest here. Note that P_{BA} is completely specified by the ABC vertex. It is also important to observe the symmetry relation:

$$P_{BA}(z) = P_{CA}(1-z) \qquad (z < 1) \qquad (54)$$

which is expected from momentum conservation: each time B is produced with fraction z, C is also produced with fraction 1-z. The condition $z < 1$ is always specified in order to recall the existence of additional $\delta(z-1)$ singularities for diagonal P functions.

As an example consider the case of $P_{Gq}(z)$. In this case we have that A and C are quarks and B a gluon and we find

$$\overline{\sum_{spins}} |V_{q\to Gq}|^2 = \frac{1}{2}\frac{4}{3} \mathrm{Tr}(\not{k}_C\gamma_\mu\not{k}_A\gamma_\nu) \sum_{pol} \varepsilon^{*\mu}\varepsilon^\nu \qquad (55)$$

where $\not{k}_C$ and $\not{k}_A$ are given in eq. (52); the factor 1/2 is from the initial quark spin average, the factor 4/3 is from color sums and average[x]

$$\frac{4}{3} = \frac{1}{3}\sum_{a=1}^{8} \mathrm{Tr}(\lambda^a\lambda^a) = \frac{1}{3}\cdot 8 \cdot \frac{1}{2}$$

x) The quark-gluon vertex and coupling constant are defined by $-i\gamma_\mu g_s \lambda^a$ with $\mathrm{Tr}\lambda^a\lambda^b = 1/2\ \delta^{ab}$

(the factor 1/3 is from the average over the initial quark colors). As discussed above the sum over the polarization of the gluon must only include physical transverse modes:

$$\sum_{pol} \varepsilon^{*\mu} \varepsilon^{\nu} \begin{cases} = \delta^{ij} - \dfrac{k_B^i k_B^j}{k_B^2} & (i,j = 1,2,3) \\ = 0 \quad \text{for } \mu \text{ and/or } \nu = 0 & \end{cases} \tag{56}$$

Then one immediately finds from eqs. (53), (55) ,and (56) with a little algebra the result:

$$P_{Gq}(z) = \frac{4}{3} \frac{1+(1-z)^2}{z} \tag{57}$$

Since P_{Gq} is a non diagonal density, this result is complete. From this expression of $P_{Gq}(z)$ we immediately find by symmetry (see eq. (54)):

$$P_{qq}(z) = \frac{4}{3} \frac{1+z^2}{1-z} \qquad (z < 1) \tag{58}$$

Similarly from the vertex of a gluon into a quark antiquark pair we find:

$$P_{qG}(z) = \frac{1}{2} \left(z^2 + (1-z)^2\right) \tag{59}$$

Finally from the three gluon vertex one obtains:

$$P_{GG}(z) = 6 \left(\frac{1-z}{z} + \frac{z}{1-z} + z(1-z)\right) \qquad (z < 1) \tag{60}$$

Note the obvious symmetry for $z \leftrightarrow (1-z)$ of $P_{GG}(z)$ and $P_{qG}(z)$. We also remark that the four gluon vertex does not contribute to this order being proportional to g_s^2 and not g_s.

The final task is to fix the $\delta(1-z)$ singularities of $P_{qq}(z)$ and $P_{GG}(z)$. This is readily done by using the constraints in eqs. (48) or (49) and (50).

Note that all moments of these functions as given by eq. (27) would be divergent at $z=1$ for $n \geq 1$. We therefore start by regularizing the factor $(1-z)^{-1}$ by reinterpreting it as a distribution $(1-z)_+^{-1}$ defined in the following way:

$$\int_0^1 \frac{dz\ f(z)}{(1-z)_+} \equiv \int_0^1 dz\ \frac{f(z)-f(1)}{1-z} \qquad (61)$$

with f(z) being any test function which is sufficiently regular at the end points. In particular we have

$$\int_0^1 dz\ \frac{1}{(1-z)_+} = 0 \qquad (62)$$

We then add, to $P_{qq}(z)$ and $P_{GG}(z)$, a $\delta(z-1)$ function with the coefficient determined by the constraints in eqs. (48), (49) and (50). We thus find

$$P_{qq}(z) = \frac{4}{3}\ \left(\frac{1+z^2}{(1-z)_+} + \frac{3}{2}\ \delta(z-1)\ \right) \qquad (63)$$

$$P_{GG}(z) = 6\ \left(\frac{z}{(1-z)_+} + \frac{1-z}{z} + z(1-z) + \left(\frac{11}{12} - \frac{f}{18}\right)\delta(z-1)\right) \qquad (64)$$

This completes the calculations of the P_{BA} functions. When these functions are inserted in the evolution equations, eqs. (42), (44), the resulting Q^2 dependence takes automatically into account the leading logarithms from diagrams of all orders in α_s. In a physical gauge the dominant diagrams are the ladder diagrams with dressed vertices and propagators, as has been shown in ref. (6,7).

6. Q^2 DEPENDENCE OF FRAGMENTATION FUNCTIONS

We now make a short digression to mention that the above method is immediately extended to obtain the differential equations for the Q^2 dependence of the fragmentation functions for a parton A to contain a given hadron h. In the naive parton model the probability of finding h in A is (Ref. 5) a function only of z, i.e. the fraction of momentum of A that is carried by h: $D_A^h\ (z)$. But in this case also, gluon corrections induce a t dependence (Ref. 15) which is ruled (Ref. 16) by the same P_{BA} functions of the previous sections. The evolution equations now read:

$$\frac{dD_q^h(z,t)}{dt} = \frac{\alpha_s(t)}{2\pi} \int_z^1 \frac{dy}{y}\ \left[P_{qq}(y)D_q^h(\frac{z}{y},t) + P_{Gq}(y)D_G^h(\frac{z}{y},t)\right] + O(\alpha_s^2(t))$$

$$\frac{dD_G^h(z,t)}{dt} = \frac{\alpha_s(t)}{2\pi}\int_z^1 \frac{dy}{y}\left(P_{qG}(y)D_q^h(\frac{z}{y},t) + P_{GG}(y)D_G^h(\frac{z}{y},t)\right) + O(\alpha_s^2(t)) \tag{65}$$

These equations have the same structure as those for parton densities apart for a transposition. For example a quark can fragment into a gluon and a quark. Then the variation with t of D_q^h is given by the probability of finding a quark in a quark times D_q^h plus the probability of finding a gluon in a quark times D_G^h. Similarly for the second equation.

This is an example of how the present diagrammatic techniques are most easily extended to situations that cannot be approached by light cone operator expansions. The arguments of refs. (7) provide a quite sound basis for eqs. (65) that now stand on the same level of eqs. (42), (44) for parton densities.

7. COMPARISON WITH THE DATA ON SCALING VIOLATIONS

We now briefly comment on the status of the comparison of QCD predictions on scaling violations with the data on electro (or muon) production and on neutrino scattering.

The qualitative pattern of scaling violations as predicted by QCD is at present well supported by the data (Ref. 17). From the explicit form of the P_{BA} functions in sect. 5 we see that moments of valence with $n > 1$ decrease with Q^2, faster with larger n. (The moment of valence with n=1 is constant due to flavor conservation). This is physically evident because a quark can only loose energy by radiating gluons and thus the x distribution of valence with increasing Q^2 becomes more and more concentrated at low x. Momentum conservation, eq. (50), then implies that since the valence contribution is decreasing, the sum of the momentum fractions carried by sea quarks and gluons must increase with Q^2. Actually both sea and gluon momentum fractions are separately increasing because their empirical values at low Q^2 are smaller than the predicted values at infinite Q^2 (for four flavors these values are 4/7 for gluons and 3/56 for each sea quark). All higher moments ($n > 2$) of sea quarks and gluons decrease with Q^2, faster with larger n. As a consequence the structure function F_2, for example, which is determined by momentum densities of quarks (times squared couplings) is predicted to decrease with Q^2 at large x and to increase at small x. The area under F_2 should slowly decrease to its fixed finite asymptotic value determined by total momentum fractions carried by sea quarks. It decreases slowly because the falling of the n=2 moment of valence (itself not very fast) is partly compensated by the raise of the n=2 moment of sea densities. The shrinking of F_2 towards small values of x is by now well established with all available lepton beams.

For a more quantitative analysis the most significant tests up to now involve the first few moments of structure functions or the behaviour at fixed intermediate values of x. For a fair judgement on their significance it is quite important not to forget a number of simple and clear facts. The QCD formulae are asymptotic predictions. They are obtained by neglecting in the first place all mass effects down by powers of m^2/Q^2, where m indicates any mass or energy scale in the theory: target or parton masses, threshold masses etc. In addition non leading terms down by powers of $\alpha_s(t)$ are also neglected. It is therefore clear that for a meaningful comparison of theory and data the values of Q^2 involved should at least be as large as 8-10 GeV^2. All attempts to exorcize this simple fact by using Nachtmann moments (Ref.18) or similar tricks pretending to include some effects of order m^2/Q^2 seem rather vain in view of the fact that terms down by powers of $\alpha_s(t)$ are in any case neglected. In my opinion the difference between taking Nachtmann or normal moments, between including or not including the elastic contribution etc. should simply be taken as a measure of the systematic error in the comparison of theory with data at a given Q^2. The non leading effects are particularly important in the large x region where resonance effects indicate the breakdown of perturbation theory. Thus the first few moments and intermediate values of x are safest.

With these words of caution in mind, we consider some examples of QCD tests. An analysis of the first few moments of F_2 for e and μ beams taken from ref. (19) is reproduced in fig. 6. The data for moments with larger n decrease faster as predicted. The corresponding values of moments of valence, sea and gluons are reasonable. However it is clear that if we insist on avoiding data at small Q^2 very few data points survive, especially for deuterium. I may refer the reader to the lectures by Williams in these same proceedings for a discussion of the quantitative importance of elastic or quasi elastic contributions in the data at low Q^2. Of particular interest is also the recent analysis of the moments of F_3 in neutrino scattering presented by the BEBC collaboration (Ref. 20). F_3 is given by a sum of quark minus antiquark differences and is therefore completely determined by valence. The simplest Q^2 dependence of valence moments implies that a double logarithmic plot of the ratio of any two moments Mn_1 and Mn_2 (versus log Mn_1 and log Mn_2) should describe a straight line with varying Q^2 with slope predicted by QCD with no parameters. The results are shown in Fig. 7 and are certainly quite impressive. However most of the data refer to small Q^2 values in this case also.

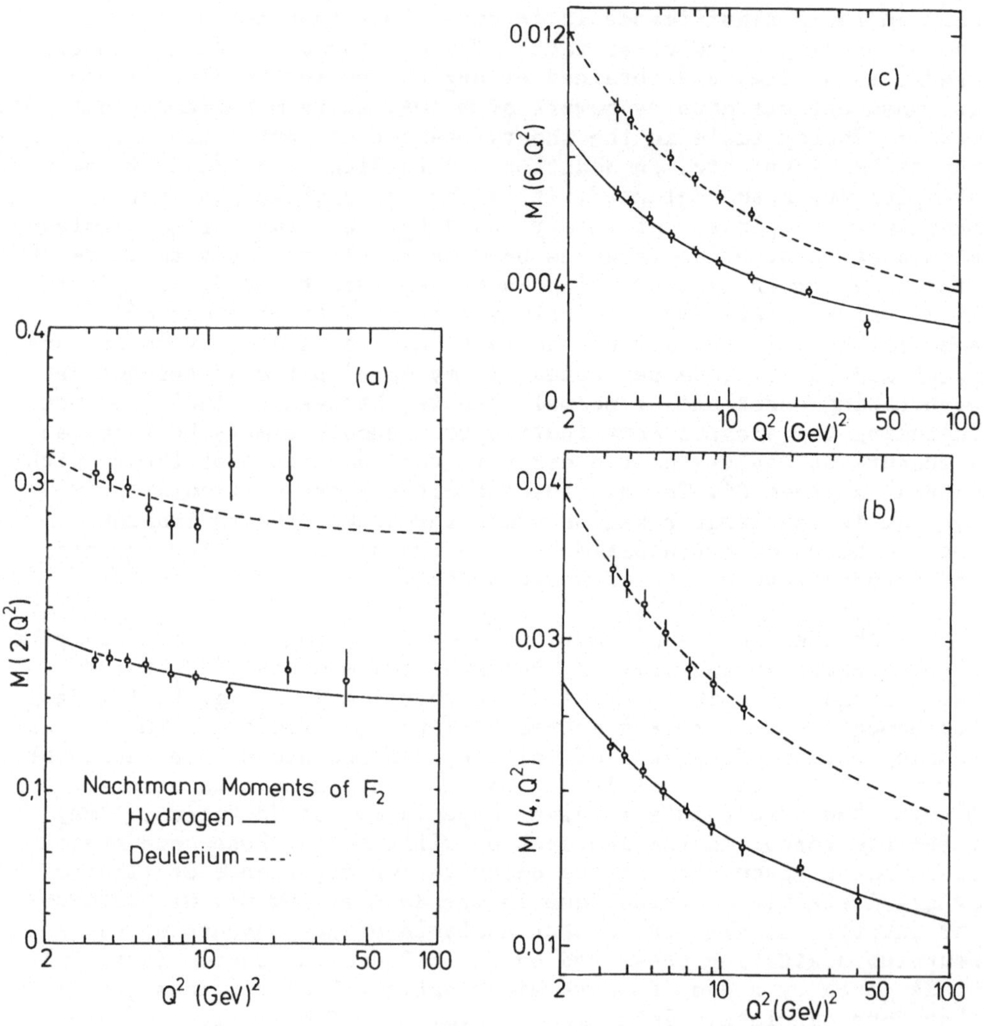
0,012
(c)
M(6.Q²)
0,004
0
2
10
Q² (GeV)²
50
100
0,4
(a)
0,3
M(2,Q²)
0,1
Nachtmann Moments of F₂
Hydrogen ——
Deulerium ----
0
2
10
50
100
Q² (GeV)²
0,04
(b)
0,03
M(4,Q²)
0,01
2
10
Q²(GeV)²
50
100

Figure 6

We finally present an example (Ref. 21) of the Q^2 dependence of data at fixed x in electroproduction. Fig. 8 is taken from Ref. (22). Note that since sea and gluons are mostly concentrated at low x, it is predominantly the Q^2 dependence of valence that is being tested here. In summary I would conclude that a) the existence of scaling violations is experimentally proved, b) the qualitative pattern and the order of magnitude of the scaling violations is in good agreement with QCD, c) the values of Λ defined in eqs. (21), (23) obtained from different independent tests show a remarkable consistency, d) within the limitations of present data QCD appears to provide a satisfactory description of scaling violations in deep inelastic phenomena.

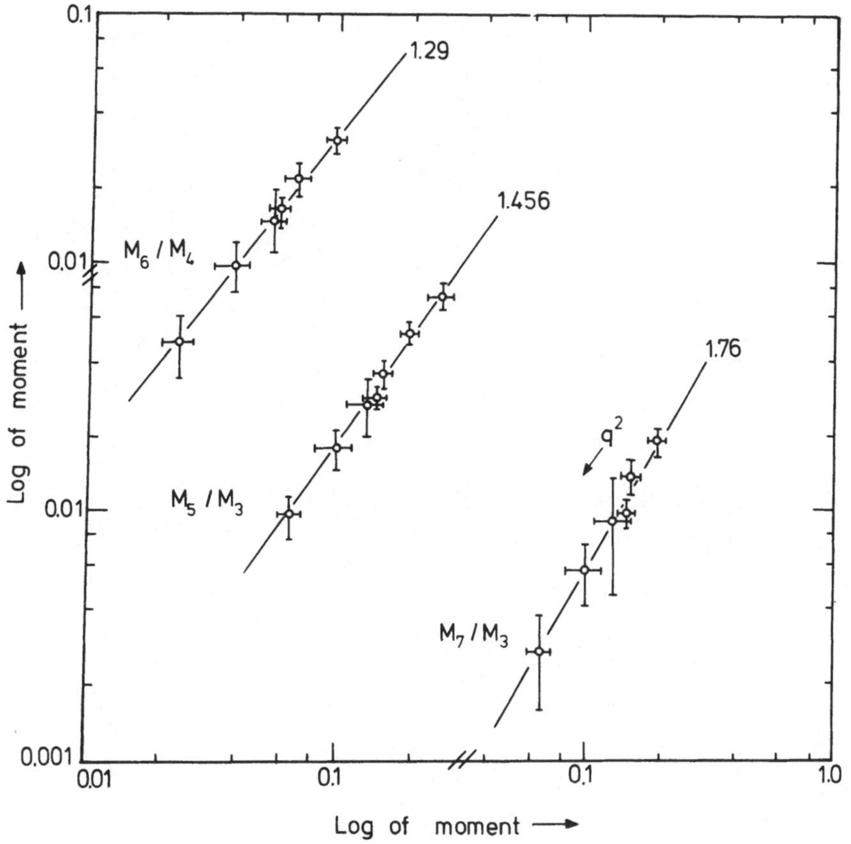

Figure 7

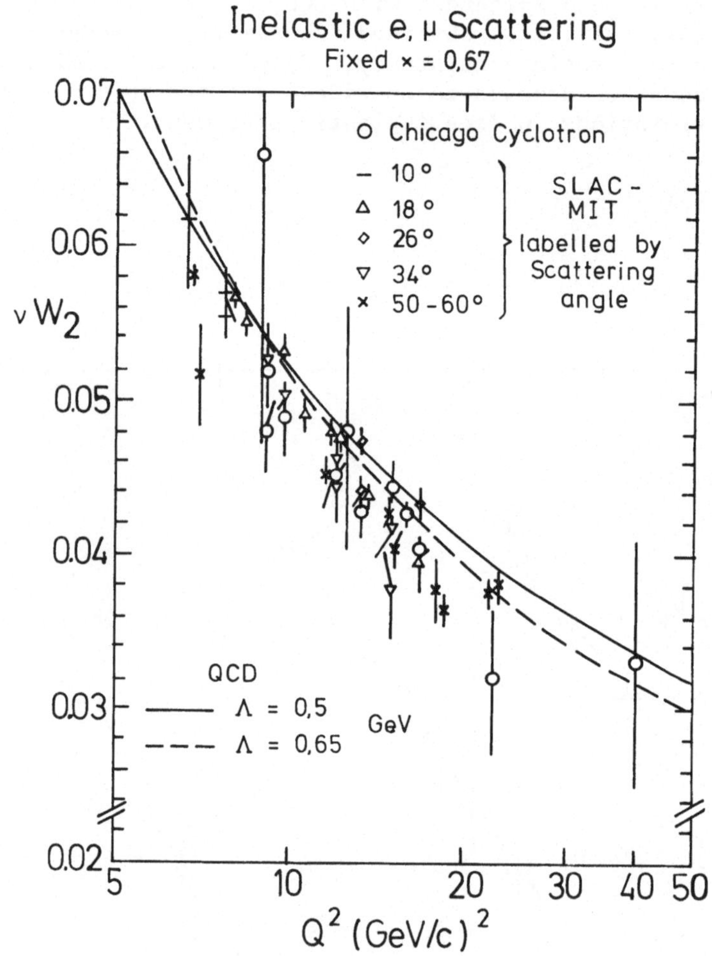

Inelastic e, μ Scattering
Fixed x = 0,67
0.07
0.06
0.05
0.04
0.03
0.02
νW2
○ Chicago Cyclotron
— 10°
△ 18°
◇ 26°
▽ 34°
× 50-60°
SLAC-MIT labelled by Scattering angle
QCD
Λ = 0,5
Λ = 0,65
GeV
5
10
20
30
40
50
Q2 (GeV/c)2

Figure 8

8. TRANSVERSE AND LONGITUDINAL STRUCTURE FUNCTIONS

We now pause for a moment in order to summarize the main points in the long argument in the previous sections.

We started from the lowest order result (naive parton model):

$$2F_1(x) = \int \frac{dy}{y} q(y)\delta(\frac{x}{y} - 1) = e^2 q(x) \tag{66}$$

(the non singlet structure function is again considered for simplicity). We then added the correction of order α_s to the point-like cross section:

$$2F_1(x,t) = \int_x^1 \frac{dy}{y} q(y)e^2 \left(\delta(\frac{x}{y} - 1) + \frac{\alpha_s}{2\pi} t\, P(\frac{x}{y}) + \frac{\alpha_s}{2\pi} f(\frac{x}{y})\right) \tag{67}$$

The necessity of including the t dependent term in $q(y,t)$ arises because $\alpha_s t$ cannot be considered as a small parameter (even if one could replace α_s with $\alpha_s(t)$). On the other hand, the independence of the P functions from the detailed structure of the probe leads to effective parton densities that only depend on t and on the target. In turn the t derivatives of the effective parton densities start with α_s and are small when α_s is replaced by $\alpha_s(t)$. By introducing $q(y,t)$, one obtains from eq. (67) (see also eqs. (16), (17)):

$$2F_1(x,t) = \int_x^1 \frac{dy}{y} q(y,t)e^2 \left(\delta(\frac{x}{y} - 1) + \frac{\alpha_s}{2\pi} f(\frac{x}{y})\right) \tag{68}$$

The possibility of implementing the above factorization to all orders is the key result of the renormalization group approach.

The most cases, for Q^2 large, we can neglect the f term and the form of the structure function as in the naive parton model is restored in terms of t dependent effective parton densities satisfying the integro-differential equations (42), (44). However, in some cases the f terms are important and we shall now discuss this matter.

First one may ask why we do not absorb the f terms in the definitions of $q(y,t)$ as well. The answer is that the f functions contrary to the P functions are not independent on the detailed properties of the current. Thus if we were to compute F_2/x instead of $2F_1$ we would have found the same P functions but a different f term. In fact the complete result for the two structure functions would be (we restored the non singlet components as well):

$$F_2(x,t)/x = \int_x^1 \frac{dy}{y} \{\sum_i e_i^2 q_i(y,t) \left(\delta(\frac{x}{y}-1)+ \frac{\alpha_s(t)}{2\pi} f_2^q(\frac{x}{y})\right) + \sum_i e_i^2 G(y,t) \frac{\alpha_s(t)}{2\pi} f_2^G(\frac{x}{y})\} \tag{69}$$

$$2F_1(x,t) = \int_x^1 \frac{dy}{y} \{\sum_i e_i^2 q_i(y,t) \left(\delta(\frac{x}{y} -1)+ \frac{\alpha_s(t)}{2\pi} f_1^q(\frac{x}{y})\right) + \sum_i e_i^2 G(y,t) \frac{\alpha_s(t)}{2\pi} f_1^G(\frac{x}{y})\} \tag{70}$$

where the index i runs over both quarks and antiquarks of all flavors. The importance of the f terms becomes manifest if we are interested in the longitudinal structure function:

$$F_L(x,t) = F_2(x,t) - 2xF_1(x,t) \tag{71}$$

In fact F_L is zero to lowest order, while to order $\alpha_s(t)$, by subtracting eqs. (69), (70) we find:

$$F_L(x,t)/x = \frac{\alpha_s(t)}{2} \int_x^1 \frac{dy}{y} \{\sum_i e_i^2 q_i(y,t) f_L^q(\frac{x}{y}) + \sum_i e_i^2 G(y,t) f_L^G(\frac{x}{y})\} \tag{72}$$

where

$$f_L^{q,G} = f_2^{q,G} - f_1^{q,G} \tag{73}$$

This is a very important result that shows that the longitudinal structure function in QCD is only expected to vanish logarithmically at $Q^2 \to \infty$. This is to be confronted with the prediction of the naive parton model of a power behaviour at large Q^2: $F_L \propto m^2/Q^2$.

We already mentioned in sect. 2 that the f functions are not well defined in that they depend on the regulator mass that makes the integral over p_t^2 in eq. (14) finite near $p_t^2=0$. But this dependence on the regulator mass cancels in the differences in eq.(73). In fact f_L^q and f_L^G can be directly related to the longitudinal structure functions for parton-virtual photon scattering and computed from the diagrams in Fig. 1b (referred to a longitudinal photon). The independence on the infrared cut off manifests itself in that the parton longitudinal structure functions are well defined even in the massless theory and one finds (Refs. 9,10,23):

$$f_L^q(z) = \frac{8}{3} z$$
$$f_L^G(z) = 2z(1-z) \tag{74}$$

For an arbitrary process (e-P or ν, $\bar{\nu}$ scattering) we then obtain the results (apart from terms of higher order in $\alpha_s(t)$):

$$F_L(x,t) = \frac{\alpha_s(t)}{2\pi} x^2 \int_x^1 \frac{dy}{y^3} \left\{\frac{8}{3} F_2(y,t)+2A(1-\frac{x}{y})yG(y,t)\right\} \tag{75}$$

$$\sigma_L(t) \equiv \int_0^1 dxF_L(x,t) = \frac{\alpha_s(t)}{2\pi} \left\{ \frac{8}{9} \int_0^1 dxF_2(x,t)+ \frac{A}{6} \int_0^1 dx\, xG(x,t)\right\} \tag{76}$$

where A is the sum of all coefficients of quarks and antiquarks in the naive parton model expression for F_2/x. For example, with four flavors beyond charm threshold, A=20/9 in electroproduction, and A=8 for ν or $\bar{\nu}$ scattering on matter.

We can easily understand why QCD leads to $\sigma_L \propto \alpha_s(t)$ and why the infrared singularities cancel in this case. In the naive parton model, (i.e. for α_s set to zero), the point-like longitudinal cross section is zero in the limit of zero p_t for the quark in the nucleon. In fact in the frame where the spacelike current carries no energy (see eq. (1)), the hit parton reverses its momentum and since the helicity is conserved (vector and axial vector interactions for massless fermions) the spin is also reversed. Thus, being the process collinear the current must carry a unit of spin component in order to conserve angular momentum and the longitudinal cross section is zero. If p_t (or m) is not exactly zero, but the p_t distribution is sufficiently damped, as one assumes in the naive parton model, then one obtains by a precise calculation:

$$\frac{\sigma_L}{\sigma_T} \simeq 4 \frac{\langle p_t^2\rangle + m^2}{Q^2} \qquad \text{(Naive parton model)} \tag{77}$$

If we now allow for corrections of order α_s as in the diagrams of Fig. 1, then the intermediate quark that actually interacts with the current has a p_t distribution that only drops as

p_t^{-2} (recall the discussion in sect. 2 and especially eq. (14)). We then have schematically in QCD

$$\sigma_T \sim 1 + \alpha_s \int^{\sim Q^2} \frac{dp_t^2}{p_t^2} \sim 1 + \alpha_s \ln Q^2 \tag{78}$$

$$< p_t^2 > \sim \alpha_s \int^{\sim Q^2} p_t^2 \frac{dp_t^2}{p_t^2} \sim \alpha_s Q^2 + \text{constant} \tag{79}$$

$$\sigma_L/\sigma_T \sim \alpha_s \int^{\sim Q^2} \frac{p_t^2}{Q^2} \frac{dp_t^2}{p_t^2} \sim \alpha_s + O(1/Q^2) \tag{80}$$

We thus expect in QCD σ_L/σ_T $< p^2 >/Q^2$ $\alpha_s(t)$. We also understand why σ_L is regular in the massless theory for partons: the singularity at $p^2=0$ is cancelled by the vanishing of the numerator for a collinear process.

We stress that the verification of this QCD prediction is of the utmost importance. Unfortunately, the present experimental situation on F_L and σ_L is confuse (Ref. 24). A precise experimental determination of these quantities would also be a most direct way of measuring the gluon content of the nucleon.

9. JETS IN LEPTOPRODUCTION AND THEIR p_t-DISTRIBUTION

A strictly related set of consequences following from the above discussion is about the p_t distribution of jets in electroproduction. Here we deal with another example of those problems recently attacked in QCD, which were not solved by the renormalization group method. As we shall see the relevant QCD predictions are quite naturally derived by the present approach.

In the naive parton model, neglecting the p_t spread of quarks in the nucleon, which is damped by the wave function and does not increase with Q^2, we expect two hadronic jets opposite in rapidity, arising from the framentation of the quark parton on one side (current region) and from the debris of the nucleon on the other side (target fragmentation). This is a two jet process with small p_t. The p_t of hadrons in the jets arises from two effects: the p_t of the quark parton and the p_t with respect to the parton axis from the fragmentation of the quark into hadrons.

We have seen that in QCD the existence of gluon interactions provides the intermediate quark (and therefore the final quark because the current carries no p_t) with a hard tail in its p_t distribution. This tail is small (because it arises from a fraction of events of order $\alpha_s(t)$) but extends up to $p_t^2 \propto Q^2$. The leading term of $< p_t^2 >$ in eq. (79) is due to this effect and can be predicted perturbatively, while the constant term depends on wave function effects and cannot be evaluated. One thus expects a fraction of events of order $\alpha_s(t)$ with the topology of three jet events: the target fragmentation jet with small p_t and two jets with large and almost opposite p_t from the quark and gluon fragments (or from quark and antiquark fragments for parton processes started by a gluon in the nucleon).

For a quantitative discussion (Refs. 25,26,27) suppose we want to compute a moment of the p_t distribution for the parton jets for three jets events in electroproduction, for example $< p_t^2 >$ as a function of Q^2, x and y. y is defined by:

$$y = \nu/E_\ell \tag{81}$$

where E_ℓ is the laboratory energy for the incoming electron. As well-known y enters because it rules the proportion of longitudinal and transverse photons according to:

$$\frac{d\sigma}{dxdy} = \left(1 - y + y^2/2\right) \; (2xF_1) + (1-y)F_L \tag{82}$$

with F_L defined in eq. (71). (The generalization to neutrino production is straightforward and has been considered in refs. 28,29). From the previous discussion, it follows that we can compute the leading term at large Q^2 of $< p_t^2 >$(i.e. the one proportional to Q^2). We saw that the physical origin of σ_L and of $< p_t^2 >$ are related. Their computation is also very similar. Just as σ_L is obtained by computing σ_L for partons in lowest non vanishing order, i.e. order α_s, then convoluting the result with t dependent parton densities and finally replacing α_s with $\alpha_s(t)$, the same strategy is followed in evaluating $< p_t^2 >/Q^2$. We evaluate the angular distributions $d\sigma/dp_t^2$ off transverse and longitudinal photons for both an initial quark (antiquark) or a gluon. We then obtain the relevant moment. In particular for $< p_t^2 >$ we define:

$$\frac{\alpha_s}{2\pi} T^{L,T}_{q,G}(z) = \int \frac{p_t^2}{Q^2} \left(\frac{d\sigma}{dp^2}\right)^{L,T}_{q,G} dp_t^2 \tag{83}$$

with the integration from zero to the maximum allowed p_t of order Q^2. We finally convolute with t dependent effective parton densities and replace α_s with $\alpha_s(t)$. The final result has the form (Ref. 27):

$$\frac{\langle p_t^2 \rangle}{Q^2} = \alpha_s(t) g(x,y,t) = \frac{\alpha_s(t)}{2\pi} \frac{x}{F_2(x,t)} \Big\{ \int_x^1 \frac{du}{u^2} \Big(F_2(u,t) T_q^T\!\left(\frac{x}{u}\right) + \\ + \sum_i e_i^2\, uG(u,t) T_G^T\!\left(\frac{x}{u}\right)\Big) + \frac{1-y}{1-y+y^2/2} \int_x^1 \frac{du}{u^2} \Big(F_2(u,t) T_q^L\!\left(\frac{x}{u}\right) + \\ + \sum_i e_i^2\, uG(u,t) T_G^L\!\left(\frac{x}{u}\right)\Big) \Big\} \qquad (84)$$

The T functions are explicitly evaluated in Ref. (27). As is clear from eqs. (14), (83), they are well defined even in the massless theory and consequently no infrared cut off dependence affects them. Note that the t dependence of the function g in the previous equation only arises from the effective parton densities.

We stress once more that this result cannot be derived from the light cone expansion and the renormalization group technique. The factorization to the leading logarithmic approximation of the same effective parton densities as for the structure functions has to be separately demonstrated. This has been done over the last year by a number of authors (Refs. 6,7), and we can by now safely state that the above result is as well founded and rigorous as the more classical predictions of QCD.

One finds that the result is fortunately not very sensitive to the badly known gluon density in the nucleon (this is not the case for σ_L/σ_T). The naive parton model prediction in eq. (71) (with m=0) is found not to be true quantitatively, but only as an order of magnitude relation. As for the properties of g(x,y,t) in eq. (84) the y dependence is generally negligible, while the x dependence is very important, but can almost completely be taken into account by saying that $\langle p_t^2 \rangle$ is not proportional to Q^2 but rather to W^2, the total hadronic invariant mass squared:

$$W^2 = (P + q)^2 \qquad (85)$$

with P the nucleon momentum. For almost all x, we roughly obtain (Ref. 27) (by inserting canonical fits for the parton densities):

$$\langle p_t^2 \rangle \simeq \alpha_s(t) W^2/32 \qquad (86)$$

Recall that to this linearly rising (in W^2) component, one should add a constant contribution (apart from logs) originating from the intrinsic p_t of partons in the nucleon. At small W^2 the constant component is expected to be dominant while at sufficiently large W^2 the first term takes over. Note however that the slope in W^2 of the linearly rising term is small, as apparent from eq. (86). Thus is not a surprise that at present energies ($W^2 \lesssim 200$ GeV) the presence of the hard component is still not clearly visible. In some recent data (Ref. 30) by the BEBC collaboration an increase of $<p_t^2>$ with W^2 is observed in ν scattering while the same effect is not visible in $\bar{\nu}$ scattering where the statistics is more limited.

Note that most of the available data refer to $< p_t^2>$ of one given hadron in the jet, while the QCD result in eqs. (84),(86) concerns the $< p_t^2>$ of the parton that produce the jet. The p_t distribution of the hadrons in the jet is also dependent on the fragmentation properties of the parton into hadrons which cannot be calculated in perturbation theory. By summing the $\vec{p}_t$ of the hadrons in the jet, one can in principle reconstruct the $\vec{p}_t$ of the parton and be freed from the dependence on the fragmentation functions.

Note also that the increase of $<p_t^2>$ with W^2 arises from the same physical mechanism (hard gluon emission) that originates the scaling violations in QCD. The two effects are therefore physically equivalent.

Further tests of QCD involving three jet events in leptoproduction concern azimuthal correlations between the final lepton and the jet axis (Refs. 31,32). Such correlations are not easy to disentangle from the data because the same sort of effect can be also produced by the presence of an intrinsic p_t of partons in the nucleon, whose effect cannot be computed preturbatively.

10. JETS IN e^+e^-

Gluon corrections to naive parton results have also been studied for e^+e^- collisions. In zeroth order in α_s (Ref. 5) (naive parton model) one expects a total cross section into hadrons compared to that into $\mu^+\mu^-$ corresponding to the famous ratio $R = 3\Sigma_i Q_i^2$ (where 3 is from color and the sum is over the squared charges of each quark flavor) . Furthermore an angular distribution $1+\cos^2\theta$ is predicted for the quark jets (Ref. 5). The inclusion of both virtual and real gluon corrections to order α_s immediately leads to subleading corrections to the total cross section so that the corrected value (Ref. 1) for R reads $R=R_0(1+\alpha_s(t)/\pi)$.

While in leptoproduction some infrared complicacies can be absorbed in the definition of parton densities, in the present case all predictable quantities (not involving fragmentation functions) must be free of infrared problems. For example the problem has been considered (Refs. 33,34), of a safe definition of two jet events including gluon corrections. This is obtained by classifying two jet events as those where a fraction $1-\varepsilon$ of the total final state energy is contained within two cones of half opening δ with axis at an angle θ with the beam. δ and ε must be chosen so that $\alpha_s(Q^2)\ln\delta, \varepsilon < 1$ and $\delta > \langle p_t^2\rangle/Q^2$ where p_t is the intrinsic transverse momentum of hadrons in the jet. Similarly the choice of a suitable variable for characterizing "jettiness" has been discussed. Values of the old variable sphericity (Ref. 35) are not calculable perturbatively because of infrared singularities and new variables (spherocity (Ref. 36), thrust (Ref. 37), etc.) have been introduced and their behaviour with Q^2 studied. General criteria for good quantities to measure have been put forward (Ref. 38).

Finally when the emitted gluon is hard we have three jet events in e^+e^- (Refs. 39,40) (with transverse momentum of the emitted gluon of order Q). The new high energy accelerators should show the first examples of this rather rare (order $\alpha_s(Q^2)$)events, of high significance for revealing the parton mechanism underlying hadron production in e^+e^- .

11. DRELL-YAN PROCESSES

In this section we consider lepton pair production in hadron-hadron collisions. The usefulness of the QCD improved parton language is again seen in this case which is outside the range of application of renormalization group methods.

In the naive parton model lepton pair production in hadron-hadron collisions is described by the famous Drell-Yan formula (Ref. 41). A quark in one hadron annihilates with an antiquark in the other hadron and produces a lepton pair with point like cross section:

$$\frac{Q^2 d\sigma^{DY}}{dQ^2} = \frac{4\pi\alpha^2}{9S} \int \frac{dx_1 dx_2}{x_1 x_2} \left(\sum_{j=1}^{f} e_j^2 q_j^1(x_1) q_j^2(x_2) + 1 \leftrightarrow 2 \right) \delta\left(1 - \frac{\tau}{x_1 x_2}\right) \tag{87}$$

where Q^2 is the invariant mass squared of the lepton pair, S the invariant mass squared of the incoming hadrons, $\tau = Q^2/S$, $x_{1,2}$ are the fractions of longitudinal momentum carried by the parton quark or antiquark in hadron 1 or 2, the labels 1 and 2 on the parton symbols refer to the hadron they belong and the sum over j runs on quark flavors (i.e. j=1, ... 4).

Starting from the Drell-Yan formula, we can compute the effects induced by gluon corrections in QCD, just as we did in leptoproduction where we also started from the naive parton model approximation. In particular, two problems can be considered. The first deals with the relation of the parton densities that appear in the Drell-Yan formula with those measured in leptoproduction. The result (Refs. 42,43) is that also in the Drell-Yan case t dependent effective parton densities are induced by gluon corrections which turn out to be the same, in the leading logarithmic approximation, as in leptoproduction at the same $|q^2|$ (i.e. there is no influence in this approximation of the current being spacelike in one case and timelike in the other). Recently the non leading logarithmic corrections to this result have also been studies (Refs. 10,11) in detail. The result is that the corrections are quite large and deserve a separate and complete discussion that we postpone to the next section. The second problem is the study of the p_t distribution of the produced lepton pair. In this case the p_t of the pair is directly related to the transverse momentum of the incoming partons with no hadronic fragmentation function obscuring the result. The problem is solved by computing the angular distributions for the parton processes and then convoluting with the effective parton densities and replacing α_s with $\alpha_s(t)$ (Refs. 44,45).

Gluon corrections are evaluated from the basic processes:

$$\begin{aligned} q + \bar{q} &\to G + \gamma^* \\ q(\bar{q}) + G &\to q(\bar{q}) + \gamma^* \end{aligned} \tag{88}$$

where γ^* is the virtual photon of mass $\sqrt{Q^2}$. One must also take into account virtual corrections of order α_s to the lowest order process $q+\bar{q} \to \gamma^*$. As we did for electroproduction we add to the point-like cross section appearing in the Drell-Yan formula the result for $\sigma_{\text{order}\,\alpha_s}$. Ignoring powers of m^2/Q^2, $\sigma_{\text{order}\,\alpha_s}$ for both processes in eqs. (88) has a term that increases logarithmically with Q^2 and a finite term. By omitting all obvious factors and sums the result for the corrected Drell-Yan formula is of the form

$$\frac{Q^2 d\sigma^{DY}}{dQ^2} \int_\tau^1 \frac{dx_1}{x_1} \int_{\frac{\tau}{x_1}}^1 \frac{dx_2}{x_2} \Big\{\Big(q^1(x_1)\bar{q}^2(x_2)+1\leftrightarrow 2\Big)\Big(\delta(1-z) + \frac{\alpha_s}{2\pi}\left(2t\, P^{DY}_{qq}(z) + f^{DY}_q(z)\right)\Big) + \Big((q^1(x_1)+\bar{q}^1(x_1))G^2(x_2)+1\leftrightarrow 2\Big) \cdot \frac{\alpha_s}{2\pi}\Big(P^{DY}_{qG}(z)t + f^{DY}_G(z)\Big)\Big\} \tag{89}$$

where

$$z = \frac{\tau}{x_1 x_2} = \frac{Q^2}{s} \leq 1 \tag{90}$$

with $s = x_1 x_2 S$ the invariant mass squared of the incoming partons. Note that the total cross sections for the parton processes in eq. (88) can only depend on Q^2 and s, and thus when an appropriate dimensional factor is extracted they are just functions of z.

An explicit calculation (Ref. 42) shows the identities

$$\begin{aligned} P^{DY}_{qq}(z) &= P_{qq}(z) \\ P^{DY}_{qG}(z) &= P_{qG}(z) \end{aligned} \tag{91}$$

where on the right hand side we have the familiar functions introduced for leptoproduction. If we neglect for a moment f^{DY}_{qG} in eq. (89) it is a simple matter to show that we can reabsorb the terms in $\alpha_s t$ into effective parton densities, as we did for leptoproduction:

$$\frac{Q^2 d\sigma^{DY}}{dQ^2} \int_\tau^1 \frac{dx_1}{x_1} \int_{\frac{\tau}{x_1}}^1 \frac{dx_2}{x_2} \left(q^1(x_1)+\Delta q^1(x_1 t)\right)\left(\bar{q}^2(x_2)+\Delta\bar{q}^2(x_2,t)\right)\delta(1-z) \tag{92}$$

with

$$\Delta q(x,t) = \frac{\alpha_s}{2\pi} t \int_y^1 \frac{dy}{y} \left(q(y)P_{qq}(\tfrac{x}{y})+G(y)P_{qG}(\tfrac{x}{y})\right) +\ldots \tag{93}$$

We thus obtained the result that at the leading log approximation the same effective parton densities describe both leptoproduction and Drell-Yan processes when the same $Q^2=|q^2|$ is considered. By taking the f^{DY}_{qG} terms also into account and replacing α_s with $\alpha_s(t)$ we then end up with the modified Drell-Yan formula:

$$\frac{Q^2 d\sigma^{DY}}{dQ^2} = \frac{4\pi\alpha^2}{9S} \int_\tau^1 \frac{dx_1}{x_1} \int_{\frac{\tau}{x_1}}^1 \frac{dx_2}{x_2} \left\{\left(\sum_{j=1}^{f} e_j^2 q_j^1(x_1,t)\bar{q}_j^2(x_2,t)+1 \leftrightarrow 2\right)\cdot\right. \tag{94}$$

$$\left.\cdot\left(\delta(1-z)+\frac{\alpha_s(t)}{2\pi} f_q^{DY}(z)\right)+\left(2F_1^1(x_1,t)G^2(x_2,t)+1\leftrightarrow 2\right)\frac{\alpha_s(t)}{2\pi} f_G^{DY}(z)\right\}$$

where F_1^1 is the electroproduction structure function F_1 for hadron 1 (that originates from $\Sigma_j e_j^2(q_j+\bar{q}_j)$ in eq. (89)).

The non leading corrections from $f_{qG}^{DY}(z)$ are a measure of how good is the approximation of including in the Drell-Yan formula the same densities as measured in leptoproduction. Note that the second term arising from gluons is certainly important in PP collisions at moderate Q^2, because the leading term is small being proportional to the sea densities. However the functions $f_{qG}^{DY}(z)$ as they stand in eq. (94) are not well defined because of the already mentioned problems of regularization of infrared singularities and of renormalization procedure. The problem of going beyond the leading logarithmic approximation will be discussed in detail in the next section.

There are several levels of possible tests of the above theory of lepton pair production. First one can test the naive parton model result, disregarding the t dependence induced by gluon corrections. After all there is no complete argument to prove that the uncorrected Drell-Yan formula is a suitable starting point to be implemented by QCD refinements. Tests of the Drell-Yan mechanism in itself involve checking the linearity in the number A of nucleons in a target nucleus, the approximate scaling of $Q^3 d\sigma^{DY}/dQ$ at fixed τ, the absolute order of magnitude of the cross section (which includes testing the famous factor of 1/3 induced by color) and the ratios of cross sections among channels with valence-sea (PP for example) or valence-valence (like π^-P , π^+P or $P\bar{P}$) dominant annihilation mechanism. All these tests are reasonably extensive and successful (Ref. 46), so that at present the Drell-Yan mechanism appears as quite well supported by the data, provided the lepton pair mass is above 3 ÷ 4 GeV (say above the ψ). At a second level of accuracy one is willing to test the specific form of the t dependence, i.e. the QCD scaling violations in the Drell-Yan context. This is done by taking the most detailed PP data (Ref. 46), inserting the rather well known quark densities of electroproduction and extracting the sea densities from the data. The results are compared with the sea densities obtained from neutrino scattering. Neglecting the scaling violations the Drell-Yan sea densities show a steeper x dependence than those obtained from neutrino scattering. The agreement improves considerably (Ref. 46) by including the Q^2 dependence of densities and this is the best evidence for the presence of the predicted t dependence. The remaining discrepancy is presumably due to the effect of non leading corrections to be discussed in the next section.

The presence of gluon corrections can also be demonstrated by a study of the p_t distribution of the produced lepton pair. We shall now briefly discuss this very important signature for

QCD effects.

As for the p_t of jets in leptoproduction, one obtains from QCD a detailed prediction for the hard component of the p_t spectrum. That is to say that for example we can predict the coefficient of S in the expression of $\langle p_t^2 \rangle$. This is of order $\alpha_s(t)$ and is a function of τ :

$$\langle p_t^2 \rangle = \alpha_s(t) S\, G(\tau, \alpha_s(t)) + \ldots \tag{95}$$

Apart for terms of order $\alpha_s^2(t)$, the dots stand for terms with no factor of S, which cannot be predicted because they contain the information about the intrinsic component of p_t from the nucleon wave function. At sufficiently large S the first term dominates, but the magnitude of the term in the dots at present energies is still non negligible, because the linearly rising term has a small derivative or order $\alpha_s(t)$.

One starts by computing the angular distributions for the parton processes in eqs. (88). From these one obtains the moments of the p_t distribution:

$$\int \left(\frac{p_t}{S}\right)^n \frac{d\sigma^{DY}}{dQ^2 dp_t^2}\, dp_t^2$$

that are well defined and convergent even in the massless theory (for $n > 1$). Convoluting with the effective parton densities and replacing α_s with $\alpha_s(t)$, one ends up with the result (for $\langle p_t^2 \rangle$ for example) (Ref. 44):

$$\langle p_t^2 \rangle \frac{Q^2 d\sigma^{DY}}{dQ^2} = \frac{\alpha^2 \alpha_s(t)}{27} \int_\tau^1 \frac{dx_1}{x_1} \int_{\frac{\tau}{x_1}}^1 \frac{dx_2}{x_2} (1-z)^3 \Big\{ \Big(\sum_{j=1}^{f} e_j^2 q_j^1(x_1,t) \bar{q}_j^2(x_2,t) \tag{96}$$

$$+ 1 \leftrightarrow 2 \Big) \cdot \left(\frac{16}{3} + \frac{16z}{(1-z)^2} \right) + \Big(2F_1^1(x_1,t) G^2(x_2,t) + 1 \leftrightarrow 2 \Big) \cdot$$

$$\cdot \left(\frac{3}{2} \frac{1}{1-z} + \frac{1-z}{4} - 2z \right) \Big\}$$

Note that $Q^2 d\sigma^{DY}/dQ^2$ is proportional to S^{-1} at fixed τ (see eq. (94)) so that $\langle p_t^2 \rangle$ is indeed proportional to S.

The most important test of QCD in this connection is the comparison of $< p_t^2>$ at fixed τ and different S. By combining the FNAL data on PP collisions at $\sqrt{S} \simeq 20 \div 27$ GeV with the recent preliminar ISR data at $\sqrt{S} \simeq 64$ GeV a clear indication (Ref. 46) of the linear rise of $< p_t^2>$ with S at fixed τ is found. The measured slope has the correct order of magnitude. Note that at present it is not possible to make the theoretical prediction for the slope in PP scattering very precise, because of the importance of the badly known gluon contribution in eq. (96).

If one tries a direct comparison of the calculated hard component of $< p_t^2>$ with the measurements one finds that the perturbative component amounts for only about one half of the effect in the FNAL data (Refs. 46,47). That means that the $< p_t^2>$ due to the intrinsic parton transverse momentum is still quite sizeable. The inclusion of the intrinsic component by a reasonable but model dependent parameterization (Ref. 44) leads to a good overall fit to the data with a $< p_t>$ intrinsic per parton of $0.5 \div 0.6$ GeV. This is a rather large value and it is interesting to look in other processes for a confirmation of this result (Ref. 48).

12. BEYOND THE LEADING LOGARITHMS

In this section we consider the problem of going beyond the leading logarithmic approximation in leptoproduction and Drell-Yan processes. We have seen that at the leading logarithmic approximation the naive parton model results are reproduced in terms of Q^2 dependent parton densities satisfying evolution equations with known kernels. Corrections of order $\alpha_s(t)$ to the naive parton model formulae and of order $\alpha_s^2(t)$ to the evolution equations have been neglected in most cases in the previous discussion. Although very much related, as we shall see in a moment, in principle the two types of corrections are relevant for two separate problems. The corrections of order $\alpha_s(t)$ to the naive parton model formulae are important at a fixed Q^2 for a detailed comparison of different structure functions in a given process or for relating different processes (for example leptoproduction and Drell-Yan processes). On the other hand the corrections to the derivatives of parton densities are important for a precise extrapolation from one value of Q^2 to another. For an introduction to this problem we go back to the renormalization group result for moments of non singlet structure functions in Sect. 3 (eqs. (31) and (32)). First of all the running coupling constant can be more precisely computed by also taking into account the next order term in the expansion of $\beta(\alpha)$ in eq. (33). The coefficient b' which is independent of the renormalization procedure in the same way as b was explicitly computed (Ref. 49) with the result

$$b' = \frac{153 - 19f}{2\pi(33-2f)} \tag{97}$$

It is then easy to show from eq. (32) the validity of the following improved form for $\alpha_s(t)$:

$$\alpha_s(t) = \alpha_s^o(t)\left(1+b'\alpha_s^o(t)\ln\ln Q^2/\Lambda^2\right) \tag{98}$$

with

$$\alpha_s^o(t) = \left(b\ln Q^2/\Lambda^2\right)^{-1}$$

as in eq. (21), but with a different Λ^2 as compared to eq. (23):

$$\Lambda^2 = \mu^2\exp\left(-\frac{1}{b\alpha_s} - \frac{b'}{b}\ln b\alpha_s\right)$$

where, as usual, $\alpha_s \equiv \alpha_s(0)$. Thus all corrections from taking the next order approximation for the running coupling constant are included by always replacing the above form of $\alpha_s(t)$ in the following. As for the exponential factor in eq. (31) we immediately derive the following expansion in terms of the parameters defined in eqs. (33) - (35):

$$\exp\int_{\alpha_s}^{\alpha_s(t)} \frac{\gamma^n(d)}{\beta(d)}\,d_\alpha \equiv \frac{H^n(\alpha_s(t))}{H^n(\alpha_s)} \simeq \left(\frac{\alpha_s}{\alpha_s(t)}\right)^{\frac{An}{2\pi b}}\left(1+\frac{An}{2\pi b}(g'^n-b')\cdot(\alpha_s-\alpha_s(t))+\ldots\right) \tag{99}$$

As a consequence the t dependent factor in eq. (31) for $J_a^n(t)$ has the following expansion in terms of $\alpha_s(t)$:

$$c_a^n(\alpha_s(t))H^n(\alpha_s(t)) \simeq (\alpha_s(t))^{-\frac{An}{2\pi b}}\left\{1+\left(c_a'^n-\frac{An}{2\pi b}(g'^n-b')\right)\alpha_s(t)+\ldots\right\} \tag{100}$$

(The constant term $H^n(\alpha_s)$ can be included in the unknown factor q_o^n). It is important to stress that while $c_a'^n$ and g'^n are not separately well defined, because they depend on the renormalization procedure and the regularization chosen for the infrared singularities, their combination that appears in eq. (100) is indeed well defined (Ref. 8). Note that the ambiguities in $c_a'^n$ are independent of the index a (because g'^n does not depend on a) and therefore the differences $c_a'^n - c_b'^n$ are well defined. Note also that since $\alpha_s(t)$ starts with α_s while $\alpha_s - \alpha_s(t)$ starts with α_s^2 in ordinary perturbation theory, $c_a'^n$ and g'^n appear in different orders in a perturbation expansion in α_s but contribute to the same order in $\alpha_s(t)$. In particular $c_a'^n$ are computable from the quantities $\sigma_{\text{order}\,\alpha_s}$ often considered in the previous sections and are in fact related[8] to the moments of the functions $f_a(z)$ appearing for example in eqs. (69), (70). The reader certainly recalls that we repeatedly stated that the functions $f_a(z)$ are not well defined,but their differences, as appearing for example in the longitudinal structure function eq. (72), (73), (74) are instead well defined. On the other hand the calculation of g'^n (and the analogue quantities for the singlet sector) is far more involved, being a second order problem. The latter calculation was attacked and completed by the authors of ref. (8).

With this general discussion in mind we can now resume the more familiar language of the previous sections. We start by improving the definition of effective parton densities. We are free to include in the Q^2 dependent parton densities not only the leading logarithmic series generated by the P functions, but also some convenient non leading term. We noticed that in general the $f_a(z)$ functions cannot be all included in the parton densities because they are different for different structure functions (i.e. they depend on the index a). But we may decide to include one specified convenient set of them, the same for all cases. We make use of this freedom by defining the quark densities in such a way that the structure function F_2 is fixed to its parton model expression to all orders in $\alpha_s(t)$ (Refs. 10,11).:

$$\frac{F_2(x,t)}{x} = \sum_i e_i^2 (x,t) \tag{101}$$

This is certainly possible. To first order it amounts to change eq. (31) into:

$$\begin{aligned}\Delta q_i(x,t) = \frac{\alpha_s}{2\pi}\{t\int_x^1 \frac{dy}{y}\left(q_i(y)P_{qq}(\tfrac{x}{y})+G(y)P_{qG}(\tfrac{x}{y})\right) + \\ + \int_x^1 \frac{dy}{y}\left(q_i(y)f_2^q(\tfrac{x}{y})+G(y)f_2^G(\tfrac{x}{y})\right)\end{aligned} \tag{102}$$

with f_2^q and f_2^G taken from eq. (69). More in general, in the notation of eq. (31) this criterium amounts to also include the coefficient $c_2^n(\alpha_s(t))$ into the definition of quark densities:

$$\sum_i e_i^2 \int_0^1 dx x^{n-1} q_i(x,t) = c_2^n(\alpha_s(t)) q_o^n \exp \int_{\alpha_s}^{\alpha_s(t)} \frac{\gamma^n(\alpha)}{\beta(\alpha)} d\alpha \qquad (103)$$

This formula is only appropriate for the non singlet quark densities but it can be directly generalized. The advantage is that the product on the right hand side of eq. (103) is well defined, while the single factors are not.

Note that the evolutions are not modified to order $\alpha_s(t)$ as in apparent from eq. (102).

Some remarks are now in order. First why F_2 ? F_2 is given a special role because it satisfies the Adler sum rule (Ref. 50). This sum rule which follows from the algebra of charges is exact and not simply an asymptotic sum rule. In particular when powers of m^2/Q^2 are neglected, it can be written in the form (to all order in $\alpha_s(t)$):

$$\int_0^1 \frac{dx}{x} \left(F_2^{\nu P}(x,t) - F_2^{\nu N}(x,t)\right) = A_o \qquad (104)$$

where A_o is a fixed number depending on the flavor content of the theory. By defining quark densities as measured from F_2 we automatically ensure that the valence sum rules in eqs. (48) are preserved from corrections of order $\alpha_s(t)$. With the present definition of quark densities they are in fact equivalent to the Adler sum rule.

The gluon density is not fully specified by the above prescription, because F_2 in the naive parton model is completely determined by quark densities. The freedom to modify the gluon density by non leading corrections can be used in order to also guarantee the validity of the momentum conservation sum rule in eq. (50), by demanding for example that (Ref. 10)

$$\sum_i q_i^{old} + G^{old} = \sum_i q_i^{new} + G^{new}$$

This amounts in first order to add a term proportional to $\alpha_s(t)$ to the previous definition of the gluon density. Working to order $\alpha_s(t)$ this has no effect in practice because the gluon density already appears in order $\alpha_s(t)$ in all structure functions and thus the additional term would be of second order. In summary with the above improved definition of parton densities to all orders in $\alpha_s(t)$ F2 is given by its naive parton model expression and the effective parton densities satisfy the charge and momentum sum rules of the naive parton model at all Q^2.

An immediate advantage of this definition is that when the other structure functions are expressed in terms of effective parton densities as measured from F_2 the non leading corrections are all well defined. In fact to each correcting function f_a we now have to subtract the corresponding function f_2. Thus we have, for example, for V-A currents and unit coupling:

$$F_2(x,t) = 2x\left(q(x,t) + \bar{q}(x,t)\right) \tag{105}$$

$$F_1(x,t) = \int_x^1 \frac{dy}{y}\left\{\left(q(y,t)+\bar{q}(y,t)\right)\left(\delta\left(\frac{x}{y}-1\right)+\frac{\alpha_s(t)}{2\pi}\left(f_1^q\left(\frac{x}{y}\right)-f_2^q\left(\frac{x}{y}\right)\right)\right) + 2\,\frac{\alpha_s(t)}{2\pi}\,G(y,t)\left(f_1^G\left(\frac{x}{y}\right)-f_2^G\left(\frac{x}{y}\right)\right)\right\} \tag{106}$$

$$xF_3(x,t) = 2\int_x^1 \frac{dy}{y}\left(-q(y,t)+\bar{q}(y,t)\right)\left(\delta\left(\frac{x}{y}-1\right)+\frac{\alpha_s(t)}{2\pi}\left(f_3^q\left(\frac{x}{y}\right)-f_2^q\left(\frac{x}{y}\right)\right)\right) \tag{107}$$

where, besides eqs. (74) we also have:

$$f_2^q(z) - f_3^q(z) = \frac{4}{3}(1+z) \tag{108}$$

and all these results are independent of definition ambiguities.

The quality of the available data on leptoproduction is at present not sufficient for a determination of the parton densities to the level of accuracy implied by the retention of the next order terms in $\alpha_s(t)$. The corrections are in fact of the order of the longitudinal structure function which is badly known at present. Some testable consequences refer to corrections of current algebra sum rules. For example the Gross-Llewellyn-Smith sum rule is modified into:

$$\int_0^1 dy\left(F_3^{\bar{\nu}P}(x,t) + F_3^{\nu P}(x,t)\right) = L_o\left(1-\frac{\alpha_s(t)}{\pi}\right) \tag{109}$$

where L_o is the predicted asymptotic value. Some recent experimental determinations (Ref. 20) appear to indicate an effect of correct sign and magnitude. Another phenomenologically important consequence has to do with the extraction from the data of the sea densities. Typically the sea densities are measured from quantities that in the naive parton model are proportional to antiquarks. An example is the right-handed cross section from V-A currents. Including non leading corrections, terms of order $\alpha_s(t)$ times quark or gluon densities also appear in the right handed cross section. The effect of these terms of order $\alpha_s(t)$ will be comparatively large at moderate Q^2 because the sea densities are themselves small. Thus they cannot be neglected for a quantitative determination of sea densities. The reader may find further details on this problem in ref. 10.

Going to the Drell-Yan processes, the total cross section including terms of order $\alpha_s(t)$ with effective parton densities defined through F_2 at the same $|q^2|$ is now given by (that replaces eq. (94)):

$$\frac{Q^2 d\sigma^{DY}}{dQ^2} = \frac{4\pi\alpha^2}{9S} \int \frac{dx_1 dx_2}{x_1 x_2} \Big\{ \Big(\sum_i e_i^2 q_i^1(x_1,t)\bar{q}_i^2(x_2,t) + 1 \leftrightarrow 2\Big) \Big(\delta(1-z) + \frac{\alpha_s(t)}{2\pi}\,\theta(1-z)\big(f_q^{DY}(z) - 2f_2^q(z)\big)\Big) + \frac{\alpha_s(t)}{2\pi}\,\theta(1-z)\,\Big(2F_1^1(x_1,t)G^2(x_2,t) + 1 \leftrightarrow 2\Big)\Big(f_G^{DY}(z) - f_2^G(z)\Big)\Big\} \tag{110}$$

The differences:

$$f_G^{DY}(z) - f_2^G(z) = \frac{1}{2}\left(z^2 + (1-z)^2\right)\ln(1-z) + \frac{9}{4}z^2 - \frac{5}{2}z + \frac{3}{4} \tag{111}$$

$$f_q^{DY}(z) - 2f_2^q(z) = \frac{4}{3}\left(\frac{3}{1-z_+} - 6 - 4z + 2(1+z^2)\left(\frac{\ln(1-z)}{1-z}\right)_+ + \left(\frac{4\pi^2}{3} + 1\right)\delta(1-z)\right) \tag{112}$$

are now perfectly well defined (Refs. 10,11,51,52).

It is simple to realize that contrary to what happens in leptoproduction where the corrections of order $\alpha_s(t)$ are always small the above corrections to the Drell-Yan formula are quite large at present values of Q^2. The gluon corrections are as expected. Namely they are important in PP collisions where the lowest order term is small being proportional to the sea densities. But they are negligible in all cases where the leading term is proportional to valence quark densities as for example is the case for $P\bar{P}$ or πP collisions. What was not expected is that the corrections of order $\alpha_s(t)$ from the quark term (eq. (112)) are quite large. Note that these corrections are universal, i.e. they are of the same relative importance in $P\bar{P}$ or PP collisions because they consist of a multiplicative factor in the leading term. The large terms are the last two in eq. (112). They are indeed independent of our particular choice of F_2 as reference structure function and would be the same had we chosen F_1 or F_3 or any suitably normalized combination.

It must be stressed that the fact that the above corrections are large does not affect the validity of the Drell-Yan formula itself, but only the relation between the quark densities as obtained from leptoproduction and from Drell-Yan processes by fitting the corresponding naive parton model formulae. In other words we could always define suitable t dependent "Drell-Yan densities" (to this order in $\alpha_s(t)$), such that the Drell-Yan cross section has the familiar parton model expression of eq. (87). The usual tests of the Drell-Yan model like the approximate scaling of $Q^3 d\sigma/dQ$ or the linear rise of $< p_t^2 >$ with S would be perfectly valid. Only the "Drell-Yan densities" and the leptoproduction densities would not be so simply related. For example consider the comparison of sea densities as extracted from Drell-Yan PP collisions with the sea densities measured in neutrino scattering. As mentioned in sect. 11 it is an established result (Ref. 44) that if all scaling violations are neglected the Drell-Yan sea densities are much steeper in x than the neutrino sea densities. When scaling violations are included to the leading order the agreement improves, but still the Drell-Yan densities are somewhat steeper. The non leading terms have exactly the right sign and magnitude to explain this residual difference.

In conclusion there are some cases where the corrections to the leading logarithmic approximation are important and cannot be neglected. Examples are the problem of a quantitative determination of sea densities in leptoproduction and the phenomenology of Drell-Yan processes. If one is working with data at widely different Q^2 values, corrections of the same order also arise from terms of order $\alpha_s^2(t)$ in the derivatives of parton densities.

REFERENCES

1. For reviews see: H.D. Politzer, Phys.Reports 14C (1974) 129
W. Marciano, H. Pagels, Phys.Reports 36C (1978) 137

2. H. Georgi, H.D. Politzer, Phys.Rev. D9 (1974) 129
D.J. Gross, F. Wilczek, Phys.Rev. D9 (1974) 980

3. G. Parisi, Physics Letters 43B (1973) 207 and 50B (1974) 367
see also J. Kogut, L. Susskind, Phys.Rev. D9 (1974) 697 and D9 (1974) 3391; G. Parisi, Proc. 11th Rencontre de Moriond (1976) ed. by J. Tran Than Van

4. G. Altarelli, G. Parisi, Nucl.Phys. B126 (1977) 298

5. See for example: R.P. Feynman "Photon Hadron Interactions" W.A. Benjamin, New York (1972), J. Kogut, L. Susskind, Physics Reports C8 (1973) 76

6. Some early works on the diagrammatic analysis are:
V.N. Gribov and L.N. Lipatov, Sov.J.Nucl.Phys. 15 (1972) 438 675
A.H. Mueller, Phys.Rev. D9 (1974) 963
L.N. Lipatov, Sov.J.Nucl.Phys. 20 (1975) 94
A.P. Bukhvostov, L.N. Lipatov and N.P. Popov, Sov.J.Nucl.Phys. 20 (1975) 287
I.G. Halliday, Nucl.Phys. B103 (1976) 343
Yu.L. Dokshitser, Leningrad Preprint No. 330 (1977) and JETP 72 (1977) 1216

7. C.H. Llewellyn Smith, Talk at the "XVII Internationale Universitätswochen für Kernphysik", Schladming, Austria, 1978; Oxford Preprint (47/78)
Yu.L. Dokshitser, D.I. D'yakanov and S.I. Troyan, "Hard Semi-Inclusive Processes in QCD", Leningrad Preprint and Proceedings of the 13th Leningrad Winter School (1978), available in English as SLAC Trans. - 183
D. Amati, R. Petronzio and G. Veneziano, Nucl.Phys. and CERN Preprint TH-2527 (1978)
R.K. Ellis, H. Georgi, M. Machacek, H.D. Politzer and G.C. Ross Phys.Lett. 78B (1978) 281 and CALTECH Preprint 68-684 (1978)
G. Sterman and S. Libby, Stony Brook Preprint ITP SB78-41, 42 (1978)
G. Sterman, Phys.Rev. D17 (1978) 2773, 2789
A.H. Mueller, Columbia Univ. Preprint CU-TP-125 (1978)
W.R. Frazer, J.F. Gunion, Univ. of Calif. at Davis Preprint 78/3 (1978)

8. E.G. Floratos, D.A. Ross, C.T. Sachrajda; Nucl.Phys. B129 (1977) 66, B 139 (1978) 545, CERN Preprint TH-2566,2570 (1978)

9. W.A. Bardeen, A.J. Buras, D.W. Duke, T. Muta ; FNAL Preprint 74/42 THy (1978)

10. G. Altarelli, R.K. Ellis, G. Martinelli, Nucl.Phys. B143 (1978) 521

11. J. Kubar André, F.E. Paige, BNL Preprint 24794 (1978)

12. For a review see for example: G. Altarelli, Rivista Nuovo Cimento 4 (1974) 335

13. K. Wilson, Phys.Rev. 179 (1969) 1499
N. Christ, B. Hasslacher and A.H. Mueller, Phys.Rev. D6 3543 (1972)
W. Zimmermann, Lectures on Elementary Particles and Quantum Field Theory, MIT Press, Cambridge, MA (1970)

14. See for example: K.J. Kim, R. Schilcher, Phys.Rev. D17 (1978) 2800

15. H. Georgi, H.D. Politzer, Nucl.Phys. B136 (1978) 445;
T. Uematsu, Phys.Letters 79B (1978) 97

16. J.F. Owens, Phys.Letters 76B (1978) 85

17. See for example the review talk of R.D. Field in Proceedings of the Tokyo Conference (1978)

18. O. Nachtmann, Nucl.Phys. B63 (1973) 237; B78 (1974) 455;
H. Georgi, H.D. Politzer, Phys.Rev. D14 (1976) 1829;
S. Wandzura, Nucl.Phys. B42 (1977) 412

19. H.L. Anderson, H.S. Matis, L.C. Myrianthopoulos, Phys.Rev. Letters 40 (1978) 1061

20. BEBC collaboration, Nucl.Phys. B142 (1978)1

21. See also for example: A.J. Buras, K.J.F. Gaemers, Nucl.Phys. B132 (1978) 249
N. Cabibbo, R. Petronzio, CERN TH-2440 (1978)

22. G.C. Fox, talk given at the Neutrino Conference, Purdue University (1978)

23. A. Zee, F. Wilczek, S.B. Treiman, Phys.Rev. D10 (1974) 2881;
D.V. Nanopoulos, G.G. Ross, Phys. Letters 58 (1975) 1o5;
see also E. Witten, Nucl.Phys. B120 (1977) 189 Eqs. (11) and (13)

24. See for example the talks by L. Hand in Proceedings of the Hamburg Conference (1977) and by R.C. Taylor in Proceedings of the Tokyo Conference (1978)

25. E.G. Floratos, Nuovo Cimento 43A (1978) 241

26. K.H. Craig, C.H. Llewellyn Smith, Phys.Letters 72B (1978) 341

27. G. Altarelli, G. Martinelli, Phys.Letters 76B (1978) 89

28. A. Mendez, A. Raychaudhuri, V.J. Stenger, Oxford Univ. Report 63/78 (1978)

29. P. Mazzanti, R. Odorico, V. Roberto, CERN preprint (1978)

30. V.J. Stenger, Talk at the 1978 Oxford Topical Conference on Neutrino Physics at Accelerators, Oxford Univ. Preprint 59/78 (1978)

31. H. Georgi, H.D. Politzer, Phys.Rev. Letters 40 (1978) 3

32. J. Cleymans, Univ. of Bielefeld Preprint BI-TP 78/02 (1978
A. Mendez, Oxford Univ. Preprint 29/78 (1978)
G. Köpp, R. Maciejko, P.M. Zerwas, Univ. of Aachen Preprint (1978)
P. Mazzanti, R. Odorico, V. Roberto, CERN Preprint (1978)

33. G. Sterman, S. Weinberg, Phys.Rev.Letters 39 (1977) 1436

34. R.K. Ellis, R. Petronzio, CERN preprint TH-2571 (1978)
C.L. Basham, L.S. Brown, S.D. Ellis, S.T. Love, Phys.Rev. D17 (1978) 2298 and Seattle Preprint RLO-1388-759
W. Furmanski, Cracow Preprint, TPJU 11/78 (1978)

35. J.D. Bjorken, S.J. Brodsky, Phys.Rev. D1 (1970) 1416

36. H. Georgi, M.Machacek, Phys.Rev. Letters 39 (1977) 1237

37. E. Fahri, Phys.Rev.Letters 39 (1977) 1587

38. G. Parisi, Phys.Letters 74B (1978) 65
G. Tiktopoulos, CERN Preprint TH-2560 (1978)

39. J. Ellis, M.K. Gaillard and G.G. Ross, Nucl.Phys. B111 (1976) 253, Erratum - Nucl.Phys. B130 (1977) 516
A. de Rujulà, J. Ellis, E.G. Floratos and M.K. Gaillard, Nucl.Phys. B138 (1978) 387
S.Y. Pi, R.L. Jaffe, F.E. Low, Phys.Rev.Lett 41 (1978) 142
S.Y. Pi and R.L. Jaffe, MIT Preprint CTP-735 (1978)
S.S. Shei, New York Univ. Preprint 1977 (unpublished)
K. Shizuya and S.-H.H. Tye, Fermilab Preprint 78/54-THY
T.A. DeGrand, Y.J. Ng and S.-H.H. Tye, Phys.Rev.D16 (1977)3251

40. For an interesting algorithm on QCD jets see: K. Konishi, A. Ukawa, G. Veneziano, Phys.Letters and CERN Preprint TH-2577 (1978)

41. S.D. Drell, T.M. Yan, Phys.Rev.Letters 25 (1970) 316

42. H.D. Politzer, Nuclear Physics B129 (1977) 301

43. I. Hinchliffe, C.H. Llewellyn Smith, Phys.Letters 66B (1977) 281;
C.T. Sachrajda, Phys.Letters 73B (1978) 185;
J. Frenkel, M.J. Shailer, J.C. Taylor, Oxford Preprint 25/78 (1978).

44. G. Altarelli, G. Parisi, R. Petronzio, Phys.Letters 76B (1978) 351, 356

45. A.V. Radyushki, Phys.Letters 69B (1977) 245

K. Kajantie, R. Raitio, Nucl.Phys. B139 (1978) 72
K. Kajantie, J. Lindfors, R. Raitio, Phys.Letters 74B (1978) 384, Nucl.Phys. B (in press), Helsinki preprint HU-TFT-78-33
H. Fritzsch, P. Minkowski, Phys.Letters 73B (1978) 80
F. Halzen, D. Scott, Phys.Rev.Letters 40 (1978) 1117
E.L. Berger, Argonne Preprint ANL-HEP-PR-78-12
M. Gluck, E. Reya, Mainz Preprint MZ-TH 78/3

46. For a summary see the talk by L. Lederman in Proceedings of Tokyo Conference (1978)

47. S.W. Herb et al., Phys.Rev.Letters 40 (1978) 435

48. See however Y.L. Dokshitzer, D.I. D'Yakonov, S.I. Troyaa, Phys.Letters 78B (1978) 290, where a method is described for taking into account the region at intermediate p_t^2

49. D.R.T. Jones, Nucl.Phys. B75 (1974) 531
W. Caswell , Phys.Rev.Letters 33 (1974) 226

50. S.L. Adler, Phys.Rev. 143 (1965) 1144

51. In ref. 10 the coefficient of the δ function in eq. (112) is misreproduced (see Erratum, Nucl.Physics to appear). There is also sign error in ref. 11. The authors of refs. (10) and (11) now agree on eq. (112) as it is written here

52. For calculations referring to the gluon quark term in eq. (111) with a different definition of parton densities, see also J. Kripfganz, Leipzig Preprint KMU-HCP 78-12

DEEP INELASTIC LEPTON SCATTERING

W.S.C. Williams

Department of Nuclear Physics

University of Oxford

I have been asked to talk about deep inelastic lepton scattering but to concentrate on electron and muon scattering, since the results of neutrino experiments have been reviewed during a seminar in this Summer Institute.

I. FORMALISM

I.1 Kinematics

The process is

$$\ell + N \rightarrow \ell + X$$

where ℓ is an electron or a positron or a μ^+ or a μ^-. N is a nucleon target and X represents the final state hadrons. The charged lepton will interact with the target either electromagnetically or by the neutral weak current. The former predominates and the effects of the latter have only just been observed. They exist at a level which has no effect on the experiments which I shall discuss except that one described briefly in IV.1.

Since the electromagnetic interaction is understood and since all the evidence is that the charged leptons behave like point particles, the only unknown in the scattering process is the effect of the target structure and thus a study of this process serves to probe that structure.

KINEMATICS

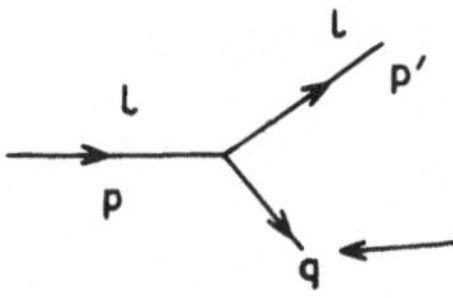

	Relativistic 4 - vector	Laboratory
Incident μ	p	$\underset{\sim}{p}, E$
Scattered μ	p'	$\underset{\sim}{p}', E'$
Energy - Momentum transfer	q	$\underset{\sim}{q}, \nu$

$$\underset{\sim}{q} = \underset{\sim}{p} - \underset{\sim}{p}'$$
$$\nu = E - E'$$

4 - momentum transfer squared	q^2	$=$	$\nu^2 - \underset{\sim}{q} \cdot \underset{\sim}{q}$
Since $q^2 < 0$ define	Q^2	$\equiv$	$-q^2$
Scattering angle			θ
If $m_l \ll E, E'$	Q^2	$=$	$4EE' \sin^2 \frac{\theta}{2}$

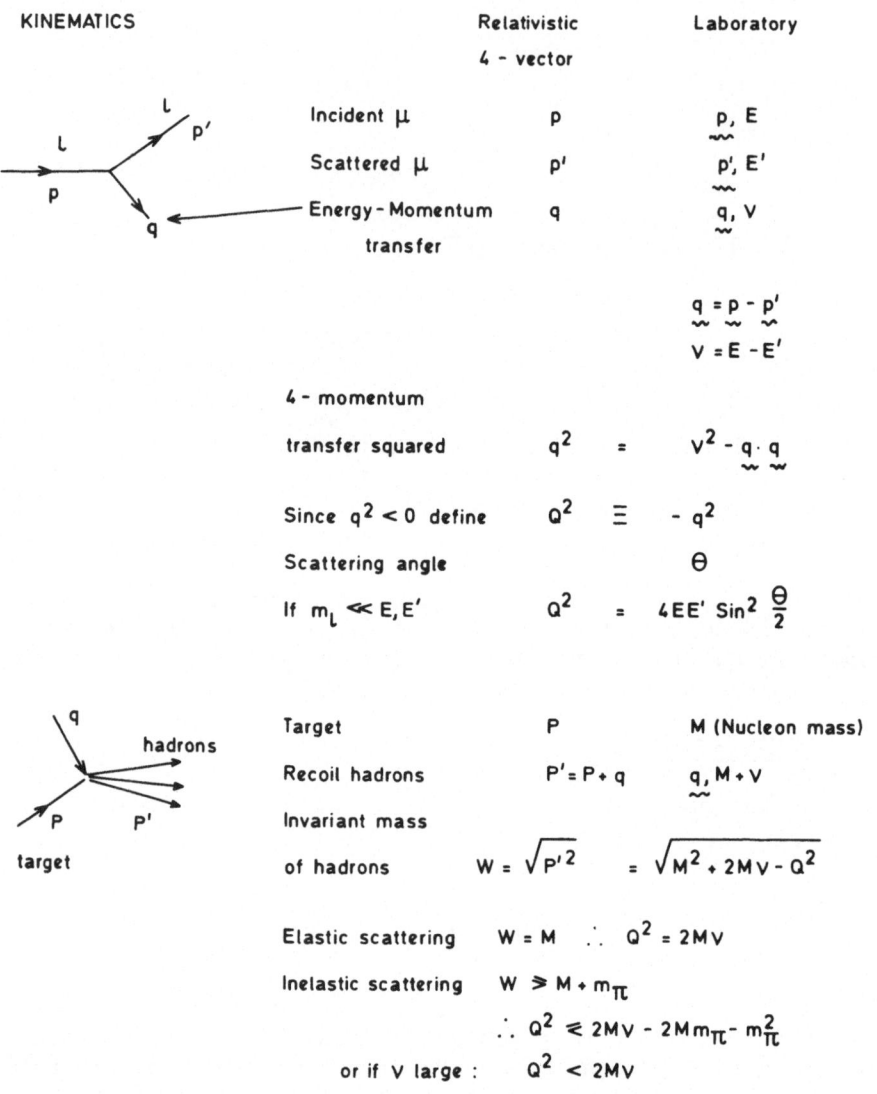

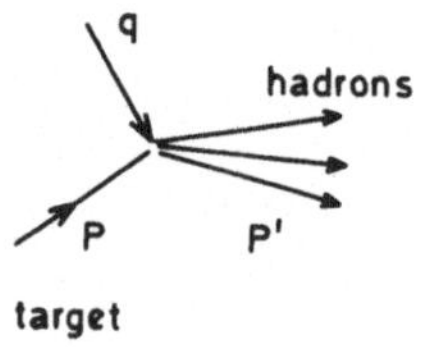

Target	P	M (Nucleon mass)
Recoil hadrons	$P' = P + q$	$\underset{\sim}{q}, M + \nu$
Invariant mass of hadrons	$W = \sqrt{P'^2}$	$= \sqrt{M^2 + 2M\nu - Q^2}$

Elastic scattering $W = M \quad \therefore \; Q^2 = 2M\nu$

Inelastic scattering $W \geq M + m_\pi$

$$\therefore \; Q^2 \leq 2M\nu - 2Mm_\pi - m_\pi^2$$

or if ν large : $Q^2 < 2M\nu$

Fig. 1 Kinematic quantitites in lepton-nucleon scattering

We shall assume muon-electron universality. This means that the muons, the electron and the positron have identical electromagnetic interactions apart from differences due to the sign of the charge or to the mass /1/.

The kinematic quantities are defined in Fig. 1. The important relativistically invariant quantities are the four-momentum transfer squared Q^2 and the scalar product $2q.P$, which is $2M\nu$ in terms

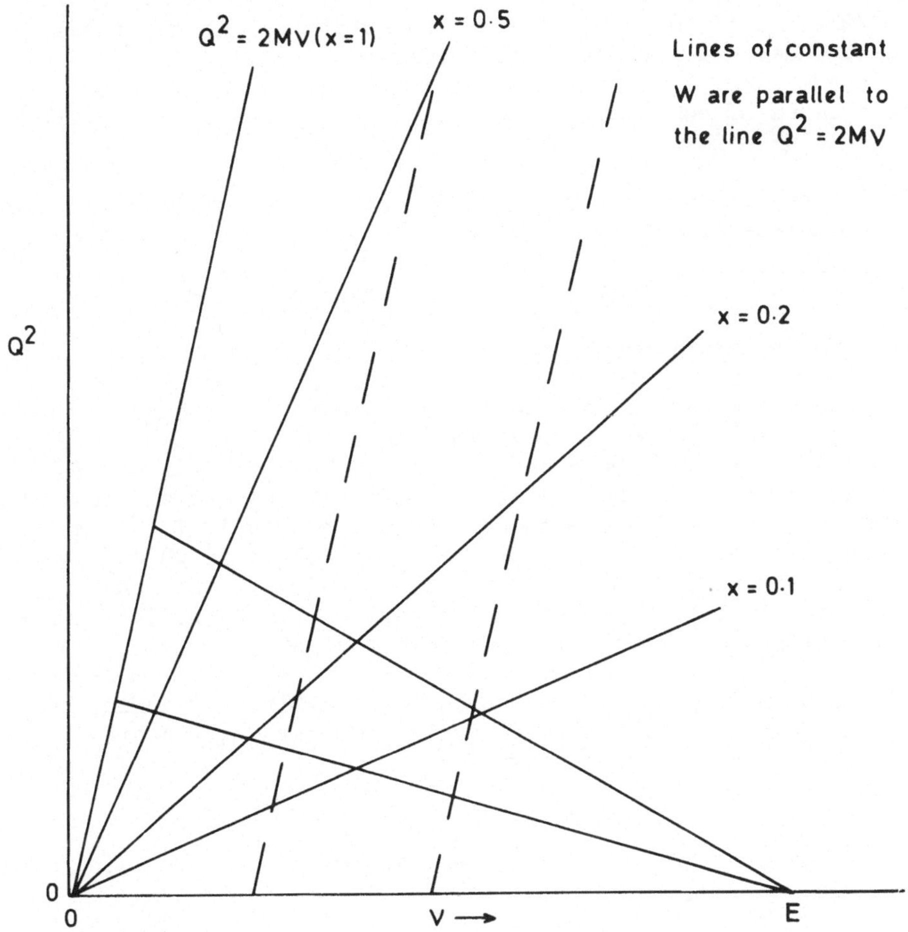

Lines of constant x radiate from $Q^2 = 0, \nu = 0$

Lines of constant Θ radiate from $Q^2 = 0, \nu = E$

Variables $x = \frac{-q \cdot q}{2P \cdot q} = \frac{Q^2}{2M\nu}$ $\quad y = \frac{P \cdot q}{P \cdot p} = \frac{\nu}{E}$

$\omega = x^{-1}$

Fig. 2 The kinematic Q^2-ν plane.

of laboratory quantities and for a stationary target of mass M. The Q^2, ν plane is shown in Fig. 2. Elastic scattering events have $Q^2 = 2M\nu$ and so inelastic scattering events lie to the right of the line defined by that equality. Also

1. Lines of constant W, the recoil hadron mass, lie parallel to $Q^2 = 2M\nu$.

2. Lines of constant $x = Q^2/2M\nu$ (the Bjorken scaling variable) radiate from $Q^2 = 0$, $\nu = 0$.

3. Lines of constant $y = \nu/E$ lie parallel to $\nu = 0$.

I.2 The Structure Functions

The mechanism of the scattering process is presumed to be due to the exchange of one or more photons between the lepton and the target - see Fig. 3. We expect the amplitudes to decrease rapidly with an increase in the number of photons exchanged and tests have shown that no effects are to be seen at the level of 1% due to the exchange of more than one photon /2/. Processes in which real photons are emitted on the lepton lines and a virtual photon loops across the lepton vertex are all a part of the radiative corrections which are invariably applied to the data.

Assuming one photon exchange, Quantum Electrodynamics can be used to show /3/ that, for an unpolarized target, the cross-section for inelastic scattering is

$$\frac{d^2\sigma}{dQ^2 d\nu} = \frac{4\pi\alpha^2}{Q^4}\frac{E'}{E}\{W_2 \cos^2\frac{\theta}{2} + 2W_1 \sin^2\frac{\theta}{2}\} \quad . \tag{1}$$

Here α is the fine structure constant and it is assumed that $m_\ell \ll E, E'$. The Structure Functions W_1 and W_2 contain all the information about the target structure. These quantities can only be

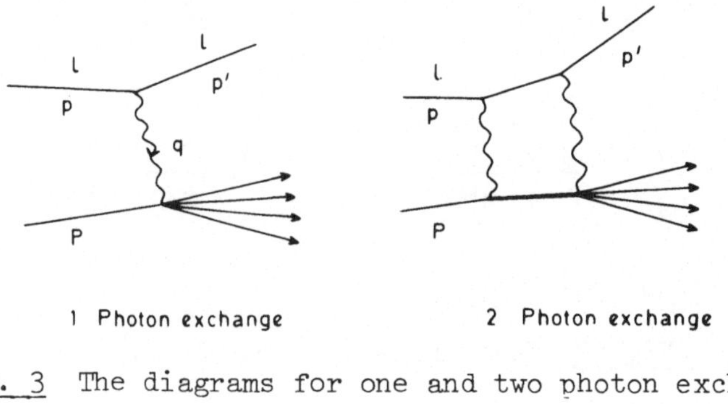

Fig. 3 The diagrams for one and two photon exchange.

functions of relativistic invariants. The only two varying with the kinematics defined by the transferred photon are q^2 and q.P. Since $q^2 < 0$, it is usual to use $Q^2 = -q^2$ and in terms of laboratory quantities $q.P = M\nu$. Thus it is usual to think of the structure functions as functions of the independent variables Q^2 and ν.

I.3 Bjorken Scaling

Bjorken proposed that in the limit $Q^2 \to \infty$ and $2M\nu \to \infty$ so that $Q^2/2M\nu = x$ remained finite, the structure functions would have the following property /4/:

$$\nu W_2(Q^2,\nu) \to F_2(x) \quad ,$$

$$2MW_1(Q^2,\nu) \to F_1(x) \quad .$$

These functions would depend only on the ratio of Q^2 to $2M\nu$ and would not be a function of the two quantities as independent variables. Incidentally, these structure functions are analogous to two of those used in neutrino physics (F_2^ν; F_1^ν) and may differ from them only by a constant factor $\sim 18/5$. The third neutrino structure function goes with parity violating terms and must be absent in the parity conserving electromagnetic structure functions.

In the very naive parton model, x is the fraction of the target nucleon's momentum carried by the parton which absorbs the single virtual photon transferred in the scattering process /5/. It then follows that, if the partons have spin 1/2,

$$F_1(x) = \sum_i e_i^2 f_i(x)$$

and

$$F_2(x) = \sum_i e_i^2 \, x \, f_i(x)$$

where: $\sum_i$ means sum over all partons;

$f_i(x)dx$ is the probability of finding parton i with momentum fraction in the range, x, x + dx;

e_i is the charge in units of electron charge of the ith parton.

Some simple sum rules can be derived from these expressions. Thus:

Conservation of probability gives $\int_0^1 f_i(x)\, dx = 1$

Conservation of momentum gives $\sum_i \int_0^1 x\, f_i(x)\, dx = 1$.

If the number of partons (η) is constant $\sum_i \int_0^1 x\, f_i(x) dx = \eta$

Then

$$\int_0^1 F_2(x)\, dx = \sum_i e_i^2 \int_0^1 x\, f_i(x)\, dx$$

If $f_i(x)$ is the same for all partons and the number of partons is constant, this equation becomes

$$\int_0^1 F_2(x)\, dx = \sum_i e_i^2/\eta$$

$$= \text{mean square charge of partons.}$$

Also we see that $F_2(x) = xF_1(x)$;

That is $Q^2 W_1 = \nu^2 W_2$,

which is the Callan-Gross relation for spin 1/2 partons.

Thus we see that measurement of the structure functions is a tool in the elucidation of the charges, momentum distributions and spins of the partons.

I.4 R and the Problems of its Measurement

Looking at Fig. 1 we can see that if E is fixed then at a given Q^2 and ν, θ is determined. Then a measurement of $d^2\sigma/dQ^2 d\nu$ at that point cannot give separated values of W_2 and W_1. It is necessary to measure at two or more different θ and that means at different E. To see how this is done we consider a different way of looking at the differential cross-sections. The charged lepton scattering creates a flux of virtual ($Q^2 \neq 0$, space-like) photons and the target presents a cross-section to these photons.

We define

σ_T = target total cross-section to transverseley polarised photons,

σ_L = target total cross-section to longitudinally polarised photons,

$R = \sigma_L/\sigma_T$.

Of course, σ_L and σ_T are functions of Q^2 and ν.

Then

$$\frac{d^2\sigma}{dE'd\Omega} = \frac{EE'}{\pi}\frac{d^2\sigma}{dQ^2d\nu} = L\,(\sigma_T + \varepsilon\,\sigma_L) \quad . \qquad (2)$$

The photon properties are

$$L = \text{flux of photons} = \frac{\alpha}{4\pi^2}\,\frac{2M\nu - Q^2}{Q^2}\,\frac{E'}{E}\,\frac{1}{1-\varepsilon} \quad ,$$

$$\varepsilon = \text{polarization parameter} \ ,$$

$$= \left[1 + (1 + \frac{\nu^2}{Q^2})\tan^2\frac{\theta}{2}\right]^{-1} ,$$

$$\simeq \frac{2EE'}{E^2+E'^2} \quad .$$

The parameter ε is relative intensity of longitudinal photons to transverse photons but it also describes the transverse polarization /6/.

The target properties of Eq. 2 are contained in cross-sections. For real photons, there is no longitudinal polarization, and gauge invariance requires

$$\lim_{Q^2 \to 0} \sigma_L = 0 \quad .$$

The relations between the structure functions and the cross-sections are given by

$$W_1 = \frac{2M\nu - Q^2}{8\pi^2\alpha M}\,\sigma_T \qquad (3)$$

$$W_2 = \frac{2M\nu - Q^2}{8\pi^2 \alpha M} \frac{Q^2}{Q^2+\nu^2} (\sigma_T + \sigma_L) \quad . \tag{4}$$

So that

$$W_1 = W_2 \left(1 + \frac{\nu^2}{Q^2}\right) \frac{1}{1+R} \quad .$$

In the Bjorken Limit $Q^2 \to \infty$, $\nu \to \infty$ the Callan-Gross relation means $R = 0$.

It is clear we can talk about and measure at any Q^2, ν the pairs of quantities (R, W_1), $(W_1, \nu W_2)$, or $(\nu W_2, R)$. The last is the usual. Then the cross-section becomes

$$\frac{d^2\sigma}{dQ^2 d\nu} = \frac{4\pi\alpha^2}{Q^4} \frac{E'}{E} W_2 \left\{\cos^2 \frac{\theta}{2} + 2\left(1 + \frac{\nu^2}{Q^2}\right) \frac{\sin^2 \frac{\theta}{2}}{1 + R}\right\} \quad . \tag{5}$$

At a fixed Q^2, ν the prescription is to vary θ (vary E). This is now equivalent to measuring $\sigma = \sigma_T + \varepsilon\sigma_L = \sigma_T(1 + \varepsilon R)$ at different ε. See Fig. 4.

The measurement of σ at different ε means that the extraction of R is very succeptible to systematic errors. The changes in in-

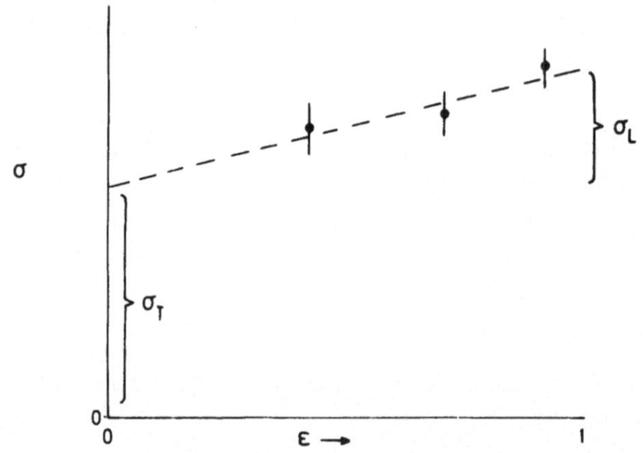

Fig. 4 Fictional plot of $\sigma = \sigma_T + \varepsilon\sigma_L$ against ε. The errors on the measured values of σ clearly allow a good git by a range of straight lines which in turn give values of σ_T and σ_L with large errors.

cident energy and angle necessary to effect a change in ε is an experimental situation which is full of the possibility of systematic errors. Even in the absence of such errors we find that cross-section measurements made to a statistical accuracy of a few percent may still lead to 100% errors on R.

The converse situation is that if it is impossible to measure R and a value is assumed, then the value of νW_2 derived from $d^2\sigma/dQ^2d\nu$ is very insensitive to this assumed value, if θ is small. We shall see this effect in results to be shown later.

II. REVIEW OF DATA AS IT EXISTED IN 1972

II.1 Review of Experiments

By 1972 the most extensive data had been obtained at the Stanford Linear Accelerator Center by a collaboration of researchers from SLAC and the Massachusetts Institute of Technology (MIT). The incident beam was electrons at energies up to 20 GeV and with an energy spread of about 0.2 to 0.5%. Two detector systems were used /7/. Each was a spectrometer capable of moving in angle and containing dipole and quadrupole magnets arranged to focus particles entering their apertures onto suitably arranged electronic detectors. Thus these spectrometers could measure for detected particle their vector momentum after scattering at the target. The 20 GeV spectrometer operated at small angles ($< 10^o$) and the 6 GeV at larger angles. The targets used were of liquid hydrogen or deuterium. Data were taken at 6°, 10°, 18°, 26° and 34° and at incident energies spanning 4.5 to 19.5 GeV.

The first noteworthy result from these measurements was that the cross-sections were much larger than could have been expected, and Fig. 5 shows this effect. The observed cross-section divided by that expected if the target were a point charged fermion is plotted against Q^2. There are two cases: one is elastic electron-proton scattering and it is seen that the ability of the nucleon to take up momentum transfer and remain unchanged, decreases very rapidly with Q^2. The second curve has W = 20 GeV. (The nucleon recoils into a hadronic system of mass 2 GeV.) The cross-section is lower and declines slowly with Q^2 compared to the elastic case. It is behaving as if in inelastic collisions the nucleon is behaving like an assembly of point-like objects each of fractional charge.

II.2 The Value of R

It is best to first consider the values of R. As we have noted, the value can only be obtained at Q^2, ν points more limited than the kinematic limits imposed by the maximum beam energy. Fig.

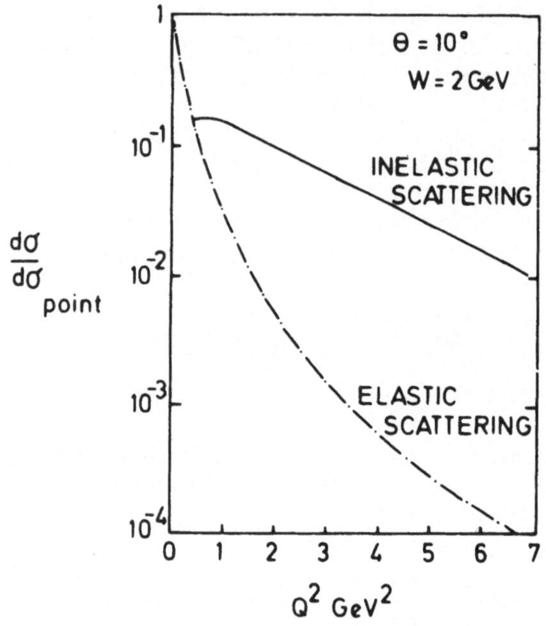

Fig. 5 The ratios of the proton inelastic and elastic scattering cross-sections to the point fermion cross-section as a function of Q^2. The inelastic cross-section corresponds to W = 2 GeV.

6 shows the region in which cross-section measurements were made and the region in which the separation of R could be done. (Note the use of the variable $\omega = 1/x$ which will occur again in this talk.) The results of measurements for both hydrogen and deuterium indicated that there were, within the errors, no detectable variations of R with Q^2 or ν, inside the region of separation /10/. Therefore R was assumed to be constant and equal to the average 0.18 ± 0.10.

II.3 νW_2 and Evidence for Scaling

Remember that scaling predicts that $\nu W_2(Q^2,\nu)$ should be independent of Q^2 so the results of Fig. 7 are particularly interesting /7/. νW_2 at $\omega = 4$ $(x = 0.25)$ is plotted against Q^2 for a series of measurements covering four different angles. It is clear that these results are consistent with νW_2 being independent of Q^2 over a considerable range. (Note that νW_2 must be zero at $Q^2=0$ so that there is a region below $Q^2 \simeq 1$ GeV^2 where we expect a rapid variation of νW_2. This is seen /9/.)

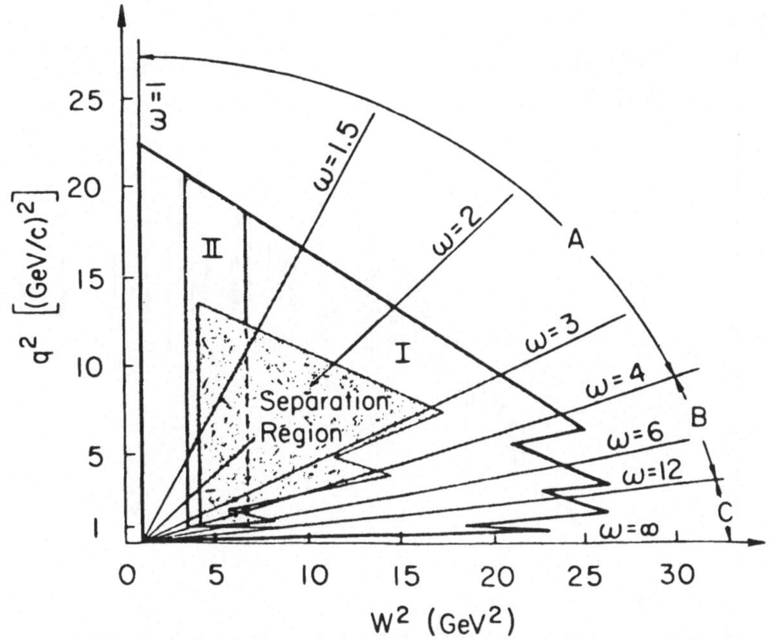

Fig. 6 The kinematic plane Q^2,W^2. The heavy line bounds all data points measured at $\theta = 6°, 10°, 18°, 26°$ and $34°$ at SLAC. In the Separation Region data at any point exist for three or more scattering angles and permit the separation of MW_1 and W_2. Taken from Miller et al. (Ref. 8).

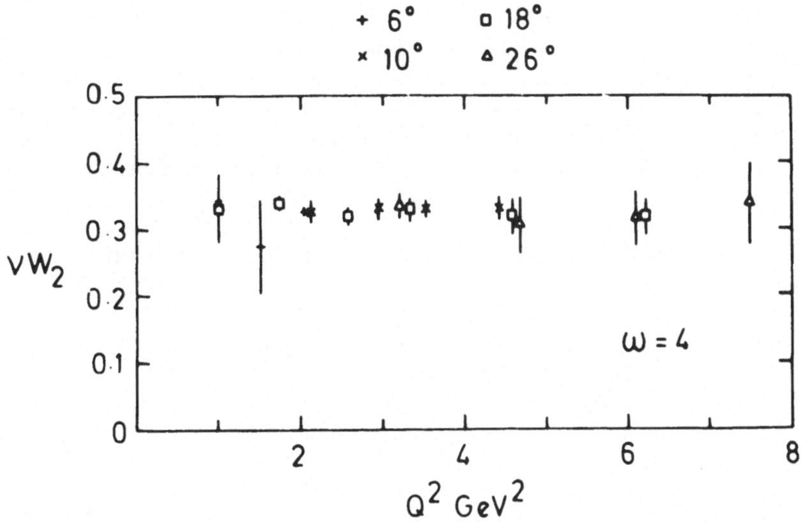

Fig. 7 νW_2 as a function of Q^2 for electron scattering data for which $Q^2 > 1$ GeV^2, $W > 2.6$ GeV, $\omega = 4$, and using $R = 0.18$. Taken from Miller et al. (Ref. 8).

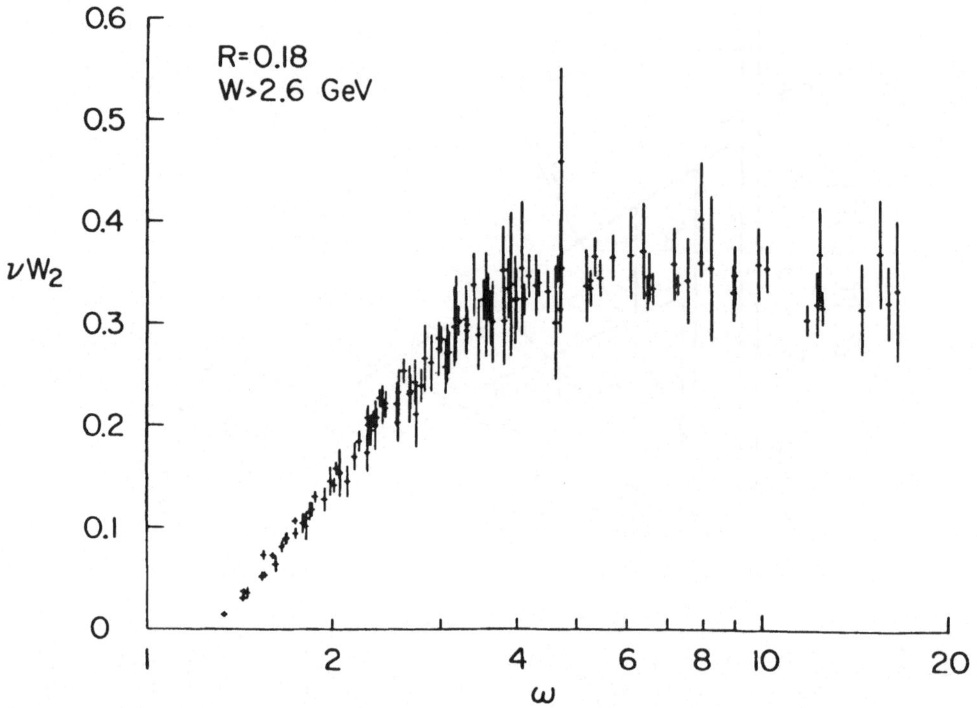

Fig. 8 νW_2 as a function of ω for electron scattering data for which $Q^2 > 1$ GeV^2, $W > 2.6$ GeV and using $R = 0.18$. Taken from Miller et al. (Ref. 8).

This is exactly as expected from scaling. It turns out $\omega = 4$ is a good place to find scaling. The independence of νW_2 of Q^2 at other ω was not quite so convincing. Nonetheless all data were assembled into a universal curve of νW_2 against ω for $Q^2 > 1$ GeV^2 and $W > 2.6$ GeV. This is shown in Fig. 8. Bjorken scaling seemed to be correct.

The conclusions of the MIT-SLAC work /8/ were that, for the proton,

1. In the region $4 < \omega < 12$, $1 < Q^2 < 7$ GeV^2 and $2 < W < 5$ GeV the structure function νW_2 scales.

2. In the region $\omega < 4$ the structure function scales if $2 < Q^2 < 20$ GeV^2 and $2.6 < W < 4.9$ GeV.

3. In the region $\omega > 12$ the kinematically accessible Q^2 was limited to values where νW_2 was still increasing from its zero value at $Q^2 = 0$. In addition the results in this region are sensitive to the value of R assumed. Thus no significant tests of

scaling were possible.

The remarkable thing was that scaling was occurring at $Q^2 >$ 1 GeV^2 and $W > 2$ GeV. The Bjorken limits had suggested that scaling might not set in until $Q^2 \gg M^2$ and $W \gg M$.

Attempts were made to extend the region of scaling by finding a new variable in place of $\omega(= x^{-1})$. The most successful was

$$\omega' \equiv \omega + \frac{M^2}{Q^2} = \frac{W^2 + Q^2}{Q^2} = \frac{2M\nu + M^2}{Q^2}$$

(equivalent to $x' = 1/\omega = Q^2/2M\nu+M^2$).

The structure function νW_2 expressed as a function of ω' was found to scale for $Q^2 > 1$ down $W = 1.8$ GeV /8,10/.

This section should be finished with a brief look at information on the neutron structure function obtained from measurements with a deuterium target. The value of R appeared, within the large errors, to be the same as for the proton, so the same (0.18) value was used to evaluate νW_2 /10/. There are some uncertainties about correctly accounting for the nuclear effects in abstracting νW_2 but these occur for $x > 0.8$. The results are shown in Fig. 9. The proton-neutron ratio is shown in (a), ($\sigma_n/\sigma_p = \nu W_{2n}/\nu W_{2p}$, if $R_n = R_p$) and the difference in (b). The structure functions are clearly different.

Thus conclusions drawn from the measurements on deuterium were that scaling held equally well but that the structure function was different, indicating the existence of a non-diffractive part to the scattering cross-section.

III. MUON SCATTERING AND FURTHER ELECTRON SCATTERING DATA

III.1 Review of Experiments

The use of muons as incident leptons in deep-inelastic scattering became important during the period 1972-76. The main reason was that muons became available at energies up to 270 GeV (compared to 20 GeV of the electron beam at SLAC). In addition, muons have two advantages over electrons:

1. The radiative corrections are less, leading to smaller uncertainties from that calculation.

2. The muon is easy to detect amongst several hadrons.

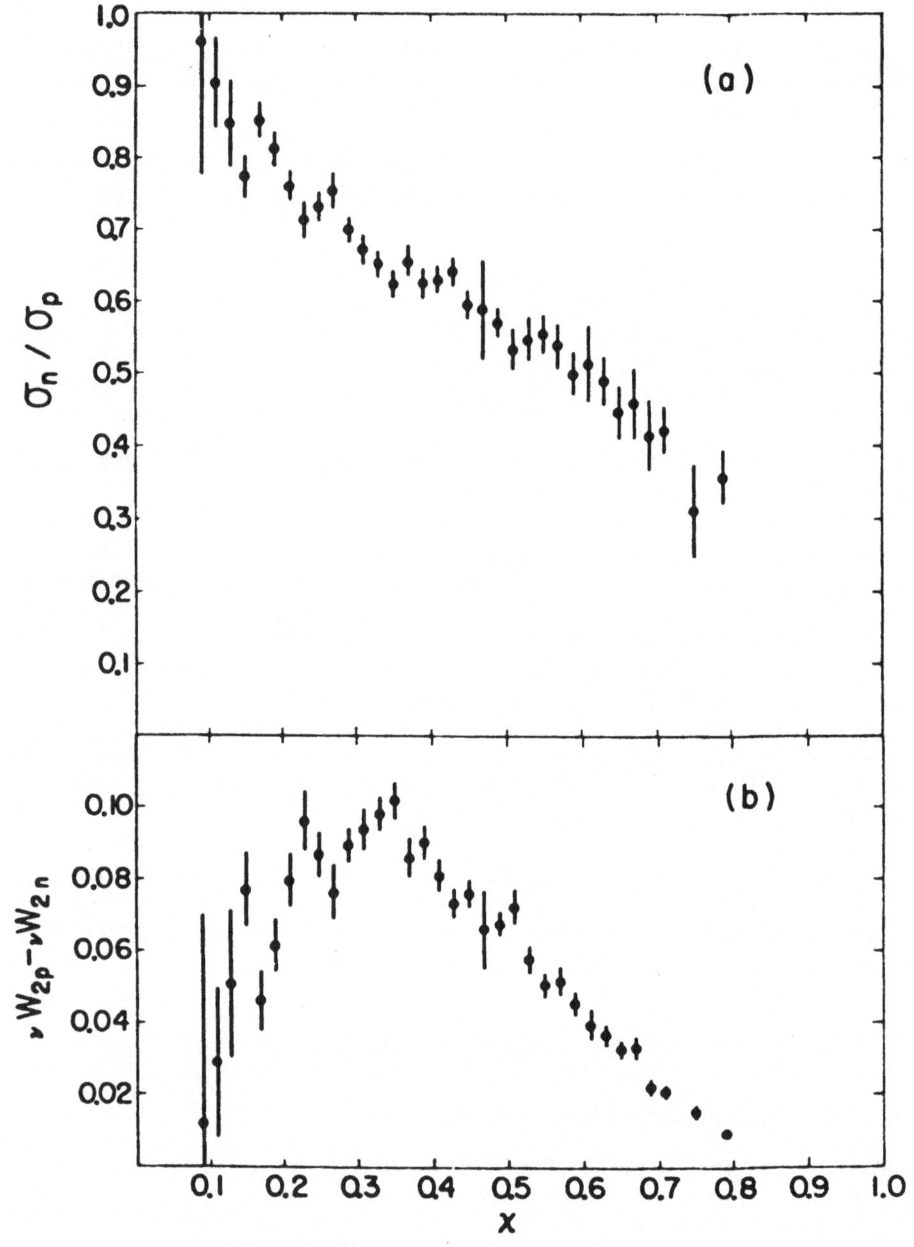

Fig. 9 (a) The ratio of neutron to proton electron inelastic scattering cross-section as a function of x. If $R_n = R_p$, this ratio is equal to the ratio of the two structure functions $\nu W_{2n}/\nu W_{2p}$.
(b) The difference of proton and neutron structure functions. Taken from Bodek et al. (Ref. 11).

The disadvantages were that:

1. The muon beams were poorly collimated and accompanied by a considerable halo, both leading to experimental difficulties.

2. The available intensities were considerably less than those at SLAC. For example:

   μ^+ Fermilab 10^5 s^{-1} at 10% duty cycle at 200 GeV,

   e^- SLAC 10^{13} s^{-1} at 0.5% duty cycle at 20 GeV.

   (Since April 1978 we have

   μ^+ CERN 10^7 s^{-1} at 10% duty cycle at 200 GeV.)

However, the most important factor is the increased kinematic range in inelastic scattering made available by the use of these high energy muon beams. Fig. 10 shows the Q^2-ν plane at E = 275 GeV. In principle Q^2 upto over 500 GeV^2 are accessible. In fact limitations of luminosity have kept measurements to below 150 GeV^2 even in the experiment with the greatest luminosity.

The experiments that have been done are:

1.	Cornell-Michigan State:	E = 56 and 150 GeV, Iron target
2.	Chicago-Harvard-Illinois-Oxford (CHIO):	E = 96 GeV, Hydrogen target E = 147 GeV, Hydrogen and Deuterium targets
2a.	CHIO + Maryland, Virginia, Michigan:	E = 219 GeV, Hydrogen target

2 and 2a will be labelled CHIO, for convenience. I was involved in these experiments and will speak of the results with some degree of possessiveness. My colleagues are listed in Reference 13.

3.	Michigan State-Fermilab	E = 270 GeV, Iron target.
4.	Lawrence Berkeley Lab, Fermilab, Princeton	E = 270 GeV, Iron target.

And I shall include in the discussion the later results from SLAC using electrons:

5.	MIT-SLAC	E = 20 GeV, Hydrogen target.
6.	SLAC	E = 20 GeV, Hydrogen target.

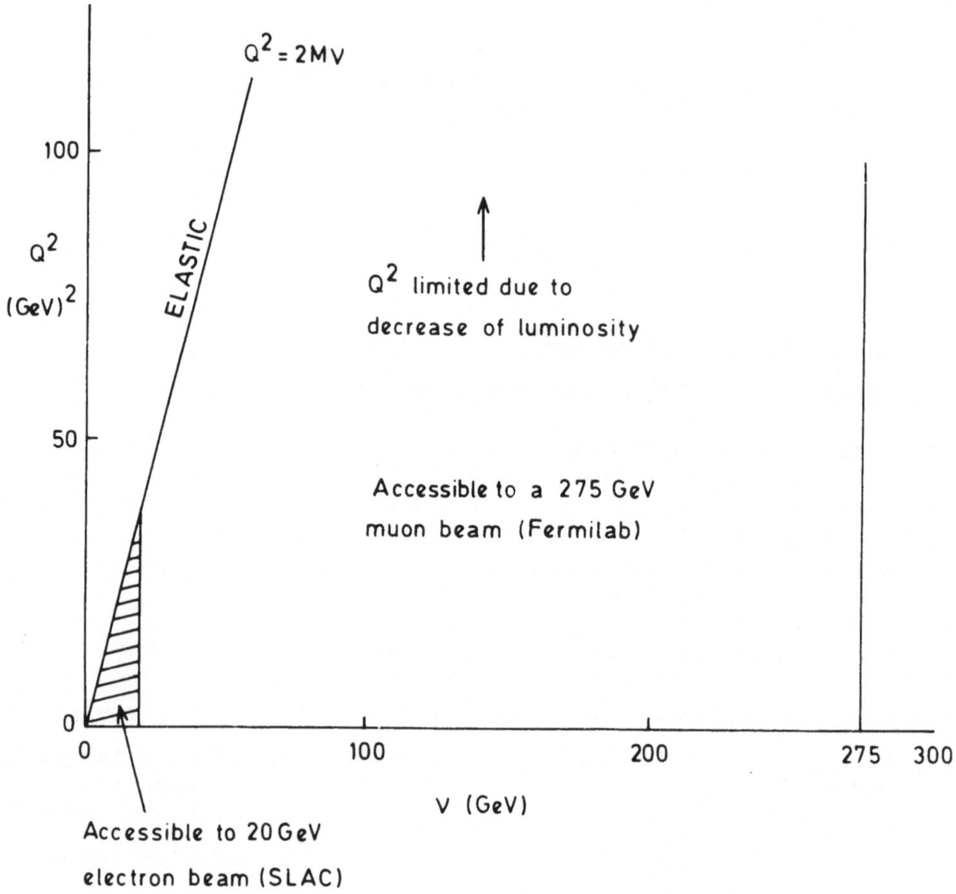

Fig. 10 The kinematic Q^2-ν plane at energies up to 300 GeV showing the regions accessible at SLAC, Fermilab and now (1978) at CERN.

III.2 Results and Scaling Violations

The results from 1, the Cornell-Michigan State experiment /12/ are shown in Fig. 11. The values of the structure function are expressed as ratios of observed to that of a Q^2 independent νW_2 from earlier SLAC-MIT results. We see immediately that the data are showing scaling violations of the kind where for $\omega < 5$, νW_2 decreases as Q^2 increases and for $\omega > 5$, νW_2 increases as Q^2 decreases. The authors of this work parametrised these effects by

$$\nu W_2(Q^2,\omega) = \nu W_2(Q^2=3,\omega)(1 + a \ln \frac{\omega}{6} \ln \frac{Q^2}{3}) \qquad (6)$$

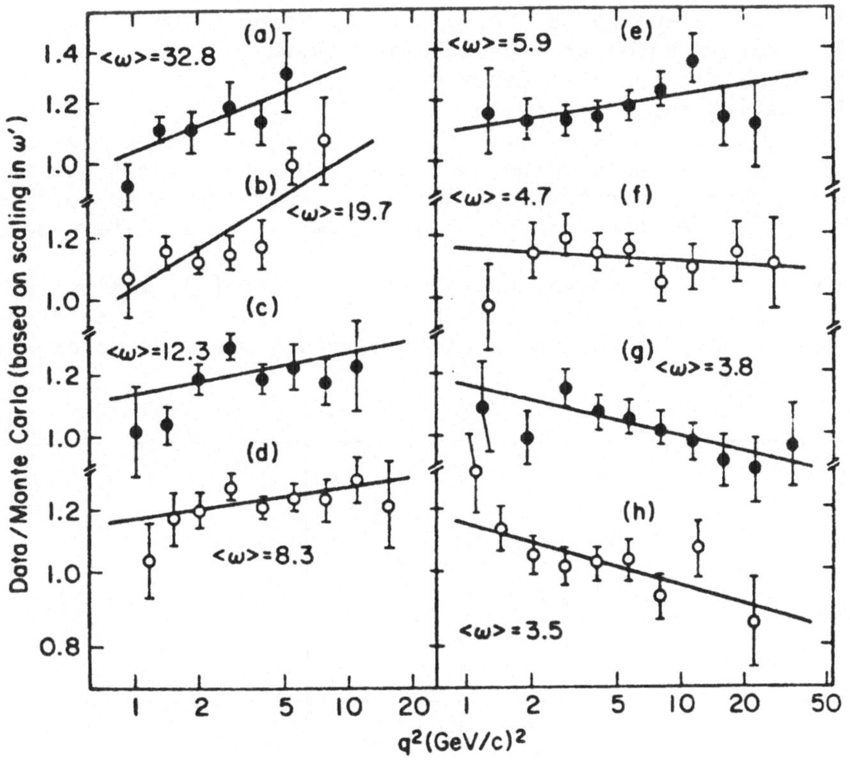

Fig. 11 Ratio of observed to simulated event rate versus Q^2 in various ω ranges for the Cornell-Michigan state experiment. In the absence of scaling violations the ratios would not vary with Q^2. Taken from Chang et al. (Ref. 12).

and found a = 0.099 ± 0.018.

The results of the experiments 2 and 2a are now available, with data from all three energies combined /13/. Since the hydrogen target data was taken at three energies a determination of R is possible, in principle. In the range

$$0.009 < x < 0.069 \quad ,$$

$$1 < Q^2 < 12 \text{ GeV}^2 \quad ,$$

we have found

$$R = 0.44 \pm 0.25 \quad .$$

The error is made up from several parts:

(a) error on energy to energy normalisation ±0.05
(b) error on acceptance calculation ±0.10
(c) track finding efficiency error ±0.15
(d) statistical error ±0.19

We have used this value to evaluate νW_2 throughout the whole kinematic range covered by our data. It is important to note that this range is greater than the range in which R was determined.

In Fig. 12 are shown the values of νW_2 ($F_2(x)$) against x (note the logarithmic scale) in bins of Q^2. The effect of changes in R are shown. Each point is plotted with a vertical error bar.

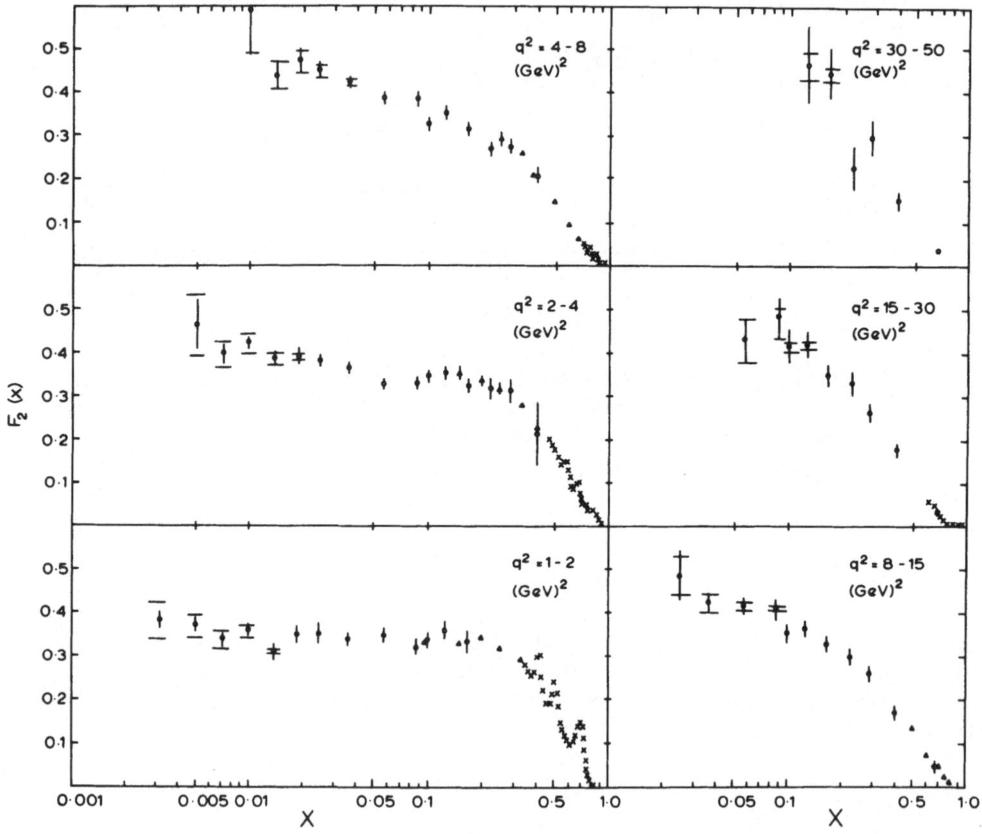

Fig. 12 The structure function $F_2(x,Q^2)$ for the proton as a function of x for various Q^2 bins using all the CHIO data (R = 0.44, open circles) and MIT-SLAC data (triangles and crosses). The errors are shown by vertical lines and are statistical only. The horizontal bars indicate the change in the value for R = 0.69 and R = 0.19 (Ref. 13).

Any significant changes in the value which occur if R is changed by one standard deviation are shown by two small horizontal bars:

	— ←		R = 0.69
Error bar {	┼ ←	νW_2 value with	R = 0.44
	— ←		R = 0.19 .

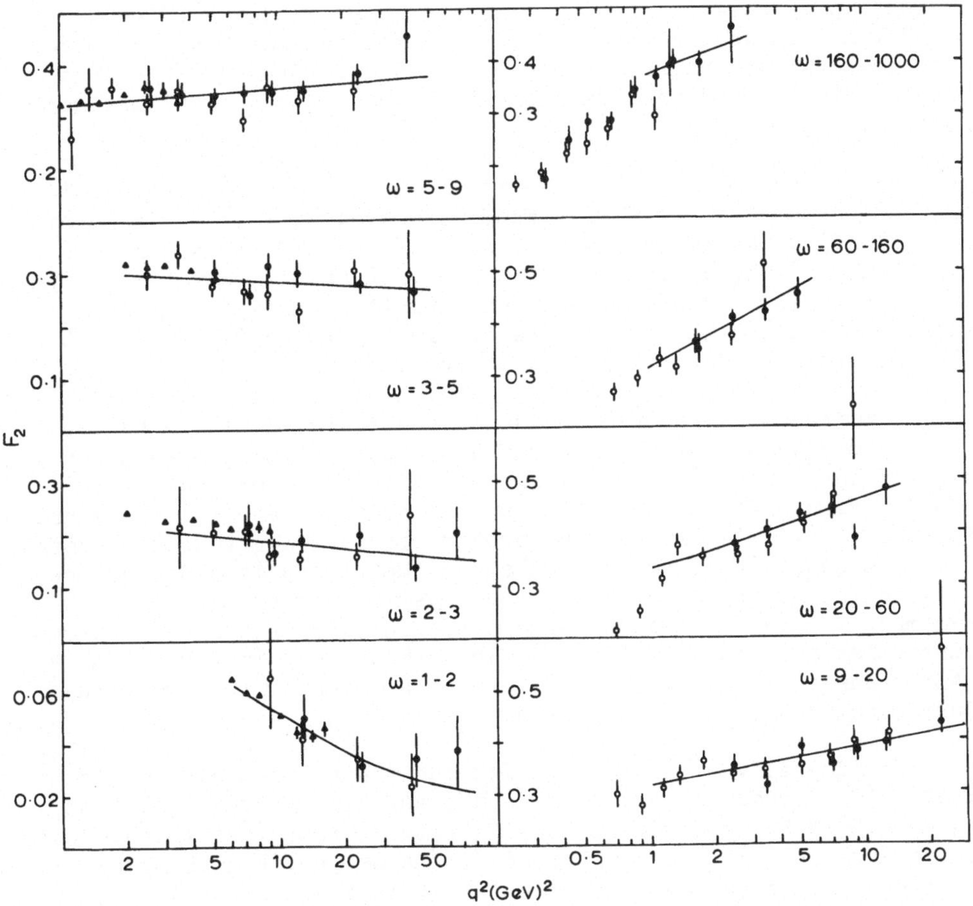

Fig. 13 The structure function $F_2(x,Q^2)$ for the proton as a function of Q^2 in various ω bins for R = 0.44 showing the 219 GeV data (closed circles), the earlier CHIO data at 96 and 147 GeV recalculated for R = 0.44 (open circles) and the MIT-SLAC (Ref. 14, triangles). Note the difference in scales and the suppressed zeroes for the ordinate. The solid lines are the fits: $F_2(x,Q^2) = F_2(x,3)(Q^2/3)^b$. (Ref. 13).

You will notice that the only data seriously affected by the "choice" of R is that at lowest x in each Q^2 range. Note also that plotted on this figure are data from an MIT-SLAC collaboration and from a SLAC compilation of data /14/. At $Q^2 < 4$ GeV^2 and $x > 0.4$ the effect of resonance electroproduction can be seen as peaks in the structure functions.

In Fig. 13 the muon data is plotted against Q^2 in bins of ω. Also plotted is the MIT-SLAC data. Here the scaling violations are clearly visibile. For $\omega < 5$, νW_2 is decreasing and for $\omega > 5$, νW_2 is increasing as Q^2 increases. Remember that $\nu W_2(Q^2 = 0) = 0$ from its definition so that there must be a region where νW_2 is decreasing towards this value, at low Q^2. Data from SLAC /9/ and from muon scattering indicate that the "turn-on" of νW_2 is almost complete by $Q^2 = 1$.

Before considering the significance of these results we will take a look at the MIT-SLAC (expt. 5, above; Ref. 14) data separately. In Fig. 14 νW_2 is shown as a function of Q^2 in separate x bins. These data, as indicated in Figs. 12 and 13, agree with the muon scattering and are clearly showing the same scaling violations. Note also that these data are in the range $0.1< x < 0.75$, whereas the muon data is mostly in the range $x < 0.3$.

Experiment 6 measured inelastic scattering at large angles and yielded values of the neutron and proton structure function MW_1 for large x /15/. This is shown in Fig. 15. The results are plotted versus Q^2 in one x bin for the neutron, and for one x and for one x' $(=1/\omega')$ bin for the proton. It is clear that the proton does not scale in either variable and that there is a difference between the neutron and the proton behaviour as a function of Q^2. Remember νW_2 (and hence MW_1) are different for proton and neutron, particularly at large x (or x'). This result may indicate that as Q^2 increases, their scaling violations are taking the two structure functions to a common value.

Incidentally we should mention the status of measurements of R at SLAC. Riordan et al. /13/ found, assuming R constant, that $R_p = 0.138 \pm 0.011$ and that $R_n = R_p$ within errors. At the Hamburg Conference in 1977, Hand reported /16/ a new average value of R obtained from a re-analysis of data. It is $R = 0.25 \pm 0.10$.

However, as far as I am aware, this has not been published elsewhere.

The scaling violations have been parametrised for the MIT-SLAC and CHIO muon data at $Q^2 > 1$ GeV^2. The form chosen to fit the data is

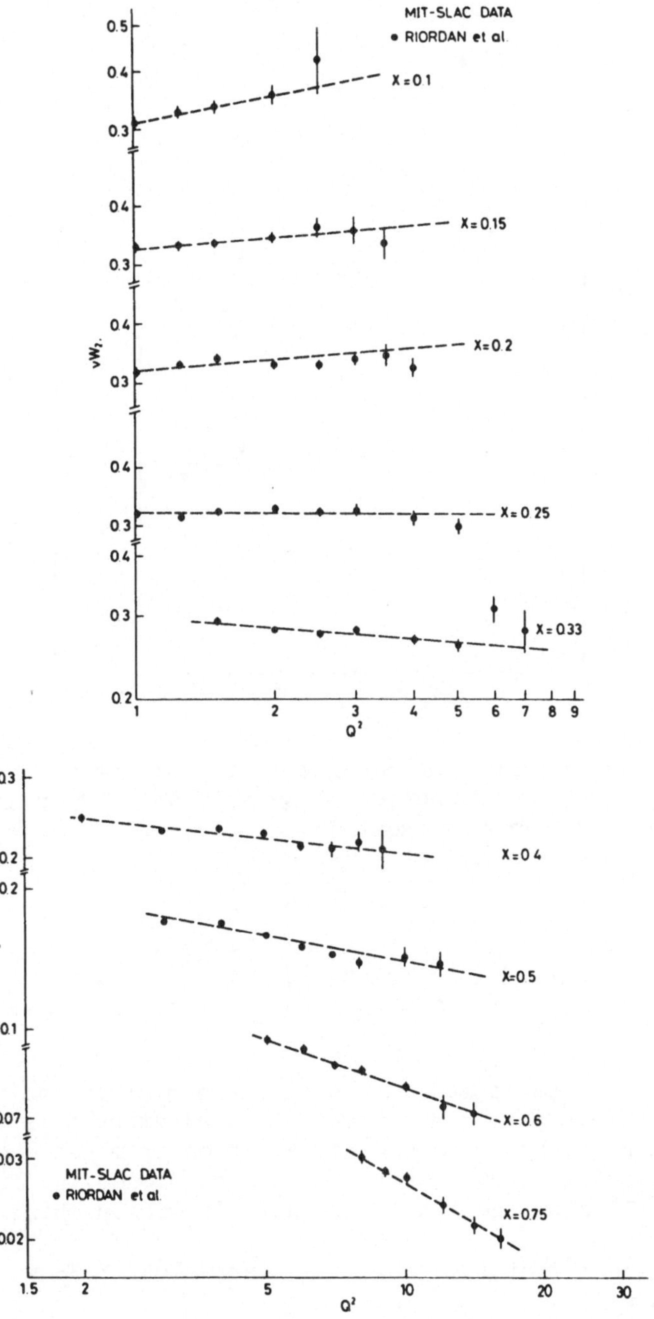

Fig. 14 The MIT-SLAC data (Ref. 14) for $\nu W_2 = F_2(x,Q^2)$ versus Q^2 in bins of x. Note the separate ordinate scales for each bin.

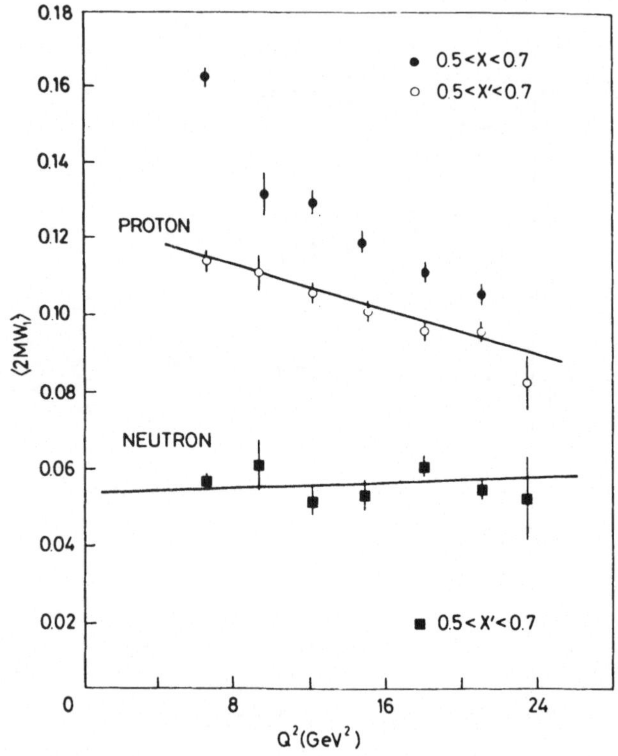

Fig. 15 The large angle SLAC results on the structure functions $2MW_1 = 2F_1$ as a function of Q^2 in bins of x or x' for the neutron and proton (Ref. 15).

$$F_2(x,Q^2) = F_2(x,Q_o^2)\left(\frac{Q^2}{Q_o^2}\right)^{b(x)} ,$$

with $Q_o^2 = 3\ \mathrm{GeV}^2$. The fitted values of b are plotted against x in Fig. 16. Again data points are given vertical error bars and the effect of changig R by one standard deviation from R = 0.44 are shown by horizontal lines, for the muon data alone. The lines drawn on Fig. 14 show the fit to the data in various bins.

It is evident that this power law variation of F_2 with Q^2 fits the data well.

The form used to fit the scaling violations observed in the Cornell-Michigan State experiment, Eq. 6, corresponds to $b \simeq -\ell n 6x$ which would be a straight line on Fig. 16. This is not in agreement with the more recent data.

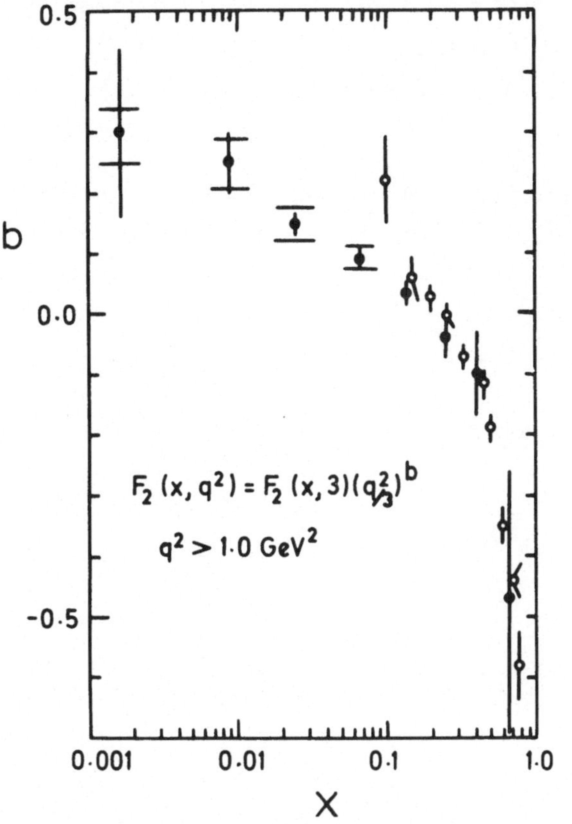

Fig. 16 The scaling violation parameter b as a function of x for all CHIO data (R = 0.44, closed circles) and for the MIT-SLAC data (Ref. 14). The upper and low horizontal bars indicate the change in b for R = 0.69 and 0.19. The errors are statistical only.

Experiment 3 may yielded preliminary data which show some big scaling violations, but I cannot discuss it here through lack of sufficient information. Experiment 4 has taken data but presumably the authours have not completed analysis.

III.3 QCD and the Momenta Analysis

You will recall that in the very naive parton model, the result

$$\int_0^1 F_2(x)dx = \text{mean square parton charge} .$$

This integral is one of a set of moments defined by

$$M(n,Q^2) = \int_0^1 x^{n-2} F_2(x)\, dx \quad .$$

At low energies, the binding energy and mass effects can be decreased by using in place of x, the Nachtmann variable /17/,

$$\xi = \frac{2x}{1 + \sqrt{1 + 4M^2x^2/Q^2}} \quad .$$

Note that $\xi \to x$ as $Q^2 \to$ large compared to M^2. The moments are redefined

$$M(n,Q^2) = \int_0^{\xi_{max}} \eta(n,Q^2)\, \xi^{n-2} F_2(\xi,Q^2)\, d\xi \quad ,$$

where $\eta(n,Q^2)$ is a function of n and Q^2 given by Nachtmann. There is not a great deal of difference between using ξ and x. For example:

$$M(2,\ 1.5\ \text{GeV}^2) = 0.215 \pm 0.006 \text{ using } x \ ,$$
$$= 0.210 \pm 0.006 \text{ using } \xi \ .$$

We use ξ as being the more respectable variable at non-asymptotic energies.

Let us first look at $M(2,Q^2)$ obtained using both MIT-SLAC and CHIO data. Fig. 17 shows this Nachtmann moment. Two sets of points have been drawn. One set includes an elastic contribution to the integral, the other does not. The first shows a decrease of $M(2,Q^2)$ with increasing Q^2, the other does not.

In Quantum Chromodynamics we expect that at very large Q^2 this moment will approach an asymptotic value of

$$M(2,\infty) = \frac{\sum_i e_i^2}{3m + 2g} \quad , \qquad (7)$$

where e_i is the charge on quark type i. The sum is to be taken

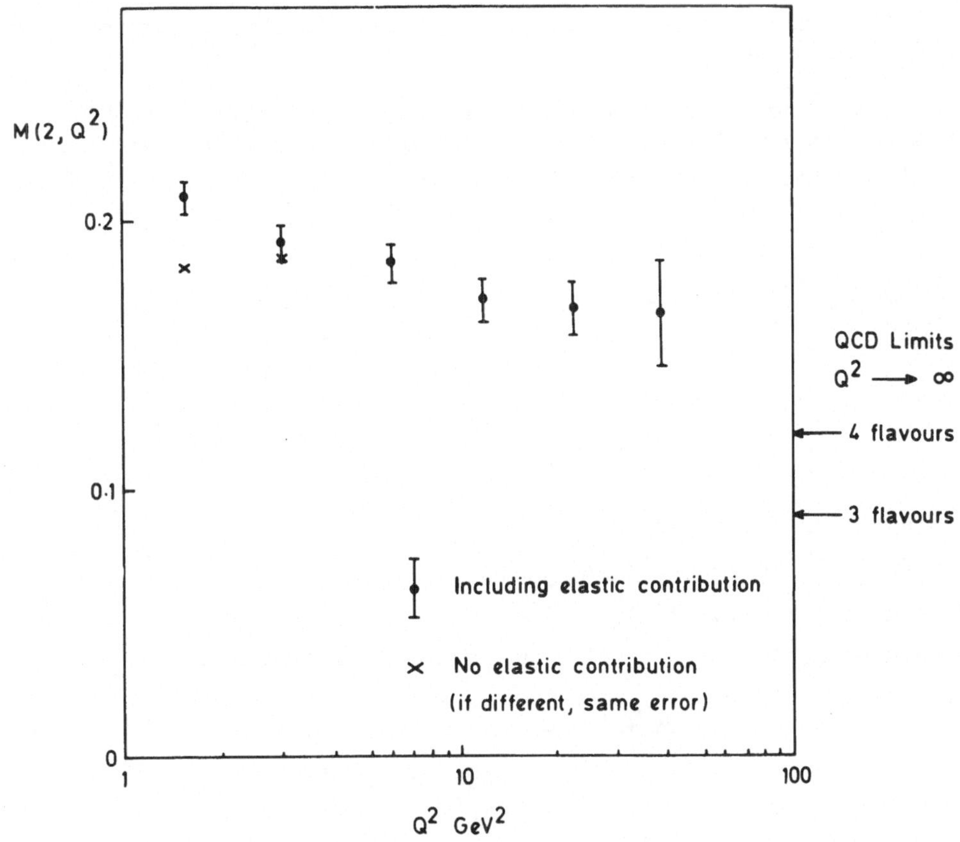

Fig. 17 The value of the Nachtmann moment $M(2,Q^2)$ versus Q^2 for the proton.

over m flavours times 3 colours. The number of gluons is g. In the usual 4-flavour, 3 colour, 8 gluon model this quantity has the value 5/42 (= 0.119).

This result may remind you of the naive quark model. Remember this moment is supposed to be the mean square parton charge. This is just what the above looks like except that the number of gluons (electrically neutral) appear to be doubly weighted in counting the number of partons and summing is done over all quarks not just valence quarks. Kogut and Susskind /18/ have given a picture of what happens as Q^2 increases. The size of the volume probed becomes smaller and instead of resolving a few partons (valence quarks), more and more partons become visible until the valence quarks become lost in a sea of $q\bar{q}$ pairs and gluons. Eq. 7 is not the mean square parton charge in this case and is a result of the Q.C.D. theory.

It is clear (Fig. 17) that if the n = 2 moment is approaching 5/42 it is doing so very slowly.

Let us now look in more detail at the predictions of QCD. This theory cannot predict the values of the structure functions but it is able to predict the variations of the moments with Q^2. The result is

$$M^N(n,Q^2) = M^N_{NS}(n,Q_o^2)\, e^{-\lambda_{NS}(n)s} + \left[M_+(n,Q_o^2)\, e^{-\lambda_+(n)s} + M_-(n,Q_o^2)\, e^{-\lambda_-(n)s} \right] \qquad (8)$$

N stands for neutron or proton.

$s = \ell n \left[\ell n(Q^2/\Lambda^2) \,/\, \ell n(Q_o^2/\Lambda^2)\right]$.

Q_o^2 = a reference $Q^2 \gg \Lambda^2$.

Λ is a parameter of the QCD running coupling constant:

$$\frac{\alpha(Q^2)}{\pi} = \frac{12}{(33-2m)\ \ln \frac{Q^2}{\Lambda^2}} \quad . \qquad (9)$$

m = number of quark flavours.

$M_{NS}(n,Q_o^2)$, $M_+(n,Q_o^2)$, $M_-(n,Q_o^2)$ are the n-1 moments of various parton momentum distribution functions at $Q^2 = Q_o^2$.

$\lambda_{NS}(n)$, $\lambda_+(n)$, $\lambda_-(n)$ are numbers (anomalous dimensions) given by QCD. For example:

$\lambda_-(2) = 0$, so that

$$\lim_{s \to \infty} M(2,Q^2) = M_-(2,Q_o^2) = \frac{5}{42} \quad ,$$

as we have indicated for m = 4, and 8 gluons. The parts of Eq. 8 inside the dotted box do not depend on whether the target is neutron or proton. This will be significant shortly (NS means non-singlet).

Anderson, Matis and Myrianthopoulos /19/ have taken all the MIT-SLAC data and the muon data and determind the n = 2,4,6 moments for the case of a proton target, as a function of Q^2. They fit these moments data with values of the $M(n,Q_o^2)$ and Λ. Assuming R = 0.25 throughout they obtain.

$$\Lambda \;=\; 600 \text{ MeV} \pm 10\% \text{ due to systematic effects.}$$

If R is increased to 0.44, Λ will decrease to about 440 MeV. However, beware! This R value is for low x and has large errors. Other values might apply to the data at large x.

The use $Q_o^2 = 30$ GeV^2. At this reference Q^2 it appears that:

u and $\bar{u}$ quarks carry about 31% of the momentum of the proton,
d and $\bar{d}$ carry about 14%,
gluons carry about 46%,
s and $\bar{s}$ carry about 9%

Another conclusion is that the average momentum of the gluons decreases with Q^2 but the total momentum of the gluons increases slightly. Therefore the number of gluons increases as Q^2 increases. This is just what Kogut and Susskind lead us to expect.

There is a similar effect, but less marked, for u + $\bar{u}$ and for d + $\bar{d}$ quarks.

Quirk at Oxford has looked in a different way at the data. If we take Eq. 8, once for the proton and again for the neutron at the same Q^2, and then take the difference, only the NS (non-singlet) term remains

$$\Delta_n = M^P(n,Q^2)-M^N(n,Q^2) = [M^P_{NS}(n,Q^2)-M^N_{NS}(n,Q^2)]e^{-\lambda_{NS}(n)s}$$

$$= \text{constant } [\ln Q^2/\Lambda^2]^{-\lambda_{NS}(n)} \qquad (10)$$

where the numbers $\lambda_{NS}(n)$ are prescribed by QCD. Therefore the difference between the proton and neutron moments should vary in this simple way with Q^2 and the observed distribution can be used to determine Λ. Quirk has taken the MIT-SLAC and the CHIO data for scattering at deuterium and hydrogen, evaluated the n = 2, 4 and 6 moments of the proton-neutron difference, and then fitted with the form of Eq. 10. This is the <u>non-singlet moments analysis</u>.

Fig. 18 shows the results for the moment difference Δ_n as a function of Q^2 and for n = 2, 4 and 6. It is difficult to identify any decrease of Δ with increasing Q^2 and the fitted value of Λ is 0.010 ± 0.100 GeV (χ^2/NDF = 61/31) assuming R = 0.53, 3 flavours

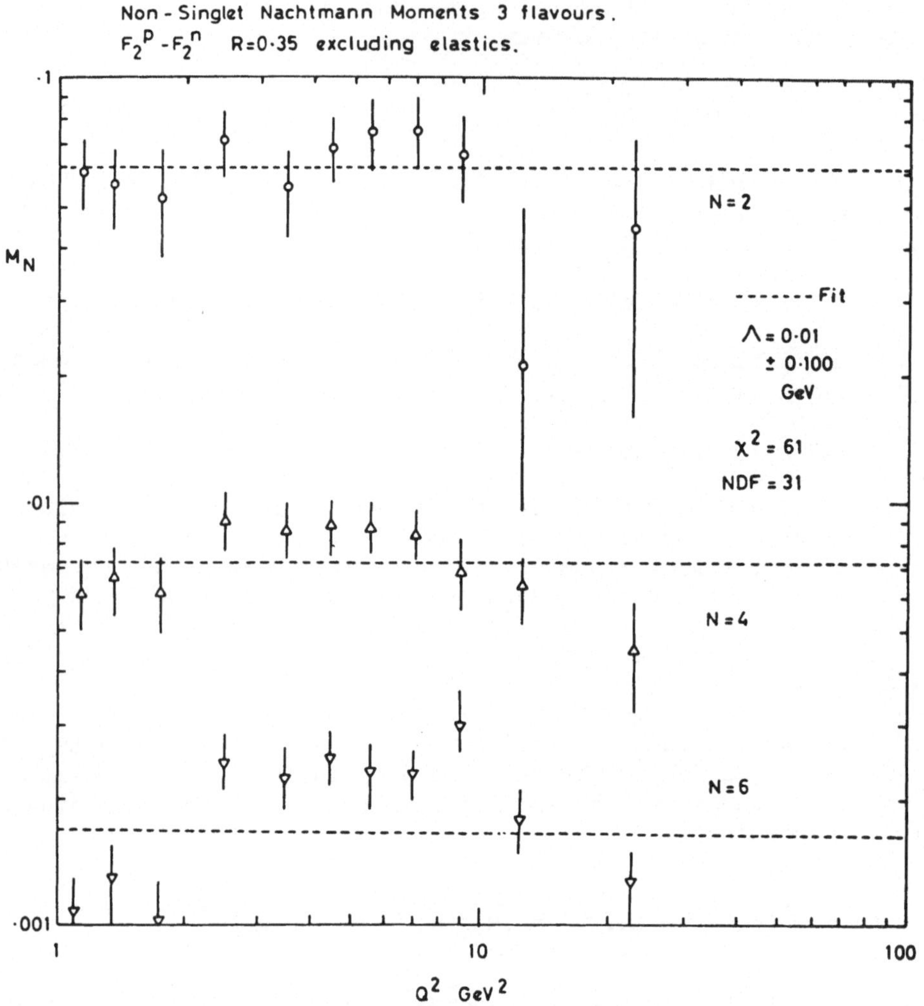

Fig. 18 The value of the non-singlet Nachtmann moments, $M^P(n,Q^2)-M^N(n,Q^2)$ for n = 2,4,6. Elastic contributions not included.

of quark and not including the elastic contribution. If the elastic contribution is included (Fig. 19), there is a clear decrease with Q^2 which gives Λ = 0.675 ± 0.100 GeV (χ^2/NDF = 22/31).

Another check on this analysis can be made. If we write down Eq. 10 for two values of n, e.g. n_1 and n_2 we find that

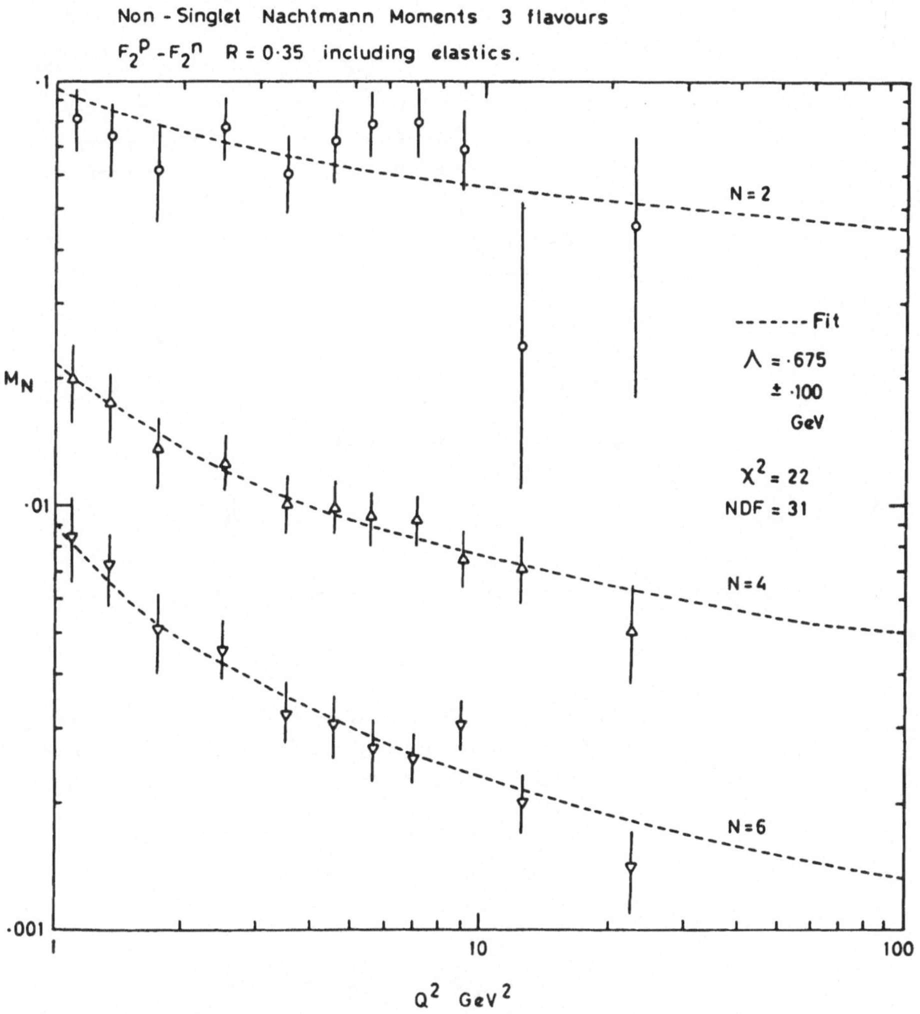

Fig. 19 The same as Fig. 18 but including the elastic contribution.

$$\ln\left(\Delta_{n_1}\right) = \frac{\lambda_{NS}(n_1)}{\lambda_{NS}(n_2)} \ln\left(\Delta_{n_2}\right) + \text{constant} \quad ,$$

so that $\ell n\Delta_{n_1}$ plotted against $\ell n\Delta_{n_2}$ should be a straight line of slope $\lambda_{NS}(n_1)/\lambda_{NS}(n_2)$.

The QCD expectations are

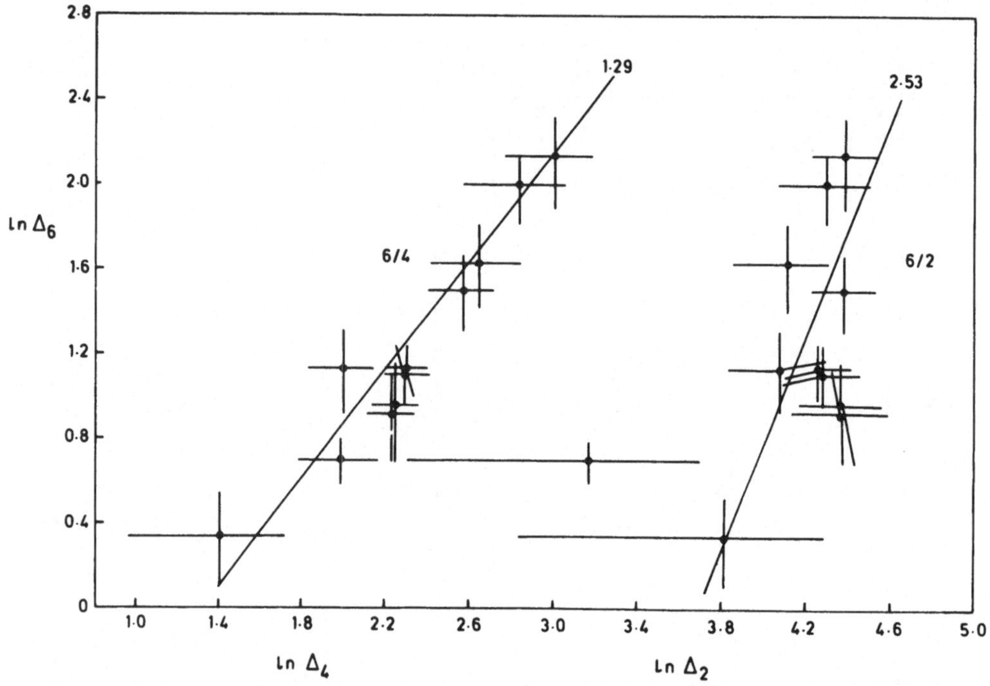

Fig. 20 The plot of $\ell n\Delta_n$ against $\ell n\Delta_m$ for n = 6 and m = 4 and 2, where Δ is the non-singlet moment. The solid lines are not fits to the data but do have the slopes expected from QCD, 1.29 and 2.53.

n	$\lambda_{NS}(n)$		
2	0.427		$\lambda_{NS}(6)/\lambda_{NS}(4) = 1.29$,
4	0.837	=>	
6	1.080		$\lambda_{NS}(6)/\lambda_{NS}(2) = 2.53$.

In Fig. 20 we have plotted $\ell n\Delta_6$ versus $\ell n\Delta_4$ and $\ell n\Delta_2$. The two lines on this figure have slopes 1.29 and 2.53 respectively. The points for 6/4 lie moderately well on the line with slope 1.29. The 6/2 points are not inconsistent with a slope of 2.53 but would naturally be fit by a straight line of greater slope. Note that the expected slopes are given by QCD. No other theory can predict these numbers. Thus the fact that the data are in agreement with these numbers is strong support for QCD. Note, however, that again this result depends on including the elastic contribution in the momenta.

Bardeen et al. /2/ have recently attempted to calculate the second order corrections to the QCD predictions. They find that if Eq. 10 is used to find Λ from the observed Q^2 variation of the nth moment, the apparent Λ found, Λ_n, will be a function of n. Fig. 21 shows how Λ_n varies with n for three values of Λ. The data of Fig. 19 has been analysed to find a value of Λ_n for each n separately and the results are plotted on this figure. The conclusion is that this approach is consistent with Λ = 400 MeV.

The conclusions of the moment analysis are:

1. From the full analysis:

$$\Lambda \ = \ 600 \text{ MeV if } R \ = \ 0.25$$

changing to

$$\Lambda \ = \ 440 \text{ MeV if } R = 0.44 \ .$$

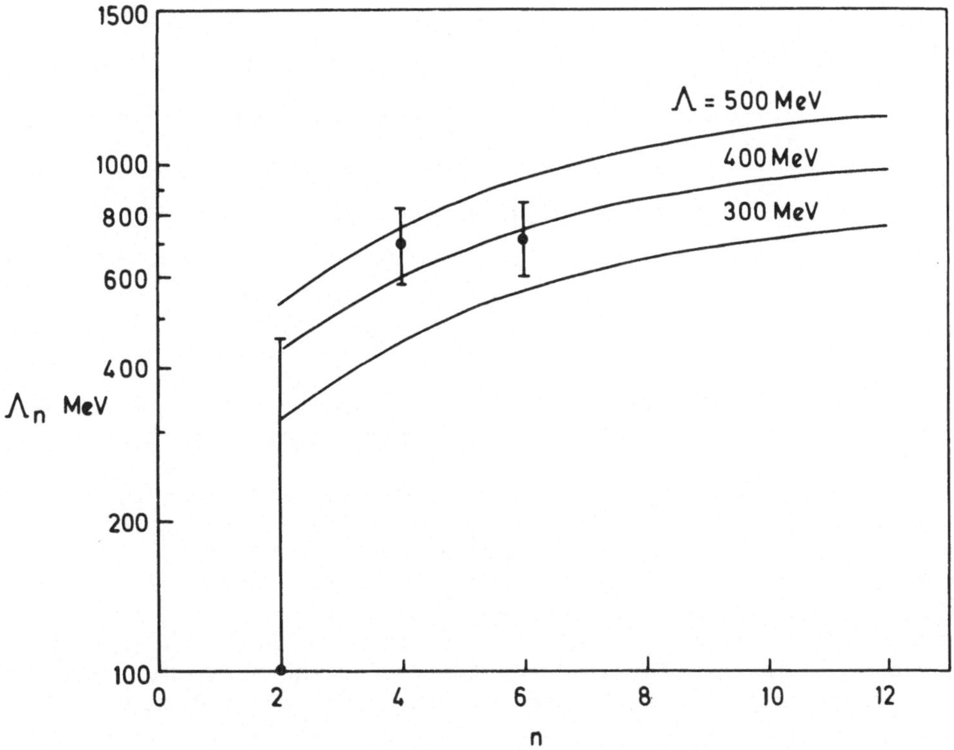

Fig. 21 The values of Λ_n plotted against n for various Λ. This shows the effects of the second order QCD calculations. The experimental points come from the separate fits to the non-singlet moments for n = 2, 4 and 6.

2. From the non-singlet analysis:

$$\Lambda = 675 \text{ MeV if } R = 0.35 \ .$$

3. Including second order corrections to the non-singlet analysis suggests

$$\Lambda \simeq 400 \text{ MeV} \ .$$

4. The Q^2 variation in the non-singlet moments which must exist if Λ is not to be very small, comes from including the elastic scattering contribution to these moments. That is from effects at $x = 1$ or $\xi = \xi_{max}$. This is just where other uncertainties exist in the QCD theory.

5. Thus rather grave uncertainties exist over the extraction of Λ from this charged lepton scattering data. The same remarks can be made about values of Λ extracted from neutrino scattering results.

III.4 Comparison with Neutrino Scattering Results

The BEBC collaboration /21/ have analysed their data and performed an analysis of the moments. They find:

$$\Lambda = 0.74 \pm 0.05 \text{ GeV}$$

which becomes

$$\Lambda = 0.66 \pm 0.05 \text{ GeV}$$

if second order QCD effects are included. In their moments calculation the contribution of the elastic scattering is included.

A more direct comparison of the structure functions may be made. In Fig. 22 we show the CERN-Dortmund-Heidelberg-Saclay neutrino scattering results /22/ expressed as a value of $F_2^\nu(x,Q^2)$ in bins of x as a function of Q^2. Also plotted are the lines from Fig. 13 which fit the CHIO data at $Q^2 > 1$, from selected x bins which match moderately closely bins in the CDHS data. The correct procedure is to take $F_2^\gamma(x,Q^2)$ for the proton and neutron, add in the correct proportions to make an iron target as used in the CDHS experiment, and divide by A (the CDHS result is F_2 for the average iron nucleon). Finally multiply by 18/5 (reciprocal of the mean of square of charges of u and d quarks) which is the expected ratio of the neutrino F_2^ν to the electromagnetic F_2^γ, for deuterium. In the absence of extensive neutron data from CHIO, I have just multiplied the proton results by 18/5. Allowing for this incorrect

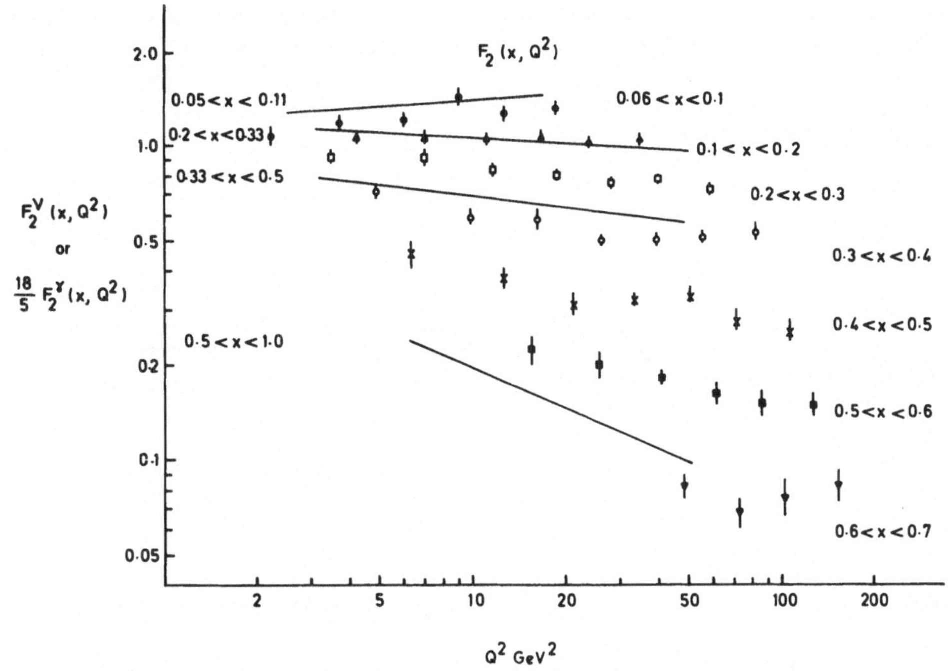

Fig. 22 The structure function from neutrino data ($F_2^\nu(x,Q^2)$) and from muon scattering data [$18/5\ F_2^\gamma(x,Q^2)$] versus Q^2. The neutrino data is from the CERN, Dortmund, Heidelberg, Saclay Collaboration and is shown as the experimental points in bins of x specified by the intervals given on the right. The muon data is given by the straight line fits from Fig. 13, in the x bins specified by the intervals given on the left.

normalisation and bin differences, the data are in good agreement. More important is the observation that the scaling violations are clearly of the same kind and in qualitative agreement.

IV. OTHER TOPICS

IV.1 Weak Interaction Effects

The neutral weak current can also contribute to the scattering of leptons. The amplitude for this process, presumably due to exchange of the Z^0 boson, interferes with the amplitude due to single photon exchange. This interference term is expected to be very small at the Q^2 values presently accessible. However, it changes sign if the electrons are polarised and if their helicity is changed from +1 to −1. Thus the observed scattering cross-sections σ_p and σ_a will be different and an asymmetry can be defined:

$$A = \frac{\sigma_p - \sigma_a}{\sigma_p + \sigma_a} ,$$

which is expected to be proportional to Q^2.

An experiment /23/ has been performed at SLAC which has measured A with a liquid deuterium target at E = 19.4 GeV, E' = 14.6 GeV, and $\theta = 40^{\circ}$ which corresponds to y = 0.24, x = 0.15, $Q^2 = 1.4$ GeV^2. They find

$$\frac{A}{Q^2} = (-0.6 \pm 0.8 \pm 0.9) \times 10^{-5} .$$

The first error is the statistical error, the second systematic.

The existence of this asymmetry immediately indicates parity non-conservation at this low level. This result gives a value of the Weinberg-Salam angle

$$\theta_W = 0.23 \pm 0.03$$

which is in good agreement with other determinations /24/.

IV.2 Hadron Production in Muon Scattering

Some preliminary remarks can be made about the hadrons produced in 219 GeV muon-proton inelastic production. The hadron distributions are given with respect to the direction of the momentum transfer $\underset{\sim}{q}$. The azimuthal angle of the hadron, ϕ, is defined with respect to the scattered muon which has $\phi = 0$.

QCD predicts /25/ that (1) the value of $\langle \cos 2\phi \rangle$ should be +ve that is hadrons are produced preferentially in the muon scattering plane, and (2) $\langle \cos \phi \rangle$ is +ve if the hadrons are produced preferentially from gluon jets and −ve if preferentially produced from quark jets.

At high z ($z = E_h/\nu$ = fraction of energy ν transferred to the observed hadron h) quark jets should dominate and hence $\langle \cos \phi \rangle$ should be −ve for large z. These effects should decrease with increasing Q^2. Table 1 gives the relevant results.

TABLE 1 Values of <cosϕ>, <cos 2ϕ> as function of Q^2 for $0.4 < z < 0.9$ in the range $W > 10$ GeV.

Q^2 GeV^2	<cos ϕ>	<cos 2ϕ>
1.5	-0.14 ± 0.03	$+0.07 \pm 0.04$
2.5	-0.04 ± 0.03	$+0.04 \pm 0.03$
4	-0.11 ± 0.04	$+0.01 \pm 0.04$
6	-0.02 ± 0.06	$+0.01 \pm 0.07$
8	-0.02 ± 0.08	$+0.04 \pm 0.10$

There is a hint of decreasing -ve values of <cos ϕ> and of decreasing +ve values of <cos 2ϕ> but the accuracy is insufficient, particularly for <cos 2ϕ> , to draw any conclusions, in favour or otherwise of the QCD predictions.

The average transverse momentum squared $\langle P_T^2 \rangle$ is also expected to vary with Q^2. The data are consistent with a 14% increase as Q^2 increases from 1.5 to 20 GeV^2 but the errors are large. The variation of $\langle P_T^2 \rangle$ with the z of the hadron is shown in Fig. 23 for two ranges of Q^2. The observed increase with z up to $z \simeq 0.6$ followed by a decrease is a widely observed effect in inclusive hadron data in all kinds of collisions.

It is interesting to compare the inclusive hadron distributions with respect to $\vec{q}$ in virtual photoproduction with the hadron distributions in e^+e^- annihilation with respect to the sphericity axis, which is supposedly the jet axis. This comparison is made to see if the forward hadrons in virtual photoproduction constitute a jet with the same properties as those of jets found in annihilation. The longitudinal variable is x_F which is defined by

$$x_F \equiv \frac{2p_H}{W}$$

where p_H is the component of the hadron momentum along the $\vec{q}$ (or jet axis), and W is the total centre of mass energy of γ_v+ proton (or of e^+e^-). The longitudinal distributions

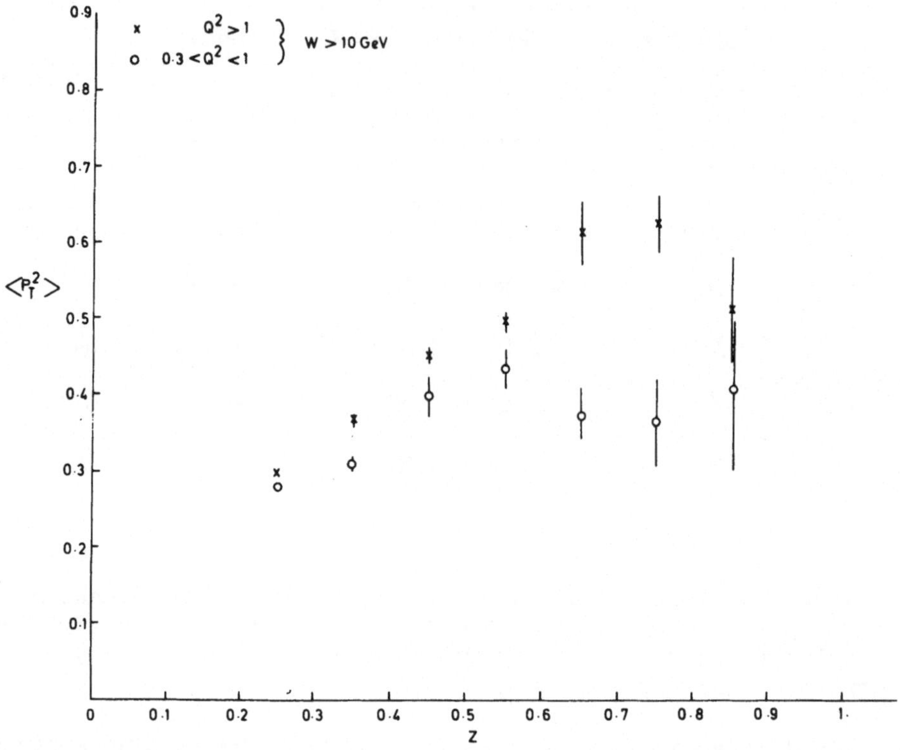

Fig. 23 The average P_T^2 versus z for hadron production in the 219 GeV muon-proton inelastic scattering.

$$\frac{x_F}{\pi\sigma} \frac{d\sigma}{dx_F}$$

agree very well once an appropriate factor for two jets in e^+e^- has been included. See Fig. 24.

The average transverse momentum $\langle p_T \rangle$ versus x_F also vary in the same way (Fig. 25) and have the same value although it is not yet clear if the distributions in p_T are the same.

The preliminary conclusions that we can draw from the hadrons are:

1. The azimuthal distributions may support QCD expectations.

2. If there are jets in e^+e^- annihilations, then the forward

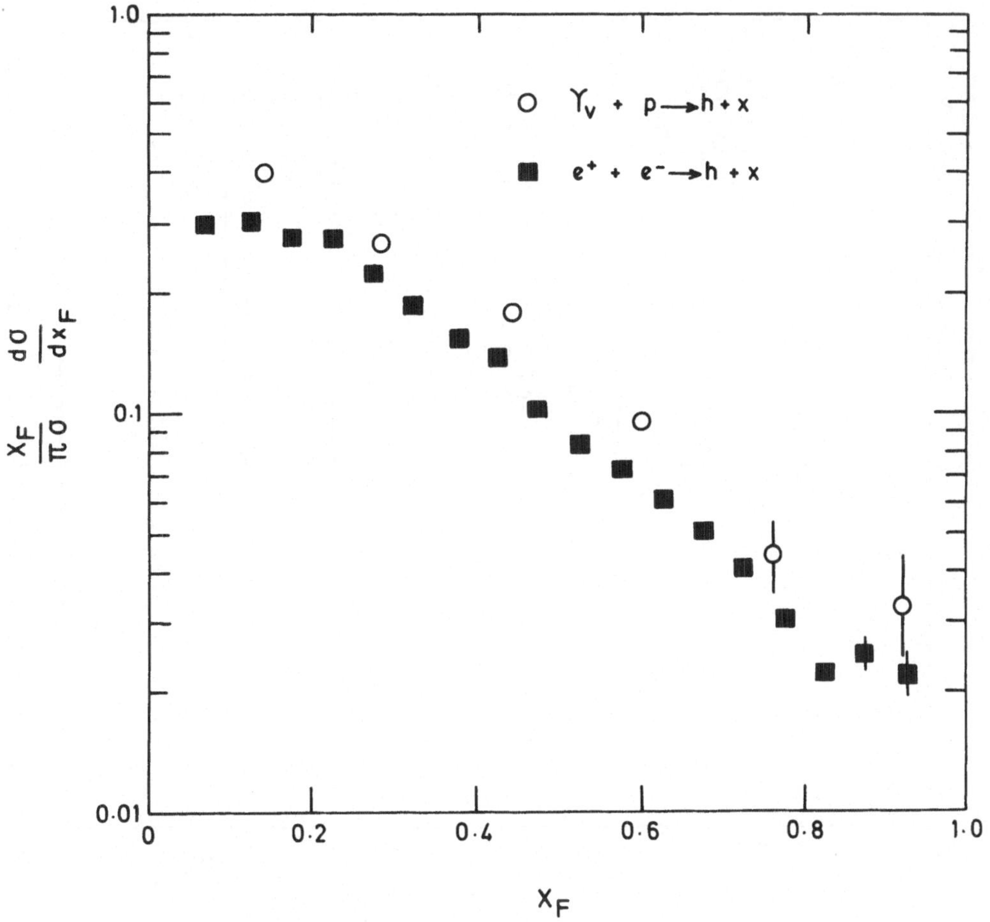

Fig. 24 The longitudinal distribution functions for hadrons in muon scattering and e^+e^- annihilation. The muon data have $W > 10$ GeV and $0.5 < Q^2 < 3$ GeV^2. The e^+e^- data is from reference 26 and has centre of mass energy squared, $s = 56$ GeV^2.

hadrons in virtual photoproduction have very similar distributions to those in the e^+e^- jets. Thus these forward produced hadrons which are, as we always expected, sharply colcollimated forwards, constitute a respectable jet.

V. NOTES ADDED AFTER DISCUSSION HOUR

During the discussion hour, one of the subjects considered was the legality or otherwise of including the elastic contribution in the evaluation of the moments. It emerged that although it

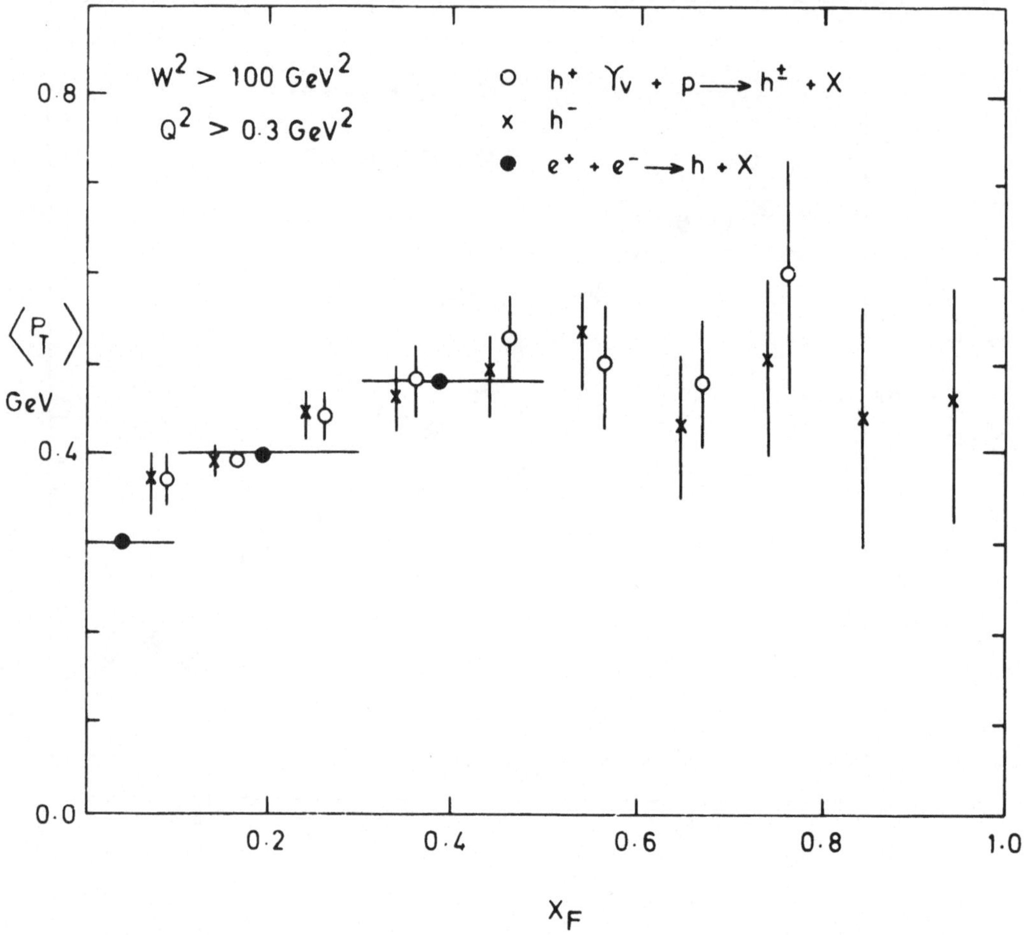

Fig. 25 The average transverse momentum of the hadrons in muon scattering and e^+e^- annihilation.

was correct, it has an effect on the moments just in that part of the Q^2 range where QCD is not reliable and that this contribution was of the order of the errors in the QCD calculations. The correct thing to do is find the moments for large Q^2, say > 10 GeV^2. However, the data is sparse in this range and for the non-singlet moment analysis it is then impossible to obtain an accurate value of Λ.

REFERENCES

1. R.E. Taylor, International Conference on High Energy Physics, Palermo (1975)

2. D.L. Fancher et al, Phys.Rev.Lett. 37, 1323 (1976); K.W. Chen, Proc. of Palermo Conference on High Energy Physics (1975)

3. S.D. Drell and J.D. Walecka, Ann.Phys. (NY) 28, 18 (1964)

4. J.D. Bjorken, Phys.Rev. 179, 1547 (1969)

5. J.D. Bjorken and E.A. Paschos, Phys.Rev. 185, 1975 (1969); R.P. Feynman, Photon-Hadron Interactions, W.A. Benjamin Inc. (1972)
Tung-Mow Yan, Ann.Rev.Nucl.Sci. 26, 199 (1976)

6. L.N. Hand, Phys.Rev. 129, 1834 (1963); N. Dombey, Rev.Mod.Phys. 41, 236 (1969)

7. W.K.H. Panofsky, SLAC-PUB-798 (1970). See also Ref. 9, and R.E. Taylor, Proceedings of Third International Symposium on Electron and Photon Interactions at High Energies, Stanford (1967)

8. G. Miller et al, Phys.Rev. D5, 528 (1972)

9. S. Stein et al, Phys.Rev. D12, 1884 (1975)

10. J.I. Friedman and H.W. Kendall, Ann.Rev.Nucl.Sci. 22, 203 (1972)

11. A. Bodek et al, Phys.Rev.Lett. 30, 1087 (1973)

12. Y. Watanabe et al, Phys.Rev.Lett. 35, 898 (1975)
C. Chang et al, Phys.Rev.Lett. 35, 901 (1975)

13. B.A. Gordon et al, Phys.Rev.Lett. 41, 615 (1978)

14. E.M. Riordan et al, SLAC-PUB-1634 (1975); R.E. Taylor and W.B. Atwood kindly made available a complete set of the electron scattering results from the MIT-SLAC experiments;
A. Bodek et al, to be published in Phys.Rev.

15. R.E. Taylor, International Symposium on Lepton and Photon Interactions at High Energies, Stanford (1975)

16. L. Hand, International Symposium on Lepton and Photon Interactions at High Energies, Hamburg (1977)

17. O. Nachtmann, Nucl.Phys. B63, 237 (1973); and B78, 455 (1974)

18. J. Kogut and L. Susskind, Phys.Rev. D9, 697 and 3391 (1974)

19. H.L. Anderson, H.S. Matis and L.C. Myrianthopoulos, Phys.Rev.Lett. 40, 1061 (1978). Also contribution to Tokyo Conference (1978)

20. W.A. Bardeen, A.J. Buras, D.W. Duke and T. Muta, Fermilab-Pub-78/42 THY (1978)

21. P.C. Bosetti et al, Nuclear Physics Laboratory, University of Oxford, 16/78

22. Data supplied by F. Dydak

23. C.Y. Prescott et al, Phys.Lett. 77B, 347 (1978)

24. W. Weinberg, International Conference on High Energy Physics, Tokyo, 1978

25. A. Mendez, QCD Predictions for Semi-Inclusive and Inclusive Leptoproduction, Dept. of Theoretical Physics, University of Oxford, 29/78

26. G. Hanson, SLAC-PUB-2118 (1978)

TOPLESS MODEL FOR GRAND UNIFICATION

Yoav Achiman and Berthold Stech

Institut für Theoretische Physik

Universität Heidelberg

ABSTRACT

A gauge model unifying weak, electromagnetic, and strong interactions is presented. It is based on the group $SU_L(3) \otimes SU_R(3) \otimes SU_C(3) \otimes P$ with a complete flavor-color permutation symmetry. All fermions are in one irreducible representation of this group. This representation contains a new heavy lepton λ and a new b-like quark h but no top quark besides u and c. The model has a global $E6 \otimes P$ symmetry before the local gauging i.e. in the limit of vanishing gauge coupling. A global SU(4) subgroup which may survive the spontaneous symmetry breaking leads to mass relations. The weak decays of b and h quarks can proceed through non-diagonal neutral currents providing for a specifice signature. The weak mixing angle is in accord with experiments.

Local gauge invariance is now generally assumed to be the basis for all elementary particles forces /1/. The difference between strong and weak interactions lies mainly in their different coupling strengths. This difference is presumably a result of a spontaneous symmetry breaking which also leads to the confining phase for quarks and gluons on the one hand and to the plasma phase in the case of color singlet hadrons, leptons, and intermediate vector bosons on the other. The minimal gauge group describing these interactions is

$$G_{minimal} = SU_L(2) \otimes U(1) \otimes SU_C(3) \qquad (1)$$

which involves 3 different gauge coupling constants.

Is is well known that the group SU(5) can be used to unite all three interactions simultaneously /2,3/. This is the minimal group for a grand unification (the flavor subgroup of SU(5) is just SU(2) ⊗ U(1)). The 5 and 10* representations of SU(5) if taken together represent particle families with 15 members, for instance

$$\underset{\sim}{5} = \nu_e, e^-, \hat{d} \qquad \underset{\sim}{10}^* = e^+, \binom{u}{d}, \hat{u} \qquad \text{electron family} \tag{2}$$

$$\underset{\sim}{5} = \nu_\mu, \mu^-, \hat{s} \qquad \underset{\sim}{10}^* = \mu^+, \binom{c}{s}, \hat{c} \qquad \text{myon family} \quad .$$

For convenience one deals in grand unified models with two component (left handed) Weyl spinor fields only. In this notation the Dirac-field for the up quark, e.g., can be written $\binom{u}{\sigma_2 \hat{u}^*}$ where $\hat{u}$ is an antiquark (left handed) Weyl spinor. If the heavy lepton τ (1.8 GeV) and the constituent of the ypsilon b (≈ 4.7 GeV) also belong to such a reducible multiplet of SU(5), a top quark t is necessary to complete the τ-family. To have all 15 particles in one irreducible representation, one needs to extend SU(5) to SO(10) /4,3/. Here the lowest non-trivial representations are 10 and 16 which decompose with respect to the SU(5) subgroup according to

$$\underset{\sim}{10} = \underset{\sim}{5} + \underset{\sim}{5}^* \qquad \underset{\sim}{16} = \underset{\sim}{1} + \underset{\sim}{5} + \underset{\sim}{10}^* \tag{3}$$

The 16 representation can thus be used to describe a lepton-quark family with one additional neutral lepton. However, it is not clear why nature favors the 16 representation of SO(10) instead of the lower 10 representation. Also, these models give no explanation why at least three such families of fermions exist.

In this talk we present an alternative to the SU(5) - SO(10) scheme. In particular, we find it attractive to perform a full unification of the flavor interactions (weak and electromagnetic) independent of the strong forces. The final grand unification will then combine the flavor and color dynamics. In this case, the new flavor group should give the correct electromagnetic charges of quarks (2/3, -1/3) and leptons (0, ±1) naturally. An obvious way to do this is to use an SU(3) group under which the quarks transform as flavor triplets and the leptons as flavor octets and singlets. It was recently shown by Okubo /5/ that this is actually the only possiblity. The minimal model of this type /5,6/ which contains the observed currents is based on the flavor group /8/

$$G_{flavor} = SU_L(3) \otimes SU_R(3) \otimes P_{LR} \tag{4}$$

P_{LR} is a parity operation which interchanges the left (L) and right (R) SU(3) groups. Because of parity invariance there is only one gauge coupling constant. The observed weak and electromagnetic currents generate the $SU_L(2) \otimes U(1)$ subgroup of (4). The charges are defined by the diagonal sum $Q = Q_L + Q_R$. The left-handed quarks are taken to be the following flavor triplets:

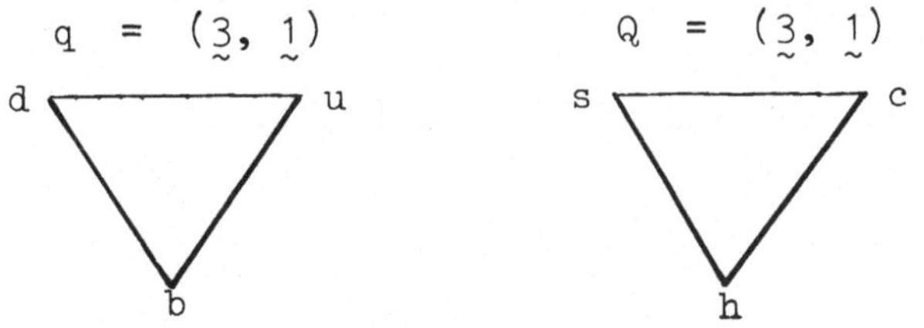

(5)

This will lead to a "topless" model with two $SU_L(2)$ singlet quarks of charge -1/3. One of these could be the constituent of the ypsilon. Let us note that the new hypercharge of $SU_L(3)$ and $SU_R(3)$ is not related to the usual hypercharge quantum number of the strange quark.
The parity transformation P_{LR} operates on the quark fields q(x) as follows

$$P_{LR}\, q(x)\, P_{LR}^{-1} = q_R(\bar{x}) \equiv \sigma_2\, \hat{q}^*(\bar{x}) \qquad (6)$$

where q_R is a right-handed Weyl spinor and $\bar{x} = (x_0, -\vec{x})$. Since P_{LR} also exchanges $SU_L(3) \leftrightarrow SU_R(3)$, the antiquarks $\hat{q}(x)$ are $(1, 3^*)$ states in the flavor space.

The gauge group that contains the strong interaction as well is now the group /9/

$$G = SU_L(3) \otimes SU_R(3) \otimes SU_C(3) \otimes P \qquad (7)$$

The color part of this group brings about a second gauge coupling constant. In view of the symmetric form (7) it is, however, very suggestive to make the three SU(3) groups equivalent before spontaneous symmetry breaking. P will then denote a generalized parity operation which includes, besides the parity operation P_{LR}, all permutations of the three SU(3) groups with the property $P^2 = +1$. With only one gauge coupling constant (7) unifies the three interactions in a remarkable flavor-color symmetric way. In the following we will explore the implications of the symmetric grand unification group (7).
Of course, as in all grand unification models, a superstrong breaking is required to explain the large difference between the actual

observed values of the strong and electromagnetic coupling constants /10,1/*. This superstrong breaking is in fact welcome since it renormalizes the weak interaction angle from the large canonical value $(\sin^2\theta_W)_0 = 3/8$, down to a value compatible with the experimental findings.

The generators of the group G and thus also the 24 gauge bosons show again the permutation symmetry inherent in (7):

$$\begin{array}{ccc} (\underset{\sim}{8}, \underset{\sim}{1}, \underset{\sim}{1}) & (\underset{\sim}{1}, \underset{\sim}{8}, \underset{\sim}{1}) & (\underset{\sim}{1}, \underset{\sim}{1}, \underset{\sim}{8}) \\ W_L & W_R & W_C = \text{gluons} \end{array} \tag{8}$$

In order that the total Lagrangian will be invariant, the fermion fields occurring in the Lagrangian must be a representation of the SU(3) groups and of this additional permutation symmetry. We will restrict ourselves to SU(3) singlet and triplets because these are the only ones needed in the color sector. From this condition and the requirement of integer charges for leptons follows the existence of 54 fermion Weyl fields:

$$\begin{array}{llll} \ell(x) = (\underset{\sim}{3}^*, \underset{\sim}{3}, \underset{\sim}{1}) & \ell_R(\bar{x}) = (\underset{\sim}{3}, \underset{\sim}{3}^*, \underset{\sim}{1}) \\ q(x) = (\underset{\sim}{3}, \underset{\sim}{1}, \underset{\sim}{3}^*) & q_R(\bar{x}) = (\underset{\sim}{1}, \underset{\sim}{3}^*, \underset{\sim}{3}) \\ \hat{Q}(x) = (\underset{\sim}{1}, \underset{\sim}{3}^*, \underset{\sim}{3}) & \hat{Q}_R(\bar{x}) = (\underset{\sim}{3}, \underset{\sim}{1}, \underset{\sim}{3}^*) \end{array} \tag{9}$$

where $P_{LR}\, \ell(x)\, P_{LR}^{-1} = \ell_R(\bar{x})$ etc.
The right-handed spinors ℓ_R, q_R, $\hat{Q}_R$ can be expressed via conjugate fields of left-handed Weyl spinors. In this way Weyl spinor fields of the corresponding antiparticles are defined:

$$\ell_R \equiv \sigma_2 \hat{\ell}^* , \quad q_R \equiv \sigma_2 \hat{q}^* , \quad \hat{Q}_R \equiv \sigma_2 Q^* . \tag{10}$$

The total Lagrangian which contains the 54 massless Weyl spinors ℓ, q, $\hat{Q}$, $\hat{\ell}$, $\hat{q}$, Q and the 24 massless gauge bosons (8) is invariant under the generalized partiy transformation P as required. Odd permutations of the three SU(3) groups involve the space reflected fields. Even permutations perform the cyclic change

*In the group G (7) the baryonic number is conserved in contrast to the situation in other unifying groups /1/.

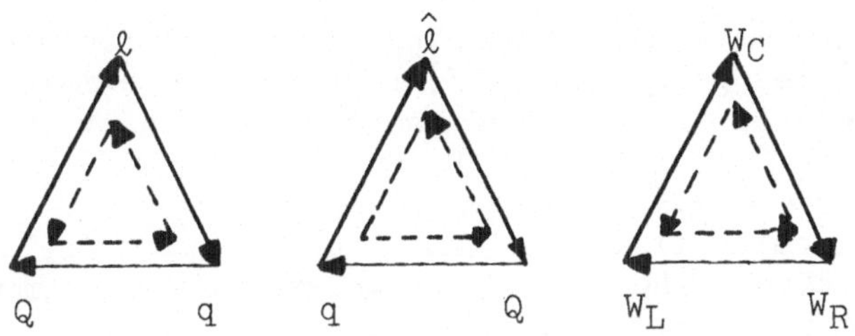

(11)

In addition to P a charge conjugation operator C which leaves the Lagrangian invariant can be defined*. Besides the gauge symmetry and the discrete symmetries P, C the Lagrangian has an additional global $(SU(2))^3$ symmetry. The corresponding symmetry transformations act on the doublets

$$\begin{pmatrix} \ell \\ \hat{\ell} \end{pmatrix} \qquad \begin{pmatrix} q \\ Q \end{pmatrix} \qquad \begin{pmatrix} \hat{Q} \\ \hat{q} \end{pmatrix} \tag{12}$$

One also finds that the color current is automatically a pure vector current. Formally, the fermion fields, the gauge fields, and the Higgs fields to be discussed later on can all be constructed from the set of elementary subquarks

$$\zeta_L = (\underset{\sim}{3}, \underset{\sim}{1}, \underset{\sim}{1}) , \quad \zeta_R = (\underset{\sim}{1}, \underset{\sim}{3}, \underset{\sim}{1}) , \quad \zeta_C = (\underset{\sim}{1}, \underset{\sim}{1}, \underset{\sim}{3})$$

and their antiparticles.

The 54 Weyl spinors can be combined into 27 Dirac-spinors consisting of a nonet of leptons with integer charges and two flavor triplets of quarks. The precise way this combination occurs will

---

*C changes particle in antiparticle fields, turns upper into lower indices and vice versa, and also interchanges L and R indices. For instance

$$C\, \ell^i{}_k(x)\, C^{-1} = \hat{\ell}^k{}_i(x)$$

$$P_{LR}\, \ell^i{}_k(x)\, P_{LR}^{-1} = \sigma_2(\hat{\ell}^k{}_i(\bar{x}))^*$$

(the first index is here an L index, the second index an R index)

of course depend on the mass matrix arising from Higgs-field couplings*. Thus, the unified gauge group (7) requires the existence of two quark multiplets as it is needed phenomenologically (see (5)).

To name the 54 fermion fields we make the following preliminary identification (before the mixing due to spontaneous symmetry breaking is taken into account)

$$q = \begin{pmatrix} u \\ d \\ b \end{pmatrix} \qquad \hat{Q} = (\hat{c},\hat{s},\hat{h}) \qquad \ell = \begin{pmatrix} N & \lambda^- & \mu^- \\ \tau^+ & \nu_\lambda & \nu_\mu \\ e^+ & \hat{\nu} & M \end{pmatrix} \tag{13}$$

$$Q = \begin{pmatrix} c \\ s \\ h \end{pmatrix} \qquad \hat{q} = (\hat{u},\hat{d},\hat{b}) \qquad \hat{\ell} = \begin{pmatrix} \hat{N} & \tau^- & e^- \\ \lambda^+ & \nu_\tau & \nu_e \\ \mu^+ & \hat{\nu}_\mu & \hat{M} \end{pmatrix} \tag{14}$$

In this presentation $SU_L(3)$ acts vertically, $SU_R(3)$ horizontally. Color indices have been suppressed. As anticipated we have no top quark apart from u and c. Instead, there are two $SU_L(2)$ singlet quarks b and h. The lepton matrix contains besides μ and e two charged leptons τ and λ with a V-A coupling to their neutrinos and a V+A coupling (via conventional W's) to a (heavy?) neutral lepton N. We note that the quarks Q have the opposite intrinsic parity compared to the quarks q and the same is true for the leptons e^-, τ^- when compared with μ^-, λ^-.

The minimal simple group which incorporates the group G of (7) is obtained by adding to the 24 generators given in (8) new ones which perform the cyclic change (see (11))

$$\ell \to q \to \hat{Q} \to \ell\,, \qquad \hat{\ell} \to Q \to \hat{q} \to \hat{\ell}$$

and vice versa. The simplest step operators of this kind have the following transformation property /9/

$$(\underset{\sim}{3}, \underset{\sim}{3}, \underset{\sim}{3}) \qquad\qquad (\underset{\sim}{3}^*, \underset{\sim}{3}^*, \underset{\sim}{3}^*) \tag{15}$$

---

* Some of the neutral leptons could also form Majorana instead of Dirac particles.

and are permutation-symmetric. The 54 new generators together with the 24 generators of the three SU(3)'s form a closed algebra. The corresponding Lie group is the exceptional group E6. E6 has been much discussed by Gürsey and coworkers /11/ as a candidate for a local gauge symmetry. In our case we must take E6 ⊗ P with the lowest 54 representation (9)(10): The fermions ℓ, q, $\hat{Q}$ form a 27 representation of E6 (without parity), the fermions $\hat{\ell}$, $\hat{q}$, Q another one. Both 27 representations transform into each other by the parity operation. The accidental global $(SU(2))^3$ symmetry is reduced to the global diagonal SU(2) acting on the doublets shown in (12).
In case one uses the group E6 ⊗ P as a local gauge group, one obtains according to (15) 54 lepto-quark gauge bosons which have to be made superheavy by a spontaneous symmetry breaking to avoid a fast decay of the proton. A parity invariant superstrong breaking with a Higgs field in the adjoint representation does this job and gives

$$E6 \otimes P \xrightarrow{H^{78}} SU_L(2) \otimes SU_R(2) \otimes U_L^Y(1) \otimes U_R^Y(1) \otimes P_{LR} \otimes SU_C(3) \quad (16)$$

The direction of breaking defines the new hypercharge Y. This breaking does not lead to superheavy fermions and gives automatically a superstrong breaking of $SU_{L,R}(3)$ which is a necessary condition for a large difference between the strong and electromagnetic fine structur constants at ordinary energies. In fact any breaking that gives mass to all lepto-quarks breaks $SU_L(3) \otimes SU_R(3)$ as well. If no further superstrong breakings occur, one finds for the renormalized weak interaction angle, using the method of Georgi, Quinn and Weinberg /10/

$$\sin^2 \theta_W = \frac{3}{8}\left(1 - \frac{1}{3}\left(1 - \frac{8}{3}\frac{\alpha}{\alpha_{strong}}\right)\right) \simeq \frac{1}{4} \quad (17)$$

If one includes an additional superstrong breaking down to the Weinberg-Salam group one obtains values for $\sin^2\theta_W$ in the range /10,3,9/

$$0.2 \lesssim \sin^2 \theta_W \lesssim \frac{1}{4} \quad . \quad (18)$$

Using Higgs fields in the adjoint representation alone is not sufficient to reduce E6 ⊗ P or G down to the Weinberg-Salam group. One needs in addition a 27 or a $(3^*, 3, 1)$ field to complete the local breaking which may also be responsable for the suppression of right-handed charged currents.

The local gauging of E6 ⊗ P is not necessary to achieve the results (16) to (18). One may just as well use the gauge group (7) with the corresponding Higgs fields. In particular, a breaking through the adjoint representation of this group leads again to the subgroup which occurs in (16) and to the result (17).
In the following we will use the gauge group (7) as before and take E6 P only as a global classification symmetry before gauging, valid only in the high energy limit where the gauge coupling constant is negligible. (Local E6 has more problems than the group (7). In particular, it is difficult to avoid sizeable neutrino masses caused by radiative corrections via lepton-quark exchanges.)

With this choice the proton remains absolutely stable. Quark masses are generated by Higgs fields (L, R) = (3^*, 3). Lepton masses are induced by Higgs fields transforming as (3^*, 3), (6, 3), (3^*, 6^*), (6, 6^*) /12/. The irreducible Higgs representations of the global E6 group which are responsible for the fermion mass terms and contain the above fields appear in the decomposition

$$(27 \quad 27')^* = (27 + 351^*)_{\text{symm.}} + 351'^*_{\text{antisymm.}} \tag{19}$$

Because the superstrong breaking via the adjoint representation is huge compared to the masses of the known fermions, the Higgs representations (19) will be influenced dynamically, and most of their neutral and color singlet members will develop vacuum expectation values which give rise to fermion masses. Simple relations between these masses will exist if there remains a global residual symmetry in the fermion sector. Suggestive candidates for such a residual symmetry are groups which leave some or all of the neutral and color singlet members of the scalar and pseudoscalar 27 Higgs representation invariant /9/. (The 27 is taken because of its relevance for the local breaking of G and because it is the classifying representation which determines the directions of charge, hypercharge, isospin, and color.) The assumption is that such a residual global symmetry is respected also by the remaining part of the Higgs Lagrangian.
The global group which leaves all the 5 neutral and color singlet member of a 27 Higgs field unchanged is a specific SU(4) subgroup of E6 /9,14/. We call it the "charge group". It is generated by the operators

$$SU(4): \; \{Q_{\text{el.mag.}}, \; (\underset{\sim}{1},\underset{\sim}{1},\underset{\sim}{8}), \; (\text{up},\text{up},\underset{\sim}{3}), \; (\text{up}^*,\text{up}^*,\underset{\sim}{3}^*)\} \tag{20}$$

and contains the locally conserved subgroup $U_Q(1) \otimes SU_C(3)$.
One finds

$$E6 \otimes P \supset SU(4) \otimes SU_L^U(2) \otimes SU_R^U(2) \otimes P_{LR} \qquad (21)$$

where U denotes the U-spin of SU(3). Decomposing the two "$\underset{\sim}{27}$" in (13), (14) one finds

$$\underset{\sim}{27} = (\underset{\sim}{6}^*,\underset{\sim}{1},\underset{\sim}{1}) + (\underset{\sim}{4},\underset{\sim}{1},\underset{\sim}{2}) + (\underset{\sim}{4}^*,\underset{\sim}{2},\underset{\sim}{1}) + (\underset{\sim}{1},\underset{\sim}{2},\underset{\sim}{2}) + (\underset{\sim}{1},\underset{\sim}{1},\underset{\sim}{1})$$

$$(u,\hat{c}) \quad ((\lambda^-,\hat{h}),\ (\mu^-,\hat{s})) \begin{pmatrix} (e^+,d) \\ (\tau^+,b) \end{pmatrix} \begin{pmatrix} \nu_\lambda & \nu_\mu \\ \hat{\nu}_e & M \end{pmatrix} \quad N \qquad (22)$$

$$(c,\hat{u}) \quad ((\tau^-,\hat{b}),\ (e^-,\hat{d})) \begin{pmatrix} (\mu^+,s) \\ (\lambda^+,h) \end{pmatrix} \begin{pmatrix} \nu_\tau & \nu_e \\ \hat{\nu}_\mu & \hat{M} \end{pmatrix} \quad \hat{N}$$

If the fermion couplings to the Higgs field are indeed left SU(4) invariant by the spontaneous symmetry breaking, one can form SU(4) invariant mass terms and obtains after diagonalization the following mass relations

$$\frac{m_d}{m_e} = \frac{m_s}{m_\mu} = \frac{m_h}{m_\lambda} = \frac{m_b}{m_\tau} = 1 \quad . \qquad (23)$$

The masses of the u and c quarks remain unrelated. A parity-invariant Higss breaking would, however, give $m_u = m_c$. Thus, the experimental mass relation $m_c \gg m_u$ is due to spontaneous parity nonconservation and may therefore have the same origin as the relation $m_{W_R} \gg m_{W_L}$ which we need in order to obtain the dominance of left-handed charged currents over right-handed charged currents at present energies.

The mass relations (23) are only valid at extreme energies where the gauge coupling constant is very small. Nevertheless, these relations can - with some optimism - be extrapolated down to ordinary energies using the renormalization group techniques /10, 3/. Estimates similar to those performed by Buras et al. /3/ for SU(5) simply change the one on the right-hand side of (23) to a larger value:

$$\frac{m_d}{m_e} \simeq \frac{m_s}{m_\mu} \simeq \frac{m_h}{m_\lambda} \simeq \frac{m_b}{m_\tau} \simeq 2.6 \quad . \qquad (24)$$

Thus, if a new heavy lepton λ is found the quark h should have a mass roughly 2.6 times as big.

As long as we do not impose further symmetries, in particular discrete ones, the remaining masses and Cabibbo-type angles are free parameters. The fact that in (22) ν_τ occurs together with ν_e makes it plausible that the mass of ν_τ is small as it is required experimentally /13/. The lepton N, on the other hand, may acquire a high mass. The specific form of the τ-neutrino decay amplitude depends on the mass of N. If N and ν_τ are massless, the τ-neutrino vertex will have a pure vector form. If N has a mass $m_N \gtrsim m_\tau$, the τ-decay proceeds by a V-A interaction.
An important difference between our scheme and more conventional models appears in the weak decays of particles which contain b or h quarks. These decays should proceed through mixings of b and h with d and s quarks. Consequently, transitions originating from non-diagonal neutral currents are expected, for instance the process

$$B \rightarrow (\pi,K) + e^+ + e^- \quad . \tag{25}$$

Here B denotes a $b\hat{q}$ meson. Processes of the type (25), if observed, would strongly support /15/ the "topless" model.

Before concluding one should point out that a very different particle spectrum from the one just discussed could arise even within the framework of the groups G or E6 ⊗ P /9/. This happens if a particular vacuum expectation value of a $\underset{\sim}{27}$ Higgs field causes a very strong or superstrong breaking. The maximal subgroup of E6 which leaves just one vacuum expectation value in $\underset{\sim}{27}$ invariant, is SO(10). E6 ⊗ P is then reduced as follows:

$$\begin{aligned} &E6 \otimes P \supset SO(10) \otimes U(1) \otimes P_{LR} \\ &\underset{\sim}{27} = \underset{\sim}{1} + \underset{\sim}{10} + \underset{\sim}{16} \end{aligned} \tag{26}$$

In this case the quarks and leptons in the 10 representation obtain very high and roughly equal masses. We cannot identify them with τ, b, ν_τ any more. The particles in the 16 representation, on the other hand, are still massless at this stage and can be identified with the conventional fermions as in (2). We see that for a suitable breaking pattern SO(10) and SU(5) are submodels of E6 ⊗ P, in which the "16" representation has the lowest energy. To incorporate the constituent of the ypsilon and the heavy lepton τ, a new fermion representation is needed, however. In this circumstance a top quark is required. Our global charge subgroup SU(4) which should now govern the residual breakings and masses leads again to the mass relations (25).

REFERENCES

1. For recent reviews see:
H. Harari, Physics Reports 42C, 238 (1978)

A. Salam, invited talk at the XIX. Int. Conf. on HEP, Tokyo 1978

2. H. Georgi and S.L. Glashow, Phys.Rev.Lett. 32, 438 (1974)

3. M.S. Chanowitz, J. Ellis and M.K. Gaillard, Nucl.Phys. B128, 506 (1977)

A.S. Buras, J. Ellis, M.K. Gaillard, and D.V. Nanopoulos, ibid, B135, 66 (1978)

4. H. Fritzsch and P. Minkowski, Ann. Phys. (N.Y.) 93, 193 (1975)

H. Georgi, Particles and Fields, 1974 (ASP/DPF Williamsburg)

5. S. Okubo, Rochester preprint UR-651

6. $SU_L(3) \times SU_R(3)$ theories with a $(3^*,3)$ nonet of leptons were first considered by: Y. Achiman, contr. paper Nr. 491 to the 16th Int. Conf. on High Energy Physics, Batavia, 1972
Y. Achiman, Phys.Lett. 70B, 187 (1977)

7. For recent $SU_L(3) \times SU_R(3)$ models with a nonet of leptons see: J.D. Bjorken and K. Lane, Prov. Int. Conf. "Neutrino 77", Elbrus (1977)

P. Minkowski, Nucl.Phys. B138, 527 (1978)

8. For earlier work on $SU_L(3) \times SU_R(3)$ models see:
R. Gatto, Nuovo Cimento 27, 313 (1963)

A. Salam and J.C. Ward, Phys.Rev.Lett. 13, 168 (1964)

S. Weinberg, Phys.Rev. D5, 1264 (1972)

9. Y. Achiman and B. Stech, Phys.Lett. 77B, 389 (1978), and paper in preparation

10. H. Georgi, H. Quinn and S. Weinberg, Phys.Rev.Lett. 33, 451 (1974)

11. F. Gürsey, P. Ramond and P. Sikivie, Phys.Lett. 60B, 177 (1976)

F. Gürsey and M. Serdaroglu, Nuovo Cim. Lett. 21, 28 (1978)

12. See P. Minkowski in ref. 7

13. G. Feldman, invited talk at the XIX. Int. Conf. on HEP, Tokyo 1978

14. In a model which unifies electromagnetic with strong interactions an SU(4) global symmetry was suggested by

M. Abud, H. Ruegg, C.A. Savoy, and F. Buccella, Lett. of Nuovo Cim. 19, 494 (1977)

15. V. Barger and S. Pakvasa, Univ. of Wisconsin preprint COO-881-64, October 1978

ASYMPTOTIC FREEDOM INFRARED INSTABILITY[×]

Peter Minkowski

Institute for Theoretical Physics, University of Berne

Sidlerstrasse 5, CH-3012 Berne, Switzerland

HEURISTIC ASPECTS OF COLOR CONFINEMENT

Electrons and nucleons bound in solids, molecules, atoms, nuclei ... can be isolated from the bound system by a finite ionization (- or breakup) energy.

The harmonic oscillator is an outstanding representative of a confined system (provided the spring does not break). Of course for small oscillations the associated normal modes describe a vast variety of phenomena from water waves to the levels of the nuclear shell model (Ref. 1)).

For such a system of two hypothetical spinless constituents of mass M (in the nonrelativistic regime) the Schrödinger equation reads

$$\left\{- \frac{1}{M} \Delta_z + \frac{\lambda^2}{M} (\vec{z})^2\right\} \psi(\vec{z}) = (E-E_o)\cdot\psi \qquad (1)$$

$$f = \frac{2\lambda^2}{M} : \text{spring constant.}$$

The associated energy values and wave functions are

$$M\cdot E = ME_o + 2\lambda n + 3\lambda \ ; \qquad n = \sum_{i=1}^{3} n_i \qquad (2)$$

$$\psi_{\underline{n}} = \left(\prod_{i=1}^{3} H_{n_i}(z_i\sqrt{\lambda}\,)\right) \exp(- \lambda\vec{z}^2/2)$$

× Work supported in part by the Schweizerische Nationalfonds

H_n: appropriately normalized Hermite polynomial.

The mean square velocity is given by

$$< v^2 >_{\underline{n}} = (n + \frac{3}{2}) \frac{\lambda}{M^2} \tag{3}$$

and of course grows without limit invalidating the nonrelativistic approximation (at least for $n \to \infty$).

The original equation considered by Schrödinger (Ref. 2) was relativistic in the form

$$\begin{gathered} (\partial_t^2 - \Delta_z + m^2(\vec{z}))\ \psi_1(t,\vec{z}) = 0 \\ \left(- \Delta_{\vec{z}} + m^2(\vec{z})\right)\psi = (E_{rel})^2\psi; \quad E_{rel} = \frac{E}{2} \end{gathered} \tag{4}$$

The harmonic potential corresponds to a space dependent mass term

$$m^2(\vec{z}) = m_o^2 + \lambda^2 \vec{z}^2 \tag{5}$$ FN1)

The eigenfunctions for eq. (4.5) are the same as in eq. (2) with the energy eigenvalues

$$E^2 = (4m_o^2 + 12\lambda) + 8\lambda \cdot n \tag{6}$$

This complex of ideas was again touched upon by Feynman, Kislinger and Ravndal (Ref. 3). They consider the spinless, fully relativistic problem involving a Bethe-Salpeter wave function of the type

$$\begin{gathered} \psi(x_1^\mu, x_2^\mu) = < \Omega | T(\varphi(x_1)\varphi^*(x_2)) | p > \\ \psi = \exp\left(-i\ P \cdot \frac{x_1 + x_2}{2}\right)\phi\ (z,P) \\ z = (\tau, \vec{z}) = x_1 - x_2 \end{gathered} \tag{7}$$

ψ is subject to two simultaneous equations

$$\begin{gathered} (- \Box_{x_1})\psi = U(P,z)\psi\ ; \ (- \Box_{x_2})\psi = V(P,z)\psi \\ U = -\lambda^2 z^2 + U_1 \ ; \ V = -\lambda^2 z^2 + V_1 \\ (- \Box_x)_{1(2)} = (\frac{P}{4} - \Box_z) \underset{(+)}{-} \frac{1}{i} P^\mu \partial_{z\mu} \end{gathered} \tag{8}$$

The oscillatorlike Ansatz

$$U_1 - c = -(V_1 - c) = if\ \lambda(P\cdot z), \quad c\text{: constant}$$

yields the following equations

$$\frac{P^2}{4}\psi = \left(\vec{a}^{+}\cdot\vec{a} - a_o^{+}a_o + 4\lambda + c\right)\psi \qquad \text{(a)}$$

$$P^{\mu}\left(\partial_{z\mu} - f\lambda\, z_{\mu}\right)\psi = 0 \qquad (9)\ \text{(b)}$$

In eq. (9) we use the oscillator variables

$$a_{\mu} = \partial_{z\mu} - \lambda g_{\mu\nu}z^{\nu};\ a_{\nu}^{+} = -\partial_{z\nu} - \lambda g_{\nu\alpha}z^{\alpha}$$

with this commutation relations

$$\left[a_{\mu},\ a_{\nu}^{+}\right] = 2\lambda\cdot(-g_{\mu\nu}) \qquad (10)$$

Consistency of equations 9a) and b) requires f = +1 or f = -1.

THE CHOICE f = +1 ↔ THE SPECTRAL CONDITIONS[4)]

It has been emphasised by Leutwyler and Stern (Ref. 4) that a Hamiltonian dynamics rooted in lightlike planes e.g. $z^+ = 0$, $z^{\pm} = z^0 \pm z^3$ determines $\Phi(z,P)$ from spacelike values of z, for which the time ordering in eq. (7) is redundant.

Thus the spectral condition for

$$A(z,P) = \left.\langle\Omega|\varphi(\tfrac{z}{2})\varphi^{*}(-\tfrac{z}{2})|P\rangle\right|_{z^+=0} \qquad (11)$$

is relevant for $\Phi(z,P)$.

$$A(z,P) = \left(\frac{1}{2\pi}\right)^3\int dq^{+}d^{2}q_{t}\ F(q;P)\left.e^{-i(q-\frac{P}{2})z}\right|_{z^+=0} \qquad (12)$$

$$q = (q^{+},\ q_{t})\ ; \quad q^{+} \geq 0$$

Combining eq. (11) and (12) with locality (for $z^2 < 0$) the spectral condition (Ref. 5)

$$\phi_1(\beta,z_t,P) = \int dz^- e^{i\beta(P\cdot z)} \phi(z,P)\Big|_{z^+=0}$$

$$\phi_1(\beta,z_t,P) = 0 \quad \text{unless} \quad -\frac{1}{2} \leq \beta \leq \frac{1}{2} \tag{13}$$

follows.

For the case of the harmonic oscillator it is essential to restrict the momentum of the intermediate states in the above argument to three momentum labels e.g. (q^+,q^x,q^y) since given (q^+,q^x,q^y) finite, q^- is infinite.

The spectral condition in eq. (13) remains unchanged if the local expression

$$\phi(z,P) = \langle\Omega|\ T\left(\varphi(\tfrac{z}{2})\,\varphi^*(-\tfrac{z}{2})\right)|P\rangle$$

is rendered gauge invariant by including a gauge compensating factor

$$G(z;\text{Path}) = P \exp\left(-\int_{\frac{z}{2}}^{-\frac{z}{2}} d\,\xi^\mu(\text{path})\ B_\mu(\xi)\right)$$

$$B_\mu = \begin{cases} ieA_\mu & \text{for an abelian gauge field} \\ ig\,V_\mu^a \frac{\chi^a}{2} & \text{for an } SU3_c \text{ gauge field} \quad a = 1,\ldots 8 \end{cases}$$

$$[\chi^a,\chi^b] = 2if_{abc}\,\chi_c,\quad Tr\chi^a\chi^b = 2\delta^{ab}$$

P : Pathordering

provided the path is taken along a straight line.

It must be stressed that the spectral condition in eq. (13) excludes spacelike momenta (P) which the associated differential equations do not eliminate. In the case of eq. (9a,b) the spectral condition implies f=1:

$$\frac{P^2}{4}\psi = \left(\vec{a}^{+}\vec{a} - a_o^{+}a_o + 4\lambda + c\right)\psi \qquad a)$$

$$(9')$$

$$P^{\mu}a_{\mu}\psi = 0 \qquad b)$$

The energy spectrum coincides with the one in eq. (6) for $m_o^2 = c$

$$E^2 = (4c + 12\lambda) + 8\lambda \cdot n \qquad (6')$$

while in the rest system of P the wave functions are given by

$$\phi_{\underline{n}} = \exp(\frac{\lambda}{2} z_o^2)(\prod_{i=1}^{3} H_{n_i}(z^i\sqrt{\lambda}))\exp(-\frac{\lambda}{2}\vec{z}^2) = \exp(\frac{\lambda}{2}\tau^2)\cdot\psi_{\underline{n}} \qquad (15)$$

Relative time (z_o) is a redundant variable (this is automatic for Hamiltonian systems (Ref. 4). For $z^o = 0(P^2 > 0)$ the wave functions in eq. (15) coincide with those in eq. (2). We retain the groundstate wave functions corresponding to eq. (15):

$$\phi_o = \exp\left(\frac{\lambda}{2}((z^o)^2 - \vec{z}^2)\right) \qquad (16)$$

ϕ_o is independent of P.

THE HARMONIC FORCE BETWEEN A (SPIN $\frac{1}{2}$) QUARK AND ANTIQUARK - THE MESON SPECTRUM

The next step is due to Leutwyler and Stern (Ref. 6) who extended to the spin 1/2 case the harmonic oscillator in eq. (7) and (8):

$$\psi(x_1,x_2) \to \psi_{AB}(x_1,x_2)$$

$$\psi_{AB} = <\Omega|T(q_A^{c'}(x_1)G_{cc'}(x_2,x_1)\bar{q}_B^{c}(x_2))|P>$$

$$G(x_2,x_1) = P\exp\left(-\int_{x_1}^{x_2} B^{\mu}(\xi,\underset{\sim}{P})d\xi_{\mu}(\text{Path})\right)$$

$\psi(x1,x_2)$ is a 16-component wave function. The physical state of polarisation of a resulting bound state depends only on four independent components out of the 16. Also the relative time variable retains its status of redundant variable.

Eq. (8) becomes

$$\gamma^\mu \partial_{x_{1\mu}} \psi = U\psi \qquad (-\partial_{x_{2\mu}} \psi)\gamma^\mu = V\psi$$

$$U\psi = -\lambda\, \gamma_5 \psi \gamma_5 \not{z} + U_1(p,z)\psi \qquad (\not{z} = z^\mu \partial_\mu) \qquad (18)$$

$$V\psi = -\lambda \not{z}\, \gamma_5 \psi \gamma_5 + V_1(p,z)\psi$$

Factorising the center of mass motion

$$\psi_{AB}(x_1,x_2) = \exp\left(-iP \cdot \frac{x_1 + x_2}{2}\right) \phi_{AB}(z,P)$$

one finds four families corresponding to four characteristic sequences of spin-orbital angular momentum configurations: (fig. 1)

(1) π-family

J^{PC}	0^{-+}	1^{+-}	2^{-+} 0^{-+}	....
particle	π	B	A_3	....
$((\text{mass})^2,\ \text{GeV}^2)$	0.019	1.52 ±0.16	2.89 ±0.49	
$(\text{Mass})^2(\text{th})$	$4c^{(\pi)}$	$4c^{(\pi)}+F$	$4c^{(\pi)}+2F$	
$(\text{Mass})^2$ chiral limit	0	F (1.2)	2 F (2.4)	no exchange forces

$$\phi^{(\pi)}_{AB} \propto (i\gamma_5)_{AB}\, \phi^{(\pi)}_{\underline{n}}(z,P)$$

$\phi_{\underline{n}}$: scalar oscillator wave function (eq. 15)

$$(P^2)_{\underline{n}} = F \cdot n + 4c^{(\pi)}$$

$$n = J + 2n_r, \qquad J = 0,1,2,\ldots \quad ; \qquad n_r = 0,1,2,\ldots$$

$$F = 8\lambda = \frac{1}{\alpha'} \simeq 1.2(\text{GeV})^2$$

(2) The ρ-family

J^{PC}	1^{--}	2^{++}	3^{--}	4^{++}	
			1^{--}	2^{++}	$\cdots$

	$\rho(\omega)$	$A_2(f)$	$g,\ \rho'$	h
Particle	0.60	1.72	2.85	4.16
(Mass^2 GeV^2)	± 0.12	± 0.13	2.56	± 0.39

(Mass^2)(th)	$4c^{(\rho)}+F$	$4c^{(\rho)}+2F$	$4c^{(\rho)}+3F$	$4c^{(\rho)}+4F$

(Mass^2)	$\frac{1}{2}F$	$\frac{3}{2}F$	$\frac{5}{2}F$	$\frac{7}{2}F$
chiral limit	(0.6)	(1.8)	(3.0)	(4.2)

Exchange force: $4c^{(\rho)} = 4c^{(\pi)} - \frac{1}{2}F$

Without exchange force: $4c^{(\rho)} = 4c^{(\pi)}$

$$\phi_{AB}^{(\rho)} \propto (\Sigma \cdot P)(\Sigma \cdot a^+)(\Sigma a)(\phi_n^{(\rho)}\ \mathbb{1}) + \cdots$$

$$\Sigma^\mu \cdot \phi = \frac{1}{2}\left(\gamma^\mu \phi + \gamma_5\ \phi \gamma_5 \gamma^\mu\right)$$

$$(P^2)_{\underline{n}} = F \cdot n + 4c^{(\rho)}$$

$$n = J + 2n_r; \quad J = 1,2,\ldots ; \quad n_r = 0,1,2\ldots$$

(3) The τ-family (A_1...)

J^{PC}	1^{++}	2^{--}	3^{++} 1^{++}	4^{--} 2^{--}
Particle (Mass^2, GeV)	A_1(D) 1.21 ±0.33			
(Mass)^2(th)	$4c^{(\tau)}+2F$	$4c^{(\tau)}+3F$	$4c^{(\tau)}+4F$	$4c^{(\tau)}+5F$
(Mass)^2 chiral limit	F (1.2)	2F (2.4)	3F (3.6)	4F (4.8)

Exchange force: $4c^{(\tau)} = 4c^{(\pi)} - F$

Without exchange force: $4c^{(\tau)} = 4c^{(\pi)}$

$$\phi_{AB}^{(\tau)} \propto (\Delta P)(\Delta a^+)(\Delta \cdot a)\ \phi_{\underline{n}}^{(\tau)} + \ldots$$

$$\Delta^\mu \phi = \frac{1}{2}\left(\gamma^\mu \phi - \gamma_5 \phi \gamma_5 \gamma^\mu\right)$$

$$(P^2)_{\underline{n}} = F \cdot n + 4c^{(\tau)}$$

$$n = J+2n_r+1\ ; \quad J = 1,2,\ldots\ ; \quad n_r = 0,1,2,\ldots$$

(4) The σ-family (δ_1 ...)

J^{PC}	0^{++}	1^{--}	2^{++} 0^{++}	3^{--} 1^{--}
Particle	$\delta(S^*)$	$\rho'(1250)$		
$(\text{Mass})^2$, $(\text{GeV})^2$	0.95 ±0.05	1.58 ±0.34		
$(\text{Mass})^2$(th)	$4c^{(\sigma)}+2F$	$4c^{(\sigma)}+3F$		
$(\text{Mass})^2$ chiral limit	$\frac{1}{2}F$ (0.6)	$\frac{3}{2}F$ (1.6)	$\frac{5}{2}F$ (3.0)	

Exchange force: $4c^{(\sigma)} = 4c^{(\pi)} - \frac{3}{2}F$

Without exchange force: $4c^{(\sigma)} = 4c^{(\pi)}$

$$\phi^{(\sigma)}_{AB} \propto \delta_{AB}\, \phi^{(\sigma)}_{\underline{n}} + \ldots$$

$$(P^2)_{\underline{n}} = F \cdot n + 4c^{(\sigma)}$$

$$n = J+2n_r+2 \;; \quad J = 0,1,2,\ldots \;; \quad n_r = 0,1,2,\ldots$$

$\phi^{(\pi,\rho,\tau,\sigma)}_{\underline{n}}$: scalar oscillator wave functions as given in eq. (15).

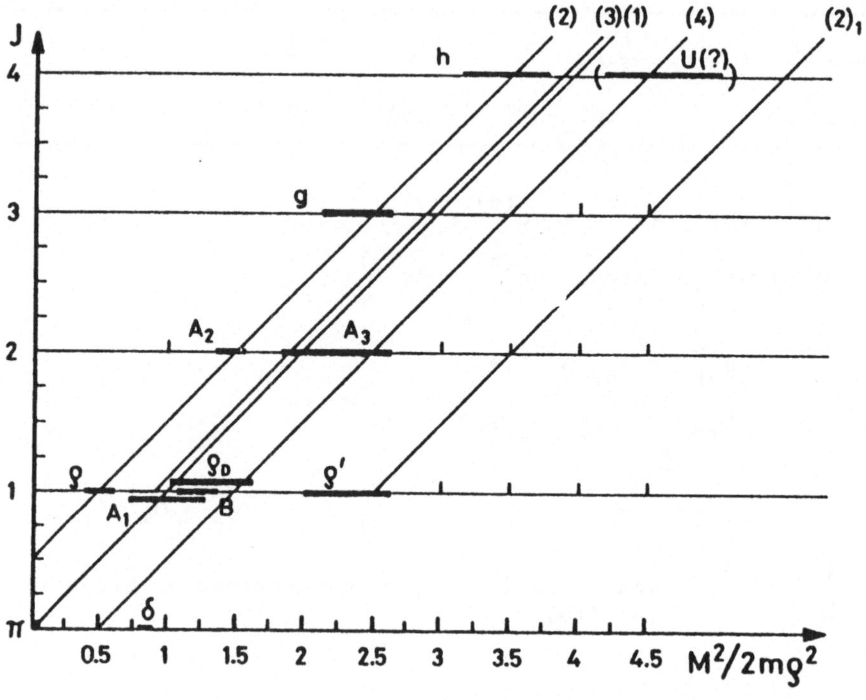

Fig. 1: Pattern of Regge families

EXCHANGE FORCES AND ABSENCE OF LS-COUPLING

In the chiral limit $(m_\pi \to 0)$ current algebra implies the well known relation

$$m_{A_1}^2 = 2m_\rho^2 \tag{19}$$

Without exchange force the above relation is satisfied. The states $(\pi;\rho,\delta, A_1, A_2)$ show the $(\mathrm{mass})^2$ pattern:

$$\begin{array}{ll} m_\pi^2 = 0 & m\delta^2 = m_{A_1}^2 = (2).F \\ m_\rho^2 = (1).F & m_{A_1}^2 = 2m_\rho^2 = (2).F \\ & m_{A_2}^2 = m_{A_1}^2 = (2).F \end{array} \tag{20}$$

The degeneracy of the spin triplet P-wave states (δ, A_1, A_2) shows the absence of LS-coupling forces in the harmonic, direct channel approximation, i.e. if exchange forces are neglected.

The combination of the chiral limit with the constraints from duality, apparent in the well known Ansatz for the $\pi^+\pi^+$-elastic scattering amplitude (Ref. 8)

$$A^{\pi^+\pi^+}(q_{\pi_1}^2, \ldots q_{\pi_4}^2; s,t,u) \propto \frac{T(1-\alpha_\rho(t)\, T(1-\alpha_\rho(U)}{T(1-\alpha_\rho(t)-\alpha_\rho(U)} \tag{21}$$

$$\alpha_\rho(t) - 1 = \frac{1}{F}\left(t - 4c^{(\rho)}\right)$$

yields the relation

$$\alpha_\rho(m_\pi^2) = \frac{1}{2} \Rightarrow 4c^{(\rho)} = 4c^{(\pi)} - \frac{1}{2}F \tag{22}$$

from the Adler consistency condition (Refs. 9,10)

$$A^{\pi^+\pi^+}(q_{\pi_1}^2 = 0, q_{\pi_2}^2 = q_{\pi_3}^2 = q_{\pi_4}^2 = m_\pi^2; S = m_\pi^2, t = m_\pi^2, U = m_\pi^2) = 0$$

As a result the pattern of eq. (20) becomes

$$
\begin{array}{ll}
m_\pi^2 = 0 & m_\delta^2 = \,? \\
m_\rho^2 = (\tfrac{1}{2}).F & m_{A_1}^2 = 2m_\rho^2 = (1).F \\
 & m_{A_2}^2 = (\tfrac{3}{2}).F
\end{array}
\tag{23}
$$

Thus the ρ-family (#2) is shifted by $(-\frac{1}{2})\cdot F$ and the τ-family) (#3) by $(-1)\cdot F$.

The broken degeneracy of the P-wave states

$$(m_{A_2})^2 - (m_{A_1})^2 = \frac{1}{2}\; F$$

is not the result of genuine LS-coupling but of an exchange force.

We still have to assign the δ (σ-family) to an exchange force shifted trajectory in the chiral limit.

To this end we consider the lightlike charges of $SU6_W$ (Ref. 11) on a plane

$$z^+ = z^0 + z^3 = \text{const.}$$

The canonical anti-commutation rules for quark fields on a plane characterised by the normal vector n^μ are

$$
\begin{aligned}
&\delta(nz)\, \not{n}\, \{q(x), q(y)\}\, \not{n} = \not{n}\, \delta^4(z)\, \delta \begin{pmatrix}\text{color}\\ \text{flavor}\end{pmatrix} \\
&z = x-y\, ; \quad n^2 \geqslant 0 \qquad n^o > 0
\end{aligned}
\tag{24}
$$

For $n^\mu = (\frac{1}{2}, 0,0, -\frac{1}{2})$; $n^2 = 0$, $n^- = 1$ the canonical ("good") components are

$$
\begin{aligned}
&\not{n}q = \gamma^0 q^{(n)} \; ; \qquad q^{(n)} = \frac{1+\alpha_3}{2}\, q \\
&q^{(n)} = \begin{pmatrix} q_1^{(n)} \\ q_2^{(n)} \end{pmatrix} = \begin{pmatrix} \frac{1-\gamma_5}{2}\; \frac{1+\alpha_3}{2}\, q \\ \frac{1-\gamma_5}{2}\; \frac{1+\sigma_3}{2}\, q \end{pmatrix}
\end{aligned}
\tag{25}
$$

The canonical components $q_{1,2}^{(n)}$ generate the algebra of colorless charges $U(2n_{fl})_W$:

$$Q_{\alpha\beta\, f_1 f_2}(z^+) = \int_{z^+=\text{const.}} dz^- d^2 z_t \sum_c (q^{(n)}_{f_1^\alpha})_c (q^{(n)}_{\beta f_2})_c$$

$$f_{1,2} = 1,\ldots, n_{fl} ; \quad \alpha,\beta = 1,2 \tag{26}$$

The charges in eq. (26) are not all conserved even in the chiral limit $m_q \to 0$, where the charges

$$Q_{11\, f_1 f_2} = Q^R_{f_1 f_2} ; \quad Q_{22\, f_1 f_2} = Q^L_{f_1 f_2} \tag{27}$$

generating the chiral group $U(n_{fl})_R \times U(n_{fl})_L$ are conserved for the axial baryon number

$$Q^{(5)} = \sum_f^{n_{fl}} (Q^L_{ff} - Q^R_{ff}) \tag{28}$$

subject to the axial current anomaly (Ref. 12).

We shall interpret the neglect of $SU(2n_{fl})$ breaking as reflecting the phenomenological fact that LS-coupling forces are quite small for the nonstrange Baryon system (Ref. 13).

This is apparent from the L = 1 SU(4)20 baryonic multiplet

	$S_{123} = \frac{3}{2}$			$S_{123} = \frac{1}{2}$	
J^P:	$\frac{5}{2}^-$	$\frac{3}{2}^-$	$\frac{1}{2}^-$	$\frac{3}{2}^-$	$\frac{1}{2}^-$
	N(1670)	N(1700)	N(1700)	Δ(1670)	Δ(1660)
				N(1520)	N(1535)
	(2I+1, 2J+1) = (2,4)			(2I+1, 2J+1) = (4,2) ⊕ (2,2)	

Thus we consider (for $n_{fl} = 2$) the scalar density

$$S_o = (\bar{u}u + \bar{d}d)$$

and the angular momentum changing charge

$$Q_1 = Q\begin{matrix}\Delta J_z = -1\\ \\ \Delta I = 0\end{matrix} = \int dz^- d^2 z_t \begin{bmatrix} U_2^{(n)*} U_1^{(n)} + \\ + d_2^{(n)*} d_1^n \end{bmatrix}$$

We study the consequence of the commutation rule

$$\left(Q_1, S_0(z)\right) = \begin{Bmatrix} \bar{u}\left(\gamma^{(x)} - i\ \gamma^{(y)}\right) u + \\ + \bar{d}\left(\gamma^{(x)} - i\ \gamma^{(y)}\right) d \end{Bmatrix} \tag{29}$$

In eq. (29) we have neglected anticommutators of canonical - and dependent components of the quark fields $q^{(n)}$, $q^{(-n)} = \frac{1-\alpha_3}{2}\ q$.

We note that it is the ($\Delta J_z = -1$) part of the isoscalar vector current which appears on the right hand side of eq. (29). Using eq. (29) we now calculate the matrix element

$$\begin{aligned} &< \Omega \left| \left(S_0(0),\ Q_1\right) \right| \omega; J_z = 1\ ,\ p> = \\ &= < \Omega \left| \left[J^{(x)} - i\ J^{(y)}\right]_{I=0} \right| \omega; J_z = 1, p> \\ &= 2F_\omega m_\omega = < \Omega \left| S_0(0) \right| Q_1(\omega, J_z = 1, p)> \end{aligned}$$

In eq. (30) $F_\omega \approx F_\rho$ denotes the decay constant of ω and ρ related to the partial width $\rho \to e^+e^-$

$$\Gamma_{\rho \to e^+e^-} \simeq 9\ \Gamma_{\omega \to e^+e^-} = \frac{4\pi\alpha^2}{3m_\rho} \frac{F_\rho^2}{2} \tag{31}$$

$$\Gamma_{\rho \to e^+e^-} \simeq 6.67 \text{ keV}, \quad F_\rho = 214 \text{ MeV}; \quad F_\pi \simeq 131.5 \text{ MeV}$$

In the preceding argument S_0 can be replaced by $S_1 = \bar{u}u - \bar{d}d$ together with $|\omega_1 J_z = 1, p> \to |\rho, J_z = 1, p>$.

Eq. (30) implies that the states

$$Q_1 \Big|^{\omega}_{\rho}\ ; J_z = 1, p^{cm}> \ ; \ p^{cm} = (m_{\rho,\omega}, 0,0,0)$$

have the quantum numbers $JP^{C_n} = 0^{++}$ and $I = \begin{Bmatrix} 0 & \omega \\ 1 & \rho \end{Bmatrix}$, i.e.

$$Q_1 \left| {\omega \atop \rho} , J_z = 1, p^{cm} \right> \simeq \left| {S \atop \delta} ; p^{cm} \right> ; \quad m_{\delta, S^*} \simeq m_{\rho,\omega} \tag{32}$$

$$\text{for } m_{u,d,s} \to 0$$

provided Q_1 is a symmetry (approximately), i.e. commutes with the Hamiltonian.

We attribute the deviation from the relation in eq. (32)

$$\frac{(m_{\delta,S^*})^2}{m_\rho^2} \simeq 1.6$$

predominantly to the strong influence of the $K\bar{K}$-threshold on the 0^{++} resonances which constitutes a large $SU3_L \times SU3_R$ breaking effect due to the strange quark mass

$$\frac{m_S}{m_\rho} \simeq \frac{1}{5} \tag{33}$$

Including in the following representative states the L=2 (first) excitation on the (S^*,δ) trajectory $\rho_D(J^{PC} = 1^{--})$ we display the pattern of the exchange force shift on trajectories in the chiral limit

$$\begin{array}{llll}
 & m_\pi^2 = 0 & m_\delta^2 = \frac{1}{2} F & \\
J^{PC}{}_n = 1^{--}: & m_\rho^2 = \frac{1}{2} F & m_{\rho_D}^2 = \frac{3}{2} F & m_{\rho'}^2 = \frac{5}{2} F \\
 & & (\sim 1.56) & (\sim 2.56) \\
 & (\simeq 0.6 \text{ GeV}^2) & m_{A_1}^2 = 1 \cdot F & \\
 & & m_{A_2}^2 = \frac{3}{2} F &
\end{array} \tag{34}$$

For the cc-system we expect two distinct effects both related to the large charmed quark mass

$$m_c \simeq (1.2 - 1.5) \text{ GeV}$$

1) Exchange forces are much reduced in strength.

2) The short range one gluon-exchange force (not an exchange force in the sense used above) generates the nonrelativistically approximated potential (Ref. 14)

$$V_c = -\frac{4}{3}\kappa\frac{1}{r} \quad ; \quad \kappa = \frac{g^2}{4\pi}$$

κ: QCD-coupling constant evaluated at a scale appropriate for the $\bar{c}c$-bound system.

Then neglecting for the sake of this discussion all effects due to 2) we obtain the $(\text{mass})^2$-pattern without exchange-force shifts (eq. (20)

$$\begin{array}{lll} m^2_{\eta_c} = K_c = 4c^{(\eta_c)} & m^2_{\psi} = K_c + (1).F & m^2_{\chi}(2^{++}) = K_c + (2)\cdot F \\ m^2_{\eta^1_c}(J^{PC} = 2^{-+}) = K_c + (2)F & & m^2_{\chi}(1^{++}) = K_c + (2)\cdot F \\ (m_{\psi'})^2 = (m_{\psi_D})^2 = (3)\cdot F + K_c & & m^2_{\chi}(0^{++}) = K_c + (2)\cdot F \end{array} \qquad (35)$$

The outstanding difference between the $(\text{mass})^2$ pattern of eq. (34) and (35) pertains to the hyperfine splitting, which is reduced by a factor 1/2 by the exchange forces giving rise to the (approximate) relations

$$\begin{array}{ccc} m^2_{\psi} - m^2_{\eta_c} \simeq & 2(m^2_{D^*} - m^2_D) \simeq & 2(m^2_{\rho} - m^2_{\pi}) \\ (1.5(\text{GeV}^2) & (1.1) & (1.2) \end{array} \qquad (36)$$

The CP = -1 $\bar{c}c$-system shows candidates (Ref. 14) for the pattern in eq. (35)

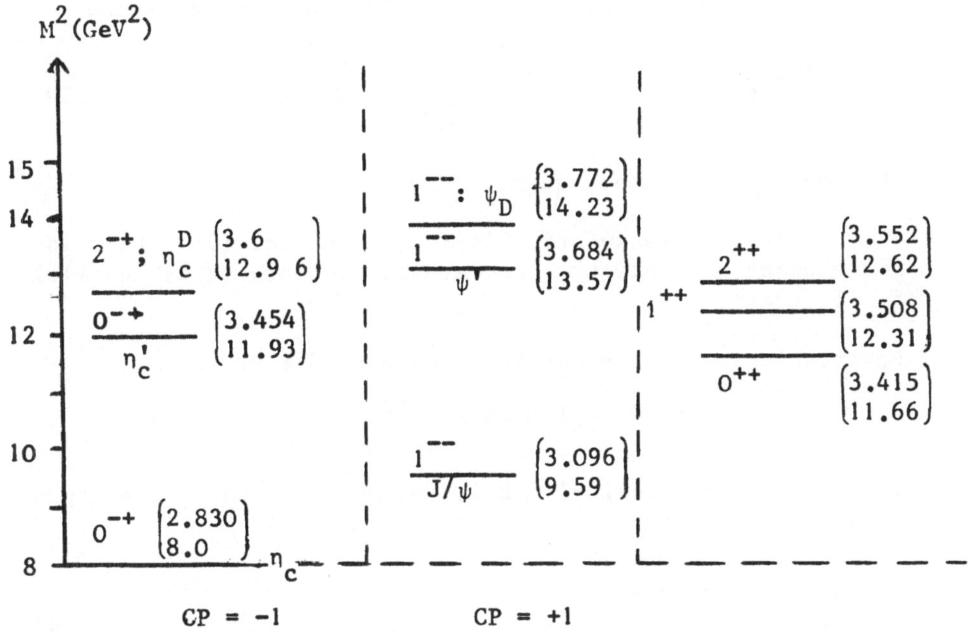

Fig. 2: Level scheme of charmonium

In parenthesis are given (mass (GeV), $(\text{mass})^2 (\text{GeV})^2$) for each level.

CYLINDRICAL GAUGE FIELD CONFIGURATIONS

What is the connection (if any) between the harmonic forces and chromdynamics (Ref. 15).

For a fictitious heavy quark-antiquark system in the static approximation the long range part of the potential in eq. (18) can be approximated by

$$\begin{aligned} U &= -\lambda\ \gamma_5 \psi \gamma_5 Z \\ U &\simeq \lambda \cdot |\vec{z}| \qquad\qquad \lambda = \frac{1}{8\alpha'} \end{aligned} \tag{37}$$

The associated Schrödinger equation would be of the form

$$\left(-\Delta_{\vec{z}} + (M_Q + U(\vec{z}))^2 \right) \psi_{\substack{\text{static} \\ QQ}} = \frac{P^2}{4} \psi \tag{38}$$

M_Q: mass of the heavy quark.

In the ultra relastivistic limit ($|U| >> M_Q$ eq. (38) which is spin independent is inadequate and should be replaced by eq. (18) of ref. (6) again.

Nevertheless we can study the region of distances

$$1 << \lambda |\vec{z}| << M_Q$$

The nonrelativistic limit (Ref. 16) corresponds to the approximations

$$\begin{aligned} &< |U(z)| >_\psi << M_Q \\ &\left| \frac{P^2}{4M_Q^2} - 1 \right| << 1 \rightarrow P^2 = (2M_Q + \varepsilon)^2 \simeq 4M_Q^2 + 4M_Q \varepsilon \end{aligned} \tag{39}$$

Then eq. (38) approximately becomes

$$\begin{aligned} &\left(- \frac{\Delta_{\vec{z}}}{M_Q} + V_{NR}(|\vec{z}|) \right) \psi_{NR} \simeq \varepsilon \, \psi_{NR} \\ &V_{NR} \simeq 2U \simeq 2\lambda |\vec{z}| = \frac{m_\rho^2}{2} |\vec{z}| = 0.30 \text{ GeV}^2 \cdot |\vec{z}| \end{aligned} \tag{40}$$

The asymptotic piece in V_{NR} agrees well with the forces used to describe the charmonium spectrum in the nonrelativistic language (Ref. 16)

$$V(r) \sim 0.25(\text{GeV})^2 |\vec{z}| - 0.76(\text{GeV}) - \frac{0.27}{|\vec{z}|} \tag{41}$$

Nevertheless the above nonrelativistic approximation is <u>not</u> adequate for the charmonium system because of the relatively large difference in mass between $m_c \sim 1.25$ GeV and $m_{D^0} \simeq 1.863$ GeV (Ref. 14).

The nonrelativistic variant of the selfconsistent harmonic potential points towards cylindrical gauge fields between the sufficiently separated QQ-system (Ref. 17).

However unlike the Hydrogen-like electromagnetic binding potential the external field is not predominatly modified vaccum polarization but rather as becomes clear from eq. (38) through polarisation of the effective mass.

The stabalizing nature of this effect can be seen in considering the classical equations

$$\frac{d}{dt}\vec{p} = -\sqrt{1-\vec{v}^2}\,\mathrm{grad}_z m$$
$$\vec{p} = \frac{m\vec{v}}{\sqrt{1-\vec{v}^2}}\,; \qquad m = m(\vec{z})\,; \qquad \vec{v} = \frac{d}{dt}\vec{z} \tag{42}$$

The nonrelativistic motion then corresponds to

$$m = m_o + V_{NR}\,, \quad \langle V_{NR}\rangle_{orbit} \ll m_o$$
$$\vec{p} \approx m_o\vec{v} \;; \quad \mathrm{grad}\; m = \mathrm{grad}\; V_{NR} = \vec{K}_{NR}$$

Eq. (42) admits a transformation to the phase space variable of a nonrelativistic system although the velocity u is not constrained to be small, following way

$$\frac{d}{dt}(\vec{p}^2+m^2) = 0 \rightarrow \vec{p}^2 + m^2 = E^2 = \mathrm{const.} \tag{43}$$

Then quantum mechanical version of eq. (43) is eq. (38) with

$$m(\vec{x}) = m_o + U(\vec{x}) \tag{44}$$

whereby velocities are not constrained to be small compared to the speed of light. The interpretation of eq. (42-44) is this: if a quark evolves in the gluon field generated by an antiquark treated as an (approximately static) external field, this field polarises the mass of the quark.

This effect can readily be compared to the evolution of a snowball rolling down a steep slope.

If $m(\vec{x})$ is spherically symmetric, angular momentum is conserved

$$\vec{J} = \vec{x}\times\vec{P} \qquad \frac{d}{dt}\vec{J} = 0$$
$$m = m(r)\;; \qquad r = |\vec{x}| \tag{45}$$

so that the classical motion proceeds in the plane perpendicular to $\vec{J}$.

The angular integration now goes in close analogy with the non-relativistic case but without any restriction on:

$$\vec{x} = r(\cos\phi, \sin\phi) = r\cdot\vec{e}$$

$$\vec{V} = \dot{r}\,\vec{e} + r\dot{\phi}\,\vec{e}_t \qquad \vec{e}_t = (-\sin\phi, \cos\phi) \tag{46}$$

$$V^2 = \dot{r}^2 + r^2\dot{\phi}^2$$

The conservation laws imply

$$\vec{x} \times \frac{m\vec{v}}{\sqrt{1-\vec{v}^2}} = E\cdot(\vec{x}\times\vec{v}) = \vec{J} \rightarrow J = E\,r^2\dot{\phi}$$

$$E^2 v^2 + m^2 = E^2 \rightarrow \left(\frac{d\rho}{d\phi}\right)^2 + \rho^2 = \frac{E^2 - m^2(r)}{J^2}, \quad \rho = \frac{1}{r} \tag{47}$$

Eq. (47) can be compared to the corresponding nonrelativistic central potential problem

$$\frac{d}{dt} m_o \vec{v} = -\text{grad } V_{NR} \rightarrow J = m_o r^2 \dot{\phi}$$

$$\left(\frac{d\rho}{d\phi}\right)^2 + \rho^2 = \frac{2m_o(E_{NR} - V_{NR}(r))}{J^2} \tag{48}$$

Of course again eq. (48) is an approximation to eq. (47) setting $m = m_o + V_{NR}$ $E = m_o + E_{NR}$ with $\langle V_{NR}\rangle \ll m_o$, $|E_{NR}| \ll m_o$.

I shall conclude this section with the remark that the harmonic potential

$$m(r) = 2\lambda r \tag{49}$$

is singled out as the only potential with classically confined and closed orbits for arbitrary initial corrections compatible with stability i.e. with $v^2 < c^2$.

REFERENCES

1) see e.g.: M. Goeppert Mayer and J. Jensen, Elementary Theory of Nuclear Shell Structure, John Wiley and Sons Inc., New York 1955

and: N. Bohr, Determination of the Surface Tension of Water by the Method of Jet Vibrations, in Niels Bohr, Collected Works, J.Rud. Nielsen ed., North Holland Publ. Co., Amsterdam 1972, p. 29

2) E. Schrödinger (1924) unpublished,
Wave Equations for a Dirac fermion corresponding to the harmonic Ansatz in eq. (5) have been studied by
C.L. Critchfield, Phys.Rev. D12, (1975) 923

3) R.P. Feynman, R. Kislinger and F. Ravndal, Phys.Rev. D3, (1971) 2706

4) H. Leutwyler and J. Stern, Ann. of Physics 112 (1978) 94
Nucl.Phys. B133 (1978) 115

5) R. Jost and H. Lehmann, Nuovo Cimento 5 (1957) 1598
F.F. Dyson, Phys.Rev. 110 (1958) 1460
S. Deser, W. Gilbert and E.C.G. Sudarshan, Phys.Rev. 115 (1959) 731

6) H. Leutwyler and J. Stern, Phys.Letters 73B (1978) 75

7) S. Weinberg, Phys.Rev. Letters 18 (1967) 507
K. Kawarabayashi and M. Suzuki, Phys.Rev.Letters 16 (1966) 255
Riazuddin and Fayazuddin, Phys.Rev. 147 (1966) 1071

8) G. Veneziano, Nuovo Cimento 57A (1968) 190
Phys. Reports 9C (1974) 199
C. Lovelace, Phys.Letters 28B (1968) 264

9) S.L. Adler, Phys.Rev. 137B (1965) 1022
Phys.Rev. 139B (1965) 1638

10) S. Weinberg, Pion Scattering Lengths, Phys.Rev. Letters 17 (1966) 616

11) see e.g.: J.L. Rosner, The Classification and Decays of Resonant Particles, Phys.Reports 11C (1974) 189

12) S.L. Adler, Phys.Rev. 177 (1969) 2426
J.S. Bell and R. Jackiw, Nuovo Cimento 60A (1969) 47
for a modern review see e.g.:
R. Crewther, CERN preprint TH. 2546 (1978)

13) H.J. Melosh, Phys.Rev. D9 (1974) 1095

14) J. Jersak, H. Leutwyler and J. Stern, Phys.Letters 77B (1978) 399

15) H. Fritzsch and M. Gell-Mann, in Proceedings of the XVI Int. Conference on High Energy Physics, Chicago, 1972, Vol. II, p. 135
H. Fritzsch, M.Gell-Mann and H. Leutwyler, Phys.Letters 47B (1973) 365
S. Weinberg, Phys.Rev. Letters 31 (1973) 494

16) for a review of the nonrelativistic approximation to the charmonium (and general -onium) system see e.g.:
J.D. Jackson, in Proceedings of the 1977 European Conf. on Particle Physics, Budapest, 1977, Vol. I, p. 603
K. Gottfried, in Proceedings of the 1977 Int. Symposium on Lepton and Hadron Interactions at High Energies, Hamburg 1977, p. 667
see also:
R. Barbieri, R. Gatto and E. Remiddi, Phys.Letters 61B (1976) 465
H. Fritzsch, Schladming Winter School Lectures 1978, CERN preprint TH-2483 (1978)

17) P. Minkowski, Structure of Cylindrical Gauge Field Configurations, University of Bern, preprint (1979)

FN1 The minimal substitution $i\partial_t \to i\partial_t - V(z)$ leads to the equation

$$[2\, E_{rel}\, V - V^2 - \Delta_z + m^2]\, \psi = (E_{rel})^2\, \psi$$

which for $|V| \to \infty$ shows the effectively repulsive nature of the dominating $- V^2$ term opposed to the harmonic Ansatz in eq. (5).

EXOTIC QUARK STATES

Peter H. Dondi

Southampton University England

FOREWORD

Much work has been done over the last few years on various unusual states. Unusual in the sense that whereas the familiar mesons and baryons are described by a quark-antiquark pair and three quarks respectively, these states have very different constituents - more than three quarks or antiquarks; gluons; new species of quarks with different colour content; or mixtures of these. At the same time, exciting new experimental results have also been presented that indicate the existence of states that may require an exotic solution.

Needless to say, in two hours it is impossible to be pedagogical, and review the whole of this very broad subject. I have therefore chosen the former course and present in some detail what I hope will be a useful introduction to some of the ideas and techniques that are being applied in this field.

In particular, I am going to concentrate on the theoretical aspects of multi-quark-antiquark states (with only passing references to other types of exotic quark states) and mention the experimental evidence presently available which is suggestive of their existence.

The lectures are divided basically into three sections. Firstly an introductory group theoretical analysis of the colour, spin and flavour content of exotic quark states. This is followed by a brief review of the experimental situation, and finally we describe how the multi-quark-antiquark states have been treated in the bag model.

Two appendices are included. One contains a short introduction to the applications of Young Tableaux in SU(N). The second lists some properties of the quadratic Casimir operators in SU(N). Both should be of some use generally, and in particular to understand calculations in the lectures and in many papers written on the subject of exotic quark states.

I. Introduction

Before the discovery of the ψ and T families, most of the "old" mesons and baryons were shown to be easily classified by a few low lying representations of the group SU(6) ⊗ O(3) [1]. In terms of an underlying quark model, the mesons are entirely described by a quark-antiquark pair and baryons by three quarks. The quarks come in three flavours (u,d,s) and transform under the action of a flavour SU(3) group. It is this flavour group combined with a quark-spin SU(2) which forms the SU(6) classification group, whilst O(3) is generated by the quark angular momentum. Since it turns out that the ground state baryon wave function prefers to be totally symmetric in space, flavour and spin, an extra dynamical symmetry, colour SU(3), is introduced so that quarks now come in three colours and, the baryon wave function can be reconciled with Fermi-Dirac statistics for the quarks, simply by being totally antisymmetric in the colour degrees of freedom. The SU(3) singlet nature of the antisymmetric colour wave function, together with the fact that the mesons can also be made into colour singlets leads to the colour confinement hypothesis that all observed states are colour singlets. Although this hypothesis means that colour is not seen directly, it can be indirectly detected in the measurement of

$$R \quad = \quad \sigma(e^+e^- \to \text{hadrons}) \; / \; \sigma(e^+e^- \to \mu^+\mu^-)$$

and the decay rate of π^0 into two photons, both of which effectively count the number of quarks and support the triplet nature of coloured quarks [2].

It should be noted however that all these arguments require only the concept of a global symmetry, and it is yet another step to the introduction of an exact local gauge theory of colour SU(3). From here on, we simply assume confinement and the existence of a set of quarks and gluons from which hadron states are built in an as yet unknown fashion, subject only to them being colour singlets (ref. 3).

The ψ family has added a considerable amount to our understanding of mesons as basically a quark-antiquark system and further results from higher mass region such as the T will no doubt only add to our confidence in that system.

So much for history. What about exotic quark states? In order to understand what I mean by exotic quark states, let us consider in some detail the basic constituents of the quark model and how colour singlet states can be constructed.

The Constituents

We shall concentrate on the old quarks (u,d,s). Then, the eighteen component quark field

$$q^{I\alpha i}$$

has baryon number 1/3 and transforms as $(3_c, 2, 3)$ under colour SU(3) (the I-index), spin SU(2) (the α-index) and flavour SU(3) (the i-index) respectively. As we have said in the Introduction, it is sometimes useful to introduce the SU(6) flavour-spin group formed by combining the SU(3) flavour with the spin SU(2) in which case the quark transforms as (3, 6) under colour SU(3) and SU(6) flavour-spin.

A similar classification can be made with respect to the SU(6) formed by combining colour SU(3) with spin such that the quark transforms as |6, 3| under SU(6) colour-spin and SU(3) flavour respectively*). We shall use all three types of classification at different times in what follows - needless to say, it is always the same quark! For completeness, the antiquark $\bar{q}_{I\alpha i}$ is in the conjugate representation and transforms as $(3^*_c, 2, 3^*)$ or equivalently $(3^*, 6^*)$ or $|6^*, 3^*|$. Together the quarks and antiquarks form the 36 basic anticommuting fermion constituents of the old quark model.

The gluon should now also be included as a possible constituent, as an addition to the old quark model coming with the arrival of Q.C.D. As a group theoretical object it transforms as $(8_c,3,1)$ under SU(3) colour, SU(2) spin and flavour SU(3) i.e. it is a colour octet, spin one boson of neutral flavour. Although massless, it has three spin components being permanently confined and presumably off-mass-shell.

---

*) In order to try and separate the classification we use (A,B) with A transforming under SU(3) colour and B under SU(6) flavour spin and |A,B| for A transforming under SU(6) colour-spin and B under SU(3) flavour. When labelling by colour, spin and flavour, the representation in the colour sector is given a suffix c.

Exotic Quark States

Bearing in mind the confinement hypothesis, we see the simplest examples of colour singlet states described in the Introduction are just the mesons

$$q^{I(\alpha i)}\bar{q}_{I(\beta j)} \quad \text{(sum over the repeated indices)}$$

in the familiar $1 \oplus 35$ representation of SU(6) flavour-spin containing the pseudoscalar (0^{-+}) nonet and vector (1^{--}) nonet, and the ground state baryons

$$\varepsilon_{IJK}q^{I(\alpha i)}q^{J(\beta j)}q^{K(\gamma k)}$$

in the 56 representation of SU(6) flavour-spin containing the nucleon octet (spin 1/2) and the Δ decuplet (spin 3/2).

Now, exotic states are states with quantum numbers outside those allowed by the above configurations, however, with quarks and gluons being taken somewhat more seriously, we add the word quark to allow us to extend our filed of study and include under the heading exotic quark states:

(i) Zero quark states - glueballs [4].
(ii) States of exotic quarks - quarks not in the basic representation of colour SU(3) [5].
(iii) Multiquark states - mesons with more than one quark antiquark pair and baryons with more than 3 quarks and antiquarks [6].
(iv) Any state containing a mixture of normal quarks, exotic quarks and constituent gluons [7].

This extended exoticism obviously includes states that are non-exotic by the original definition. Glueballs are flavour singlet. Exotic quarks are exotic by their unusual colour and may have perfectly normal flavour. Multiquark states often have some non-exotic components. What we have done really is to shift the emphasis from the flavour sector to the colour sector, as it is this sector that ultimately determines the existence or non-existence of states in the above categories, and through them to the old style exotic states (if present ideas of Q.C.D. are correct).

How then are exotic quark states built up?

In the main I want to discuss the properties and relevance of the states under subsection (iii), the multiquark states, but before passing on to this topic let me briefly review the structure of glueballs and states with exotic colour quarks.

(i) Glueballs

In order to construct colour singlets from octets of colour, we note from appendix (1) that

$$8 \oplus 8 \quad = \quad 1 \oplus 8_S \oplus 8_A \oplus 10 \oplus \overline{10} \quad 27 \qquad .$$

Thus, two gluons can form a colour singlet and this is none other than the trace over the colour indices

$$G \quad = \quad V^{\alpha}_{A} V^{\rho}_{A}$$

Furthermore, since the gluons are bosonic we can say that the ground state of two gluons must also be symmetric in spin i.e. it has six spin components which are none other than the spin 0 and spin 2 combinations from the product $3 \otimes 3$. Thus we might expect 2 ground-state glueballs with $J^{PC} = 0^{++}, 2^{++}$.

The next set of glueball states that one can envisage comes from three gluon combinations, and here, since $8 \otimes 8$ contains the two independent octets formed from the totally symmetric and totally antisymmetric, d_{ijk} and f_{ijk} invariant tensors we can form two independent colour singlets

$$G_f \quad = \quad f_{ABC} V^{\alpha}_{A} V^{\beta}_{B} V^{\gamma}_{C} \qquad \text{and} \qquad G_d \quad = \quad d_{ABC} V^{\alpha}_{A} V^{\beta}_{B} V^{\gamma}_{C} \qquad .$$

Again, using Bose symmetry, the ground state G_f boson is a $J^{PC} = 0^{-+}$ glueball, whilst G_d has 10 components which are a $J^{PC} = 3^{--}$ glueball and a $J^{PC} = 1^{--}$ glueball.

It is not easy to say what exactly the masses of these states are, but it is worth noting that the Q.C.D. Lagrangian should imply mixing through diagrams such as those shown in fig. 1., quantum numbers, angular momentum barriers and running coupling constants permitting!

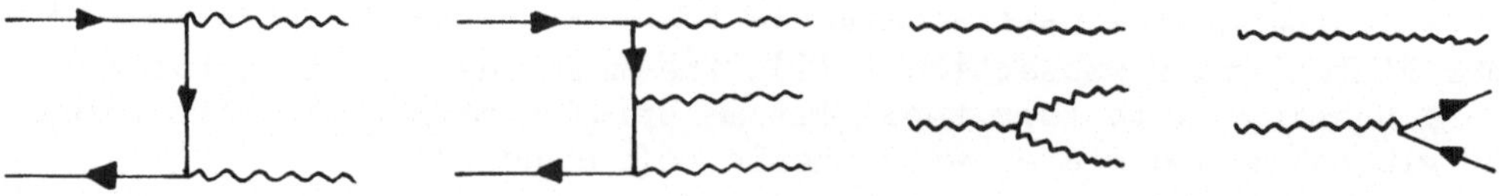

Fig. 1 Typical contributions to mixing between glueballs and quark states.

(ii) "Exotic Quark" States

Here again, it is easy to use a bit of group theory to predict the type of colour singlets that might exist given the particular colour SU(3) representation the quarks are in, but setting a mass scale for the production of these states is problematic. Nevertheless, one possible signature for production would come from a rise in R in e^+e^- experiments, if the quarks are charged.

For example, a quark in the colour 6 representation might have charge -1/3 or 2/3 so that it can combine with two ordinary quarks to produce colour singlet*, integer charged particles. Similarly a quark in the colour 8 representation might have charge 1 to combine with an ordinary quark-antiquark pair to produce again colour singlet* integrally charged particles. Then, the change in R given by

$$\Delta R = \sum_{\text{colour}} e_q^2$$

can vary from 2/3 in the first case, to a spectacular 8 in the latter example.

Finally, it may be worth pointing out that which ever representation of colour SU(3) the quarks are in, there is always a basic colour singlet boson formed by taking a quark-antiquark pair and always a colour singlet fermion formed by taking the combination of three quarks.

(iii) Multiquark States

Let us return now to the quarks we know and love, the basic flavour triplet (u,d,s). We know that this transforms as (3, 6)

*) See section (iii)

and have seen how the mesons and baryons can be constructed. Looking a bit closer at the colour structure, we find that the quark-antiquark pair transform as

$$(3, 6) \otimes (3^*, 6^*) = (1 \oplus 8, 1 \oplus 35) \quad .$$

The colour singlet is unique. From two quarks, we have

$$(3, 6) \otimes (3, 6) = (3^* \oplus 6, 21 \oplus 15)$$

and in the spacially symmetric ground state, antisymmetry of the total wave function implies the colour-flavour-spin combinations

$$(3^*, 21) \oplus (6, 15) \quad .$$

There are no colour singlet combinations, so the diquark does not exist as a physical state. Moreover, taking the combination of diquark and quark to produce the baryon singles out the antisymmetric colour 3^* since $3^* \otimes 3$ contains a singlet whilst the combination $6 \otimes 3$ does not. Thus we see that up to the baryon level, colour singlets are uniquely formed by contracting 3 with 3^*.

It is only when we look at the esoteric combinations including constituent gluons, quarks with exotic colour, or states with a high multiplicity of old quarks and antiquarks that the more unusual colour combinations begin to play a role.

Thus, the diquark can be combined with an antidiquark in two possible ways, either we have the colour singlet from

$$(3^*, 21) \otimes (3, 21^*) = (1 \oplus 8, 1 \oplus 35 \oplus 405)$$

or we can choose the combination

$$(6, 15) \otimes (6^*, 15^*) = (1 \oplus 8 \oplus 27, 1 \oplus 35 \oplus 189)$$

and pick out this colour singlet. Thus in the diquark-antidiquark system we can produce two colour singlets with different basic colour structure which, through the Pauli principle manifests itself in terms of a different SU(6) flavour-spin content in each case.

(Naturally, one does not need to stop here in the construction of colour singlets from unusually coloured objects. In the multiquark-antiquark sector we can form singlets from

$q\bar{q}(8) - q\bar{q}(8)$ where the number in parenthesis indicates the particular colour representation chosen from the preceeding quark state

$qqq(8) - q\bar{q}(8)$ $qqq(10) - \bar{q}\bar{q}\bar{q}(10^*)$
$qqq(8) - qqq(8)$ $qqqq(6^*) - qq(6)$

and so on and so on. On top of these, we also have states already mentioned in the earlier sections where constituent gluons or quarks with exotic colour are introduced to produce combinations like

$q\bar{q}(8)$ - Glue (8)
$qq(6)$ - Exotic quark (6^*)
$q\bar{q}(8)$ - Exotic quark (8)

to mention but a few.)

Notice, that in the flavour sector the diquark decomposes as

$$(6,\ 15) = (6_c,\ 3,\ 3^*) \oplus (6_c,\ 1,\ 6)$$

$$(3^*,\ 21) = (3^*_c,\ 3,\ 6) \oplus (3^*_c,\ 1,\ 3^*)$$

The 3^* is the antisymmetric combination of the two flavour indices ($qq(3^*) \sim \varepsilon_{ijk}q^i q^j$) and the 6 is the symmetric combination (see fig. 2) so that a diquark-antidiquark meson system always contains non-exotic parts (in the old sense of the word) along with the exotic components from $6 \otimes 6^*$.

So what!

In the above, the diquarks and multiquark systems have been treated as entities in their own right*), but unless we go a few

*) It is perhaps worth mentioning, that before the introduction of colour, Lichtenberg suggested that the diquark should be treated as a completely separate system so that the baryon was a diquark plus quark and the statistics would no longer be a problem. Naturally he then has the possibility of meson states made up out of diquark-antidiquark pairs. Including colour, several models of ψ were proposed in which ψ was a multiquark-antiquark system (ref. 6 - History).

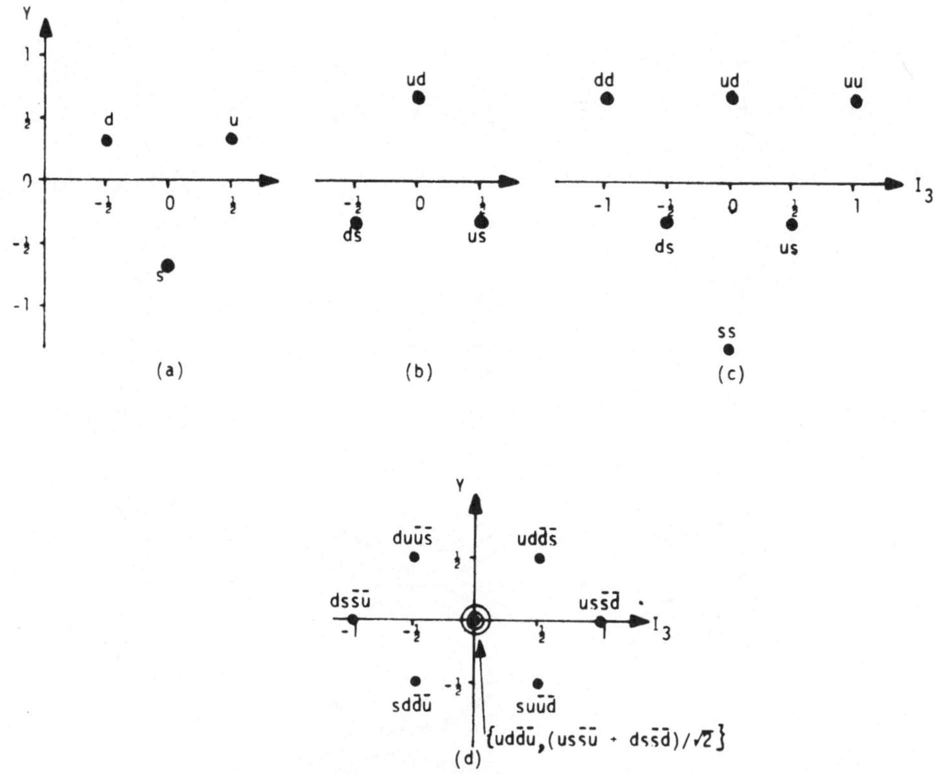

Fig. 2 SU(3) flavour weight diagrams for: (a) quark triplet (b) quark pair forming antitriplet (c) quark pair forming sextet (d) diquark-antidiquark forming nonet.

steps further and realise a dynamical situation by which we are allowed to consider them as such, we must bear in mind that as far as group theory is concerned any colour singlet state made out of quarks and antiquarks can be arranged into multiples of $q\bar{q}$, qqq and $\bar{q}\bar{q}\bar{q}$ colour singlets. We must therefore search for systems, within present day understanding of Q.C.D. that do infact have the characteristic of keeping their identity in terms of some underlying colour structure rather than just falling apart into multi-meson and/or baryon colour singlets as indicated by group theory.

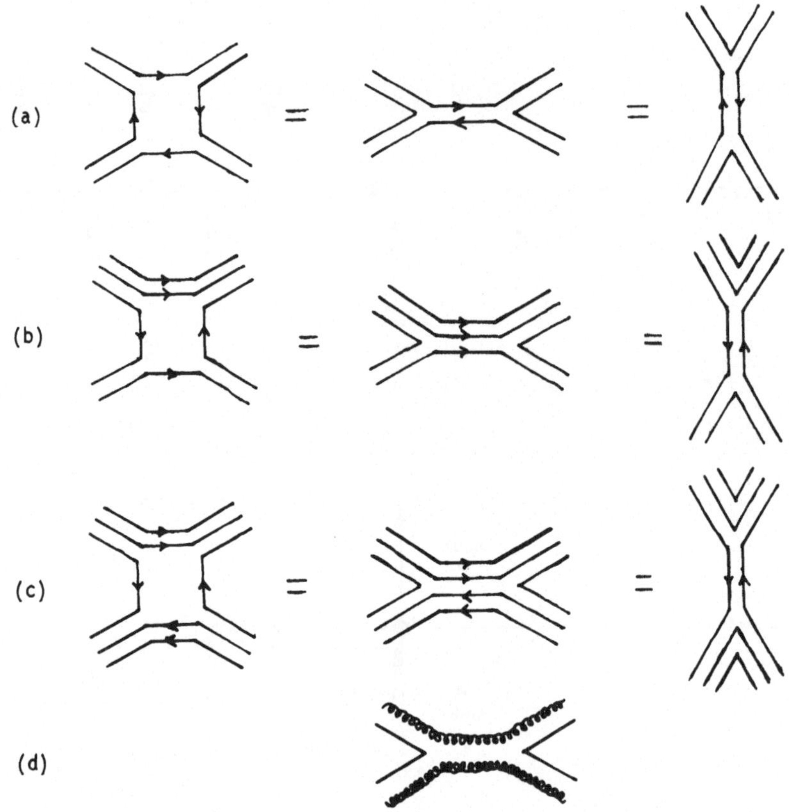

Fig. 3 Duality diagrams for (a) Meson-Meson; (b) Meson-Baryon; (c) Baryon-Antibaryon processes; (d) indicates diquark-antidiquark picture of $B\bar{B}$ scattering.

A Brief Interlude with Duality

The concept of Duality [8] i.e. the idea that Resonance formation and Regge exchange are manifestations of the same basic dynamics, the former being more appropriate to the description of low energy processes whilst the latter is applicable at high energies, has been of major importance since its introduction in the late sixties. It was however troubled by the occasional appearance of exotic quark states in channels dual to ordinary mesons, because of the lack of any experimental evidence to support their existence. This can easily be seen by glancing at the duality diagram for $B\bar{B}$ (Baryon-Antibaryon) elastic scattering and comparing it with meson-meson and meson-baryon scattering where the exchange and resonance states are perfectly reasonable (fig. 3). Using the diagram to our advantage, we might infer that if the diquark keeps its identity in $B\bar{B}$ scattering, then we can effectively replace it by a new "diquark line" and see that a signature for the diquark-antidiquark state is that it prefers to decay into a $B\bar{B}$ final state rather than a host of mesons that would be typical of an ordinary meson with an identical amount of available phase space. Fortunately, the experimental evidence for such states has improved over the last few years, and it is no longer a futile theoretical exercise to try and produce models of exotic quark states such as the diquark-antidiquark (often refered to as Baryonium or diquonium) or more general multiquark-antiquark configurations.

2. Experimental Evidence

Before analysing a particular dynamical model which can allow for the existence of such states, it seems a good time to review briefly the experimental situation from an optimist's stand point. We have seen in Professor Greco's talk that recent measurements suggest things are not as simple as they used to be in e^+e^- between about 1 - 3 GeV. If any of the bumps in this region are for example diquark-antidiquark states, we can immediately fix them as $J^{PC} = 1^{--}$ states which naturally correspond to odd angular momentum, and thus their couplings to the photon can be expected to be a good deal smaller than S-wave quark-antiquark states (by wave function at the origin arguments). We must await better information on these states, and turn to the more direct evidence available in hadronic experiments [9].

$p\bar{p}$ total cross section measurements

Has structures

Mass (GeV)	Width (MeV)	Comments
(S)1.936	4 - 8	$J^{PC} = 2^{++}$ favoured, I probably 1
(T)2.190	90	I = 1
(U)2.350	160	I = 1 and 0

$p\bar{p} \to p\bar{p}$ [10]

S, T, U, present, slightly different masses and widths quoted. All are highly elastic.

$\bar{p}p \to \pi^+\pi^-$ (Mass range 2.08 GeV - 2.58 GeV)

Branching ratios for $\pi\pi$ small - few %, however, high statistics allows amplitude analysis.

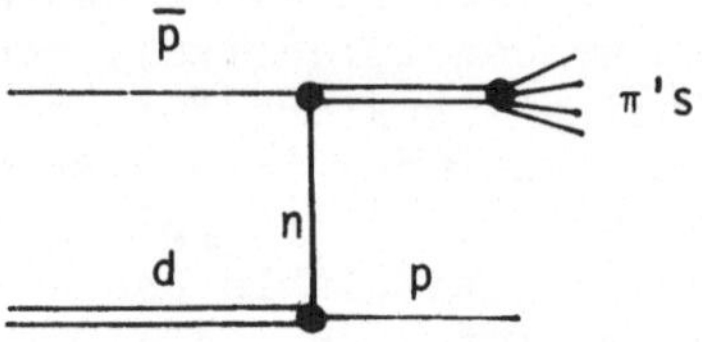

Fig. 4 Resonance production in $\bar{p}d \to pX$

Bose statistics implies J even I = 0
J odd I = 1

Breit Wigner fits give

Mass (GeV)	Width (MeV)	J^{PC} I^G
(T)2.15	200	3^{--} 1^+
(U)2.31	210	4^{++} 0^+
(V)2.48	280	5^{--} 1^+

$p\bar{p} \to K^+K^-$ 11)

Much less statistics than ππ above, but the analysis confirms $I = 1$, $J^P = 3^-;5^-$ states and $I = 0$, $J^P = 4^+$ and provides evidence for states

$I = 0$, $J^P = 3^-$ and 5^- with the same masses as $I = 1$ states

Other formation experiments showing indications of enhancements are $p\bar{p} \to \pi^0\pi^0$, $p\bar{p} \to \pi^0\eta$, $p\bar{p} \to 5\pi$.

Off Mass Shell $\bar{p}n$ experiments (fig. 4) have shown enhancements at:

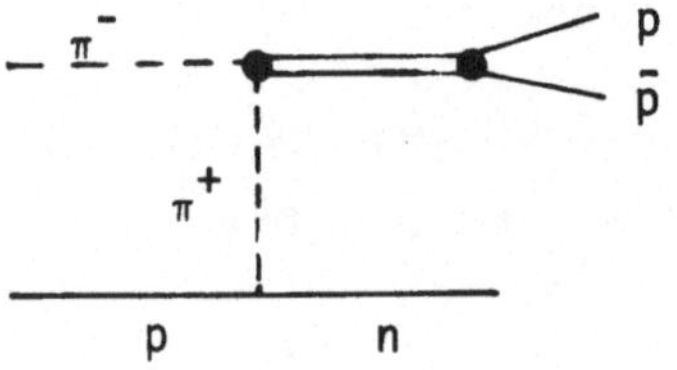

Fig. 5 Resonance production in $\pi^- p \to p\bar{p}\, n$

Mass 1.897 GeV Width ~ 22 MeV

Mass 1.932 GeV Width ~ 5 MeV

with the 1.897 not being a single state.

Production Experiments

$\pi^- p \rightarrow p\bar{p}n$ [12] (fig. 5). Using the analysis of the $p\bar{p} \rightarrow \pi^+\pi^-$ formation experiments, it was found that along with the states given above, an extra J = 1, Mass 1.95 GeV, Width 200 MeV Breit Wigner was necessary to fit the data. Further inclusion of J = 2, Mass 1.91 GeV, Width 60 MeV Breit Wigner gave an even better fit but was not truly necessary.

From all these results we can conclude that there is good evidence for resonance states coupling preferentially to $B\bar{B}$ (c.f. g(1680), h(2040), these meson states have similar mass, width and spin but decay into mesons).

However, these are not the only states that have been detected coupling to nucleon-antinucleon channels. In fact, perhaps the most exciting states are the following, which are increasingly high mass states, rather far above threshold, and yet with very narrow widths.

Searches in production experiments, $p\bar{p}$ (or $\bar{p}n$) $\rightarrow X\pi$, $X \rightarrow N\bar{N}$ or $N\bar{N}\pi$ have indicated enhancements at masses of 2.85 GeV and 3.05 GeV with widths of less than 40 MeV and 20 MeV respectively. More recently in the reaction $\pi^- p \rightarrow (p\bar{p})(p\pi^-)$ two narrow reasonances have been seen at 2.02 GeV and 2.264 GeV with widths of the order of 20 MeV (figs. 6,7) in the $p\bar{p}$ system, and with $\pi^- p \rightarrow (p\bar{p}\pi^-)X$ a very narrow state of mass 2.95 GeV, width less than 15 MeV has been detected (figs. 8,9) with signals in the $p\bar{p}$ channel of the 2.02 GeV and 2.2 GeV states and one at 2.62 GeV.

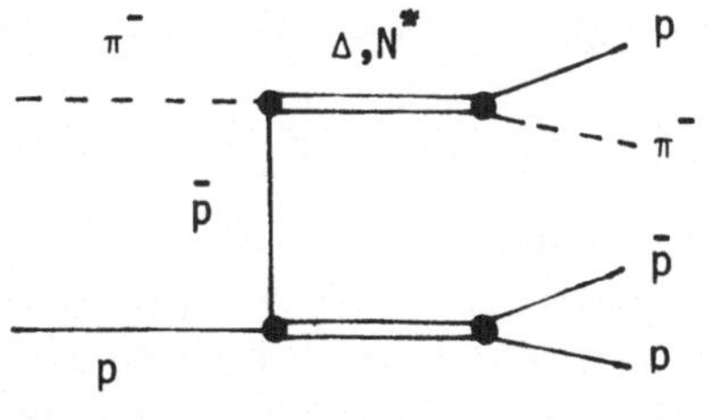

Fig. 6 Resonance production in $\pi^- p \rightarrow p\bar{p}\, p\pi^-$.

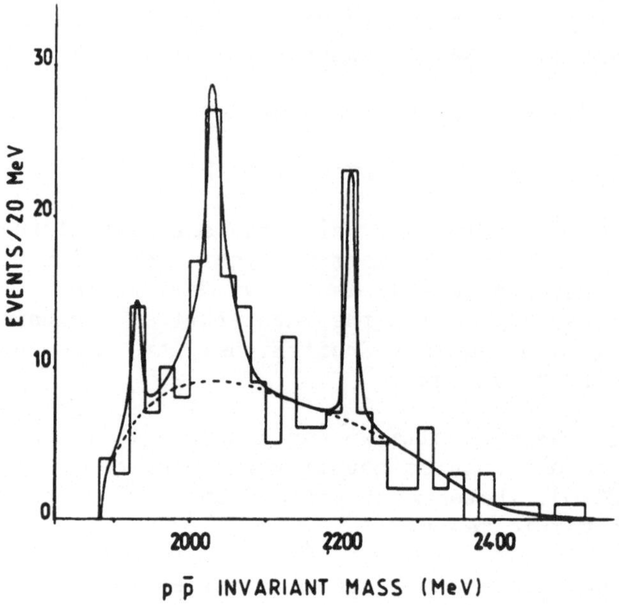

Fig. 7 $p\bar{p}$ effective mass for $\pi^- p \rightarrow p_f \pi^- p\bar{p}$ at 9 and 12 GeV/c (ref. P. Benkhieri et al., Phys. Lett. 68B, 483 (1977)).

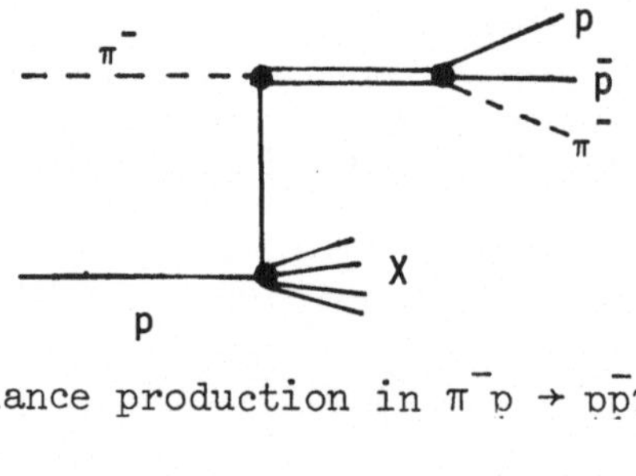

Fig. 8 Resonance production in $\pi^- p \rightarrow p\bar{p}\pi^- X$

All these states are in need of confirmation [13] but there seems to be an accumulation of good evidence for exceptionally narrow states of high mass not involving the production of new quarks.

Further evidence for narrow states in $p\bar{p}$, this time below threshold [14] has been obtained by taking measurements of monoenergetic γ-rays from a stopped $\bar{p}$ beam in a Hydrogen target. Three possible states are seen (fig. 10) corresponding to masses 1.684 GeV, 1.646 GeV and 1.395 GeV with widths less than 20 MeV. In comparison,

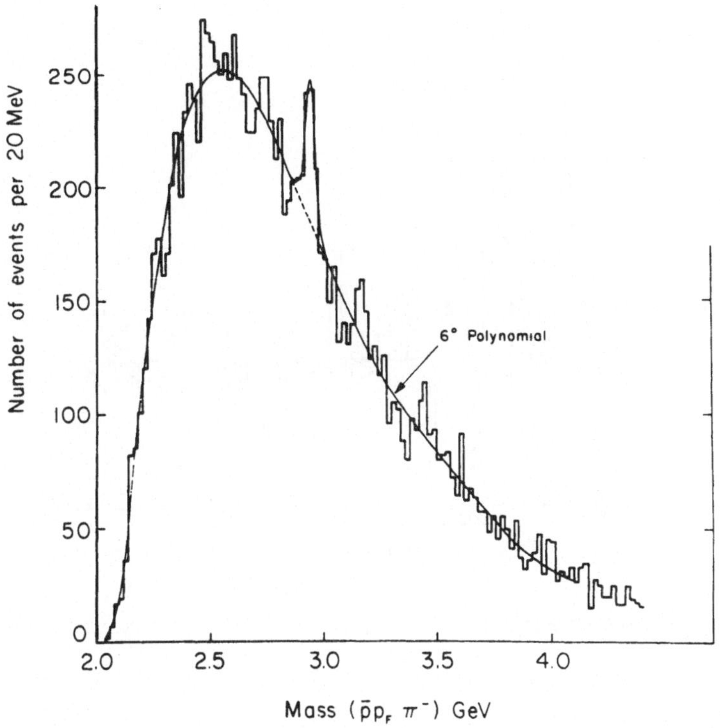

Fig. 9 Total unweighted $\bar{p}p\pi^-$ mass spectrum at 16 GeV/c (13998 events) (ref. C. Evangelista et al., Phys.Lett. 72B, 139 (1977)).

the only direct measurment of a resonance below threshold that I am aware of is in $\bar{p}d \rightarrow pX$ [15] where a state of mass 1.794 GeV in $\bar{p}n$ with width about 7 MeV has been reported.

Again, confirmation of these states must be awaited, perhaps the 3 states in the γ-ray experiment can be detected in $pd \rightarrow nX$, but an optimist can certainly conclude that there is plenty of scope to begin to believe in the diquark-antidiquark states.

There is still an outstanding question. If we believe in diquark-antidiquark states, where are the exotic states of old that surely must exist? Well, let me close this brief experimental review by showing you the results of a recent experiment [16] on $K^+p \rightarrow (\bar{\Lambda}p\pi^+)n$ (fig. 11). In $\bar{\Lambda}p\pi^+$, a charge 2, strangeness 1 combination, a signal of a resonance with mass 2.46 GeV, width less than 20 MeV has been detected. Furthermore, all measurements imply a two body Baryon-Antibaryon final state, either $\bar{\Lambda}\Delta^{++}(1232)$ or

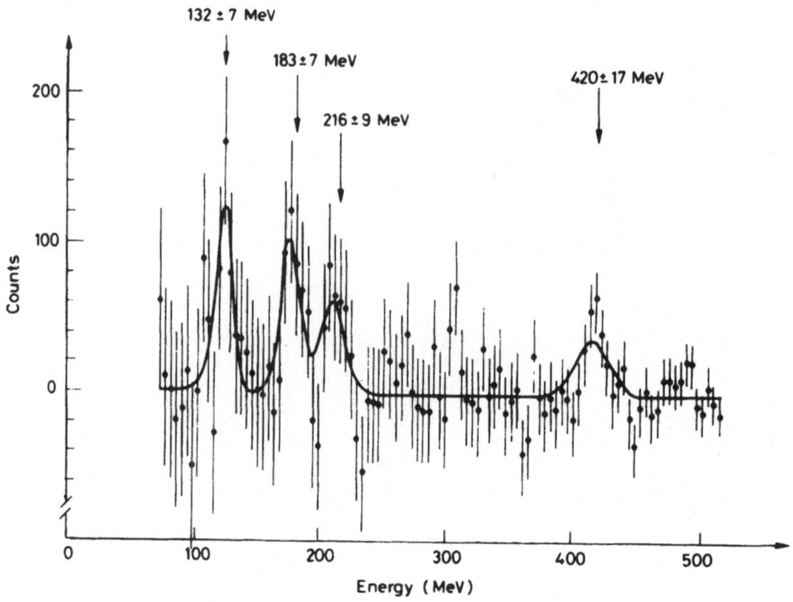

Fig. 10 γ-ray spectrum after background subtraction (ref.14)

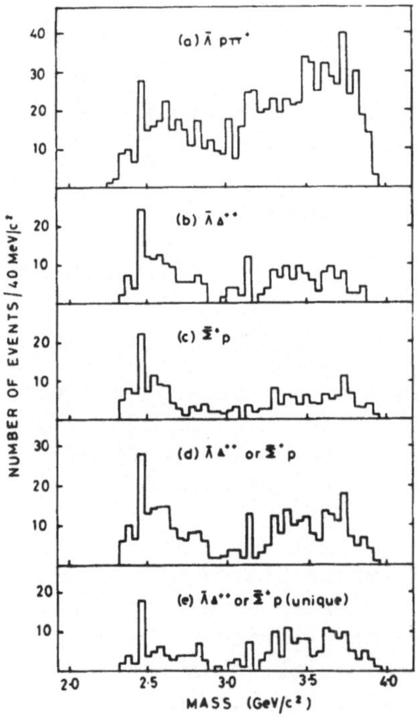

Fig. 11

$\bar{\Lambda}p\pi^+$ mass spectrum:

(a) complete mass spectrum,

(b) for events with $p\pi^+$ in the $\Delta^{++}(1232)$,

(c) for events with $\bar{\Lambda}\pi^+$ in the $\bar{\Sigma}^{++}(1385)$,

(d) for events with $p\pi^+$ in the $\Delta^{++}(1232)$ or $\bar{\Lambda}\pi^+$ in the $\bar{\Sigma}^+(1385)$,

(e) as in (d) for unique fits. (Ref. 16.)

$p\bar{\Sigma}^+(1385)$. A really good candidate for Baryonium if it can be confirmed.

As a summary of the situation, we seem to have related to the $N\bar{N}$ system:

(i) Narrow states below threshold and narrow states just above threshold.
(ii) Further above threshold are broad states which decay into $N\bar{N}$ rather than mesons.
(iii) Some very narrow resonance which are also very heavy.

Unfortunately, from all these states we only know quantum numbers of the states detected in the $p\bar{p} \rightarrow \pi^+\pi^-$ experiments. Finally, we also have one possible candidate for a truly exotic state.

3. A Model

There have been many papers written on models of the proposed mutliquark-antiquark states [6], and in this section, we shall look at some of the ingredients that have gone into some of the models and consider their relevance to experiments given the limited information available at present.

In order to find candidates to describe all these states, notice that qq pairs in baryons are always in a 3^* representation of SU(3) colour. If then we have a diquark-antidiquark formation

$$qq(3^*) \quad - \quad \bar{q}\bar{q}(3)$$

in a colour singlet state with high angular momentum separating the diquark from the antidiquark, we can begin to understand that this meson has the possibility of decaying into $B\bar{B}$ by creating a $q\bar{q}$ pair, whilst the decay into mesons is inhibited by the need to overcome an angular momentum barrier. This is succinctly described by the quark line diagrams, fig. 12.

Now, when we come to a high angular momentum diquark-antidiquark colour singlet state from the configuration

$$qq(6) \quad - \quad \bar{q}\bar{q}(6^*) \quad ,$$

meson decays are again inhibited by an angular momentum barrier, but since the diquark and antidiquark are in 6 and 6^* representations of colour SU(3), we can expect that these states do not couple strongly to $B\bar{B}$ either, and are therefore considerable nar-

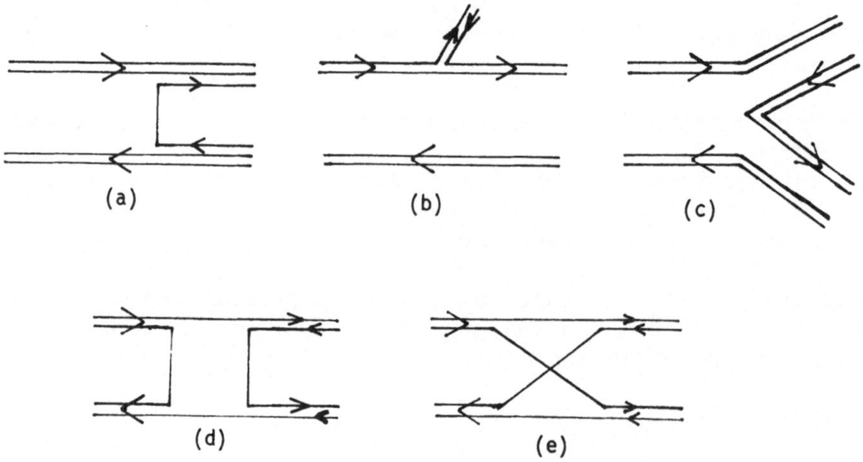

Fig. 12 Allowed - (a), (b) and (c) and Disallowed - (d) and (e), processes involving diquark-antidiquark mesons (represented by the double lines with one arrow head) of the $qq(3^*)$-$\overline{qq}(3)$ type.

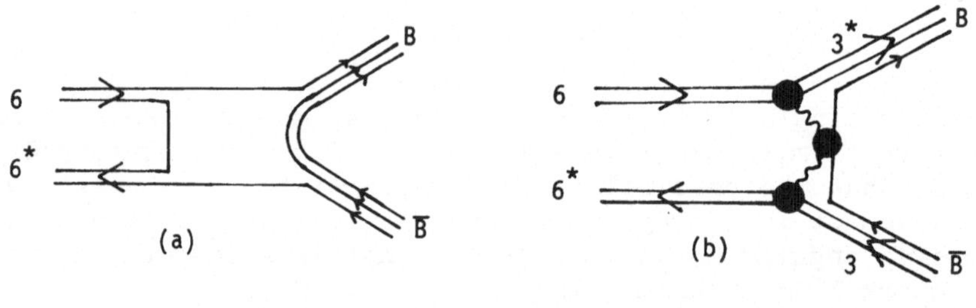

Fig. 13 Further disallowed processes for $qq(6)$ - $\overline{qq}(6^*)$ mesons.

rower (fig. 13). If these states do not couple easily to the $B\bar{B}$ system, it should not be a surprise that the high mass, narrow states that we wish to describe by this system have been found in production experiments and not in formation experiments.

In order to make a first attempt at describing the spectrum of these states we follow to a great extent the Bag model calculations of Jaffe [6.1]. The advantage of the Bag is that it treats states with any number of quarks and antiquarks on the same footing, and has many features that one might expect in any simple phenomenology of states built out of confined constituents. If you like, in its basic form it is a good analogy to the Semi-Empirical-Mass-Formula of nuclear physics since it can describe a broad range of particle states with a few odd exceptations [17]. Naturally there are problems. Solutions of bag equations have been obtained for a spherical bag with relative angular momentum zero [17] and for a highly elongated bag with relative angular momentum between quark clusters at either end tending to inifinity [18] (fig. 14). Unfortunately, the states we are presently interested in fall between these two extremes having low but non-zero angular momentum. We must therefore make a few ruthless extrapolations.

The mass of the ground state [17] (spherical bag) is given by

$$M = \frac{4}{3}\pi R^3 B - \frac{Z_o}{R} + \sum_i \frac{N_i}{R}\left[x^2(m_i R) + (m_i R)^2\right]^{1/2}$$

$$- \frac{\alpha_c}{R} \sum_{a=1}^{8} \sum_{i>j} \vec{\sigma}_i \cdot \vec{\sigma}_j \, \lambda_i^a \lambda_j^a \, M(m_i R, \, m_j R')$$

$$+ \frac{\alpha_c}{R} \sum_a \sum_i \sum_j \frac{\lambda_i^a}{2} \frac{\lambda_j^a}{2} E(m_i R, \, m_j R)$$

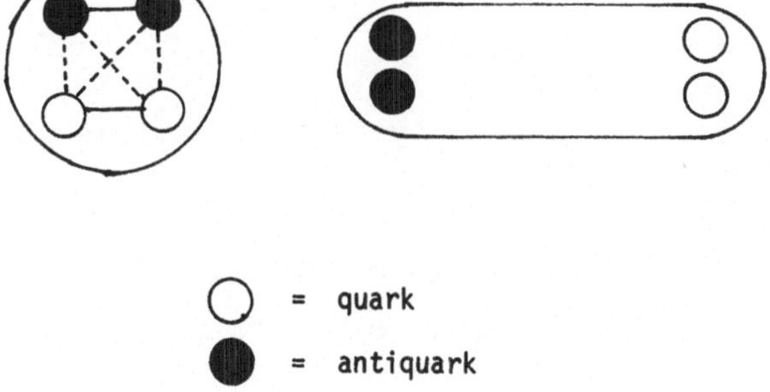

Fig. 14 Spherical and Elongated Bag. The solid lines indicate the magnetic interaction that is taken in determining M_o, the dotted lines are those artificially switched off.

The first three terms are the volume energy, zero point energy and quark kinetic energy (for N quarks of the ith variety of mass m_i) respectively whilst the final two terms are basically a one gluon exchange term (σ and λ are the spin and colour representation matrices involved) together with an extra self energy correction to satisfy boundary conditions. Fortunately, for the nonstrange states we are interested in, only the u and d quarks contribute and the quark mass parameter can be set equal to zero. The final term becomes proportional to the colour SU(3) quadratic Casimir which is zero for colour singlet states and thus drops out of the equation and we are left with three terms providing an average mass to states with the same number of quarks and one term which splits up states with differing spin

$$M = \frac{4}{3}\pi R^3 B + \frac{1}{R}\left[\sum_i N_i x(0) - Z_0 - \alpha_c M(0) \sum_a \sum_{i>j} \vec{\sigma}_i \cdot \vec{\sigma}_j \lambda_i^a \lambda_j^a\right] .$$

With M given as a function of R, the boundary condition determining the equilibrium radius reduces to finding the extremum

$$\frac{\partial M}{\partial R} = 0$$

which implies

$$4\pi R_0^2 B = \frac{1}{R_0^2}\left[\sum_i N_i x(0) - Z_0 - \alpha_c M(0) \sum_a \sum_{i>j} \vec{\sigma}_i \cdot \vec{\sigma}_j \lambda_i^a \lambda_j^a\right]$$

where R_0 is the equilibrium radius, and thus

$$M = \frac{4}{3}\pi R_0^3 (4B) .$$

4B is the effective energy density. The quantities which can be calculated in the bag are

$$x(0) = 2.04$$
$$M(0) = .177$$

whilst the arbitrary parameters

$$B^{1/4} = .146 \text{ GeV}$$

$$Z_o = 1.84$$

$$\alpha_c = .55$$

are determined by fitting the masses of ω, p and Δ.

It would seem then that we could now go ahead and use the above formula to calculate the masses of states containing two quarks and two antiquarks, and so we could. We want however to add an extra ingredient, involving knowledge of the spin splitting term. This term is expected to be a short range interaction as experience with the non-relativistic model in which $\delta^3(\vec{r})$ appears, and the $q\bar{q}$ meson spectrum 6.2) (fig. 15) suggests. At large separation (i.e. high angular momentum) between diquark and antidiquark we expect therefore that the spin splitting term acts only on the quarks in the diquark and antiquarks in the antidiquark but that the interaction between quark and antiquark has completely died out. Since the gluons are flavour singlet, this is the major term that involves a colour changing interaction between the diquark and antidiquark. Thus the hypothesis that a

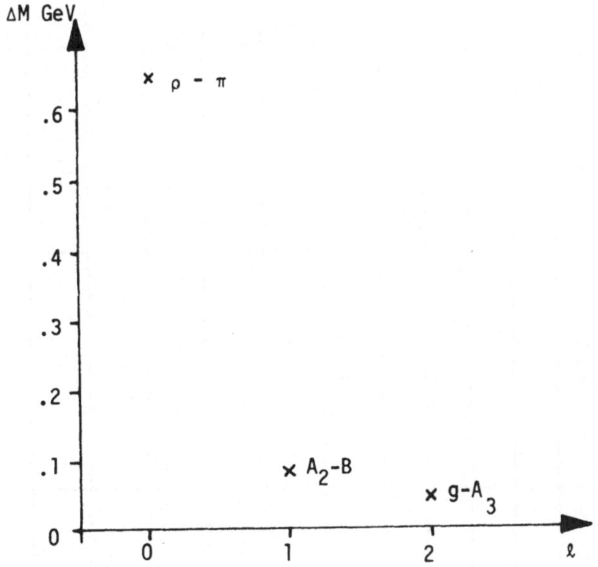

Fig. 15 Mass differences between different spin states as a function of orbital angular momentum between the quark and antiquark.

Table of Diquarks and Diquark-Antidiquark Mesons containing only u and d quarks.

Diquark Name	Representation Labelled by $SU_c(3), SU(2), SU(3)$	Representation Labelled by $SU_c(3), SU(2), SU_I(2)$	Spin	Iso-spin	$-\sigma_1 . \sigma_2 \lambda_1 . \lambda_2$
α	(6, 1, 6)	$(6, 1, 3_I)$	0	1	4
β	(3*, 1, 3*)	$(3^*, 1, 1_I)$	0	0	-8
γ	(6, 3, 3*)	$(6, 3, 1_I)$	1	0	-4/3
δ	(3*, 3, 6)	$(3^*, 3, 3_I)$	1	1	8/3

Mesons	Spin	Isospin	$(-\sigma_1 . \sigma_2 \lambda_1 . \lambda_2)$ Diquark and Antidiquark	M_o^2 GeV^2
$\beta\bar{\beta}$	0	0	-16	1.4
$\frac{1}{\sqrt{2}}(\beta\bar{\delta} \pm \delta\bar{\beta})$	1	1	-16/3	1.9
$\delta\bar{\delta}$	0, 1, 2	0, 1, 2	16/3	2.4
$\alpha\bar{\alpha}$	0	0, 1, 2	8	2.6
$\frac{1}{\sqrt{2}}(\alpha\bar{\gamma} \pm \gamma\bar{\alpha})$	1	1	8/3	2.3
$\gamma\bar{\gamma}$	0, 1, 2	0	-8/3	2.0

qq(6) - $\overline{qq}$(6*) with high angular momentum does not flip to qq(3*) - ($\overline{qq}$(3) is intimately related to the short range nature of the spin splitting term. At this stage we define a modified ground state mass equation in which the quark-antiquark spin splitting interaction is switched off and treat this as the limiting case of the situation that appears at high angular momentum. Using the formula from Appendix 2, we get a simple spin splitting interaction for the diquark-antiquark configuration

$$-\frac{\alpha_c M(0)}{R}\{\Sigma_a \vec{\sigma}_1\cdot\vec{\sigma}_2\lambda_1^a\lambda_2^a)_{quarks} + (\vec{\sigma}_1\cdot\vec{\sigma}_2\lambda_1\lambda_2)_{antiquarks}\}$$

$$= \frac{-4\alpha_c M(0)}{R}\{(C^2(\text{diquark} - 2C^2(q))(C^3(\text{diquark}) - 2C^3(q))$$

$$+ (C^2(\text{antiquark}) - 2C^2(\bar{q}))(C^3(\text{antidiquark}) - 2C^3(\bar{q}))\}$$

which leads to the masses of pure (in the sense that the Mass Equation is diagonal in diquark-antidiquark colour labels) meson states with zero angular momentum, (see table).

At high angular momentum, we have from Johnson and Thorn [18] the result that trajectories are linear and the slope is given by (fig. 14)

$$(\alpha')^{-1} = 8\pi\sqrt{2\pi\alpha_c BC^3}$$

where C^3 is the quadratic Casimir of the colour SU(3) representation of the ensemble of quarks at either end of the elongated bag, in this case the diquark or antidiquark. Since the parameters α_c and B are fixed, all states of colour 3 at one end and 3* at the other have the same slope - hence the universal slope for $q\bar{q}$ mesons and baryons. Thus, qq(3*) - $\overline{qq}$(3) has the same slope α'_3 = .9 GeV^{-2}, qq(6) - $\overline{qq}$(6*) has α'_6 = .57 GeV^{-2}, whilst α'_8 = .6 GeV^{-2} and bigger representations have flatter and flatter trajectories as shown in fig. 16. Combining angular momentum with spin we have a series of trajectories of total spin J as a function of M^2.

Although the high angular momentum bag is supposed to be an asymptotic calculation, we now take the drastic step of assuming the trajectories remain linear, as they do for ordinary mesons and baryons, down to zero angular momentum and determine the intercept by using the value of the masses of the pure states calculated in the spherical bag, i.e. we write

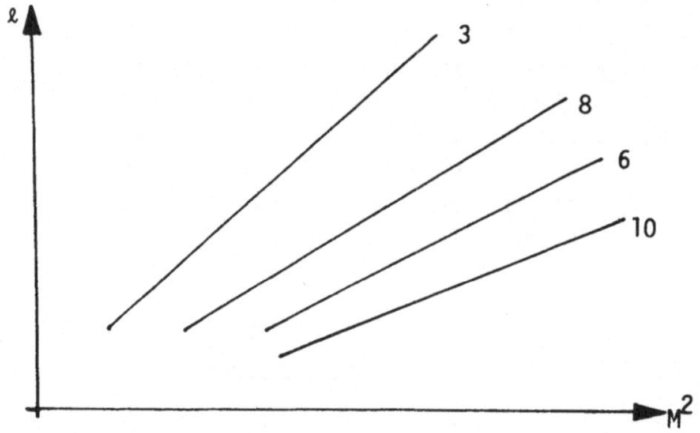

Fig. 16

Comparison of slopes of linear trajectories for various quark cluster representations.

$$\ell = \alpha_c' M^2 + \alpha_o$$

and

$$\alpha_o = -\alpha_c' M_o^2$$

where α_c' is given for each colour label, and M_o^2 is given from the table. This linear extrapolation should not be expected to work for angular momentum zero where mixing of pure states occurs through quark-antiquark interactions which are on an equal footing with the quark-quark interactions (the full interaction can in fact be handled by the mass formula [19]).

Even in the region say $\ell = 1$ or 2, where we expect the spin splitting interaction to become less important as it does for ordinary $q\bar{q}$ mesons, some mixing is probably going to occur. Similarly we should also expect some spin orbit and tensor force effects which we have not allowed for but which are known to appear in the non-relativistic one gluon exchange potential [20] and although short range, are not as short range as the spin-spin interaction. The general conclusion to be drawn is that the linearised trajectories are probably good at high angular momentum but that curvature undoubtedly sets in as $\ell = 0$ is approached.

We have then a large number of states available even in this restricted realm containing only the u and d quarks. Nevertheless several valiant attempts have been made to fit the experimental states [6.1,2,3]. Two particularly simple examples are worthy of mentioning. Firstly the state 2.95 GeV, which has yet to be confirmed, has been detected with pion cascade decays into states of mass 2.62 GeV and 2.2 GeV. Simply assuming that the pion carries away one unit of spin between each state puts all three states on a nice linear trajectory of slope identical to the $qq(6) - \bar{q}\bar{q}(6^*)$

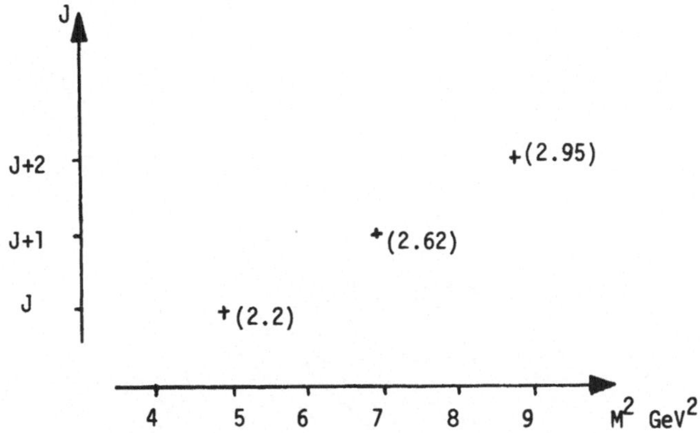

Fig. 17 Position of 2.95 Resonance and two of its cascade products assuming they differ by 1 unit of J.

configuration (fig. 17) as we might hope.

Secondly, some interesting information can be obtained on the states seen in $B\bar{B}$ formation if the $^3P_0$ model of quark pair creation is assumed to hold [21]. In particular, this model has the $q\bar{q}$ pair created in fig. 12a carrying the quantum numbers of the vacuum and having no effect on the existing quarks and antiquarks. This naturally leads to the fact that $qq(6)$ - $\bar{q}\bar{q}(6^*)$ mesons have no coupling to $B\bar{B}$ states. Using the coupling scheme of fig. 18 [6.1] leads to the system of trajectories being specified by both the angular momentum in the $B\bar{B}$ system and the $qq(3^*)$ - $\bar{q}\bar{q}(3)$ meson angular momentum. Of all the states, those analysed in $N\bar{N} \to \pi^+\pi^-$ namely T, U and V are well identified. Furthermore, the analysis indicates that $J = L - 1$, where L is the angular mo-

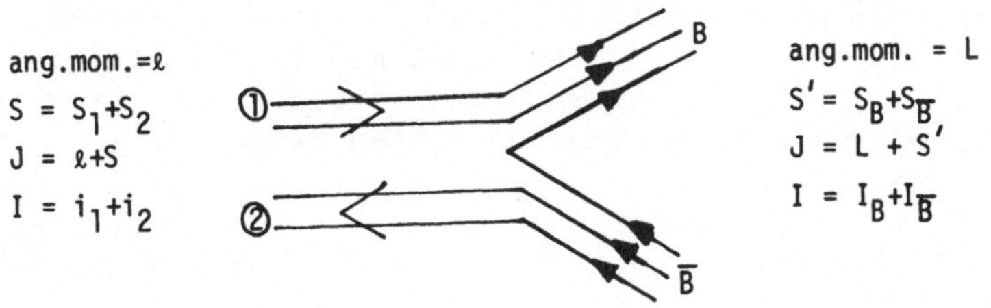

Fig. 18 Coupling of diquark 1 and antidiquark 2 to $B\bar{B}$ in $^3P_0$ model.

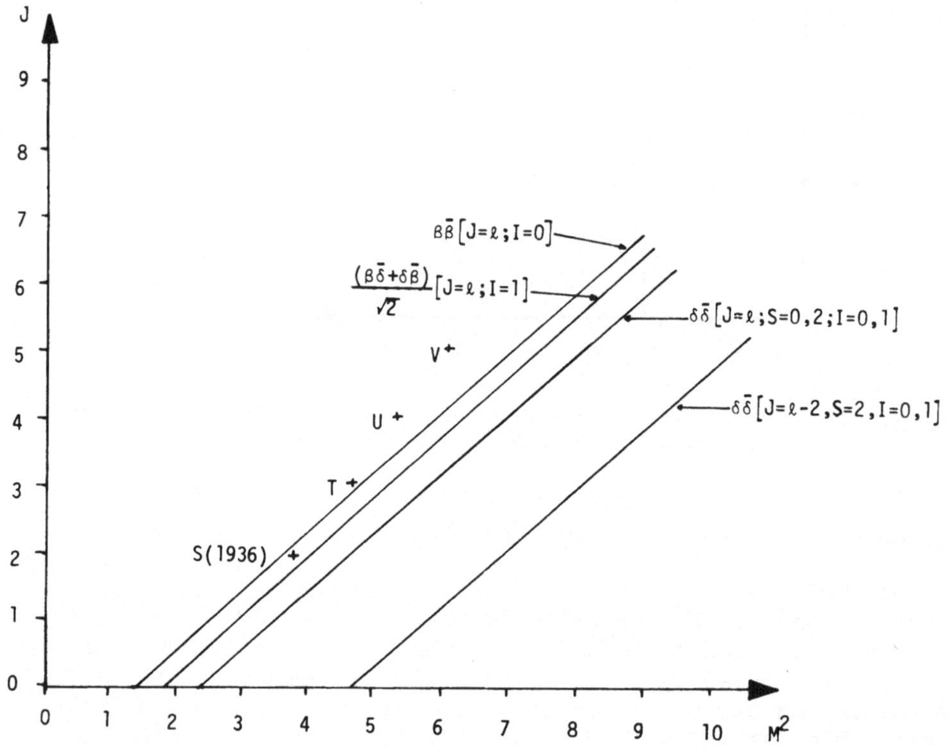

Fig. 19 Trajectories coupled to the J = L − 1 channel of the $B\bar{B}$ system. T, U and V measured in $p\bar{p} \to \pi^+\pi^-$. Favoured spin of S is J = 2.

mentum of $N\bar{N}$ pair, is favoured and it can be seen from fig. 19 that the theory is in reasonable agreement with experiment! It should perhaps be noted in passing that although these resonances fall pretty well on a straight line, the slope is greater than the maximum slope allowed by the quark model calculations, and we would prefer this to be accidental, an explanation lying in one of the many approximations that have been made.

Indeed, one may believe the bag is too restrictive and, instead of applying the bag philosophy to the full, simply abstract successful components such as the spin splitting term and the proportionality $\alpha' \sim (C^3)^{-1/2}$. The absolute mass scale of the multiquark states is then determined by fitting one of the experimental points.

Finally, a word about decay rates. The bag model which seems quite adequate for describing the spectrum of these diquark-antidiquark mesons and can generally treat $q^n\bar{q}^m$ states along the same lines, providing we are aware of the assumptions and limitations of the linear extrapolation, is very limited when it comes to describing decay rates. At present therefore we must be content with at least having the qualitative explanation of why one type of meson easily decays into Baryon-Antibaryon whilst the other doesn't and why both have small probability of decaying into mesons, whilst the most likely decay of cascading, which is allowed by the quark line rules, depends on the interplay between the amount of angular momentum lost and the available phase space for the decay to proceed.

Conclusion

Within the framework of Q.C.D. there exists the possibility that there are many particles or resonances to be found with fundamentally different constituent structure to the familiar mesons and baryons. At present there is no reason why quarks with unusual colour cannot exist and form a new basis for constructing particles. Similarly, the colour vector gluons that play an increasingly important role as the mediators of the strong interaction could equally well be confined constituents in their own right leading to the possibility of flavour singlet glueballs.

Perhaps more topical with the new experimental information that we have briefly reviewed here, is the existence of multiquark states.

As we have tried to indicate in the short time available, a major part is played by angular momentum separating clusters of quarks and/or antiquarks, themselves in relatively stable (i.e. s-wave) configurations. The colour configuration of the clusters in high angular momentum states is expected to remain constant and dominate the decay characteristics since the angular momentum barrier effectively damps out the possibility of meson decays. At low angular momentum some small mixing becomes important whilst at zero angular momentum, interaction between clusters is as important as that within clusters and the description in terms of separate clusters is really no longer valid. At this point, with no angular momentum barrier to hold them apart, quarks and antiquarks readily combine and one has "fall apart" or superallowed decays. Although we have not had time to enter into it here, the combination of two quarks and two antiquarks with zero angular momentum has been used by Jaffe [19)] to describe the low lying $J^{PC} = 0^{++}$ states. Perhaps one of the most important points to mention in this respect is that the spin splitting force actually forces the octet structure of fig. 2 to be lower than the other multiplets, and thus overcomes the old objections of low lying exotic states being in the region of 1 GeV.

Although we have treated, in the main, the properties of diquark-antidiquark mesons, the techniques are completely generalisable and I hope by now all the papers under ref. 6 are understandable.

We are just scraping the tip of the iceberg as far as exotic quark states are concerned. Present experiments seem to require some of the new ideas that we have looked at, and further results in this direction are sure to be a major testing ground for the extended quark model of multiquark states. Where should we see glueballs and "exotic quark" states? Have we already seen them? Perhaps they don't exist. If not why not?

APPENDIX 1

The Young diagram technique [23] is particularly useful for obtaining the decomposition of products of two representations of SU(N) into their irreducible components. Each representation is described by a pattern of boxes that reproduces the symmetry structure of the representation in terms of its multispinor labels. The basic spinor representation is given by a single box 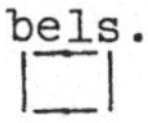 and all other representations are built up from this. In fact, there is a theorem which states that the symmetry classification is complete i.e. if we specify the symmetry of the multispinor labels uniquely then it is an irreducible representation of SU(N). A multispinor with m-indices is represented by m boxes which form a pattern, the number of boxes in the nth row ≤ the number of boxes in the $(n - 1)^{th}$ row.

Eg.

Diagram	Description	Dimension
	Basic Representation	N
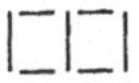	Symmetric 2-component spinor	$N(N + 1)/2$
	Symmetric 3-component spinor	$N(N + 1)(N + 2)/3!$
	Antisymmetric 2-spinor	$N(N - 1)/2$
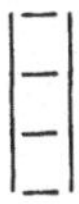	Antisymmetric 3-spinor	$N(N - 1)(N - 2)/3!$
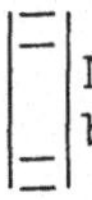 N - 1 boxes	Antisymmetric (N - 1) spinor	N
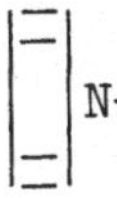 N-boxes	Antisymmetric N-spinor	1

Obviously, since the basic spinor has N independent components, a column with greater than N boxes vanishes.

When we come to representations with mixed symmetry the dimension formula is somewhat more complicated but can be simply obtained by the following prescription (see fig. A.1).

N	N+1	N+2	N+3	N+4
N-1	N	N+1	N+2	
N-2	N-1	N	N+1	
N-3	N-2			
N-4				
N-5				

10	7	5	4	1
8	5	3	2	
7	4	2	1	
4	1			
2				
1				

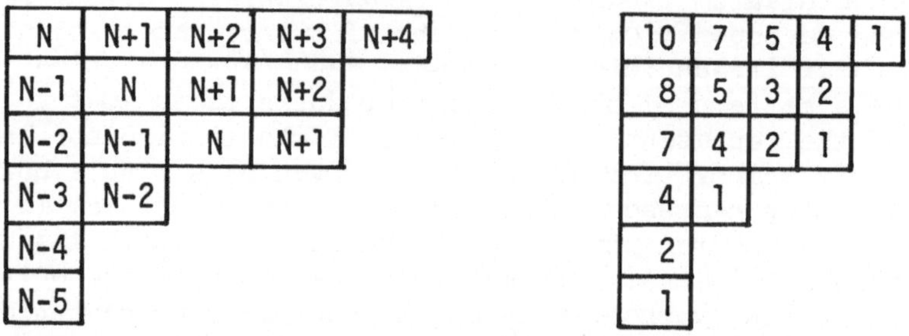

Fig. A.1   Example of numbering boxes for the
Hook Rule.

$$D_N = \frac{\boxed{\begin{array}{|c|c|c|}\hline N & N+1 & N+2 \\\hline N-1 & N \\\cline{1-2} N-2 \\\cline{1-1}\end{array}}}{\boxed{\begin{array}{|c|c|c|}\hline 5 & 3 & 1 \\\hline 3 & 1 \\\cline{1-2} 1 \\\cline{1-1}\end{array}}}$$

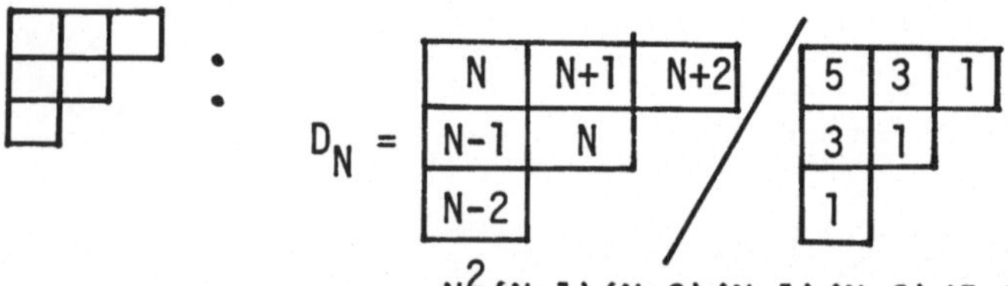

$$= N^2(N+1)(N+2)(N-1)(N-2)/5.3^2$$

for SU(3) $D_3 = 8$ and for SU(4) $D_4 = 64$

Fig. A.2   Example of the Hook Rule.

(1)   (2)

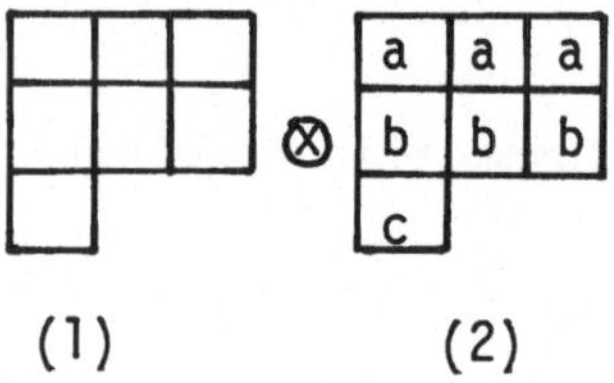

Fig. A.3   Starting point for multiplying Young Tableaux.

	a	a	b
	b	c	
a			

$\sim$ baacba

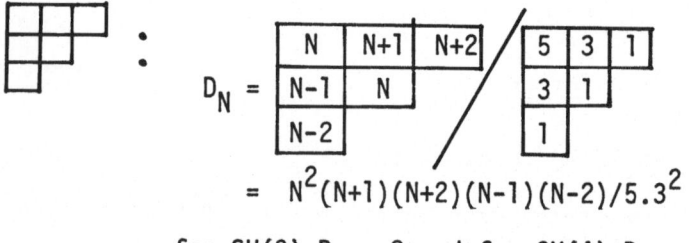

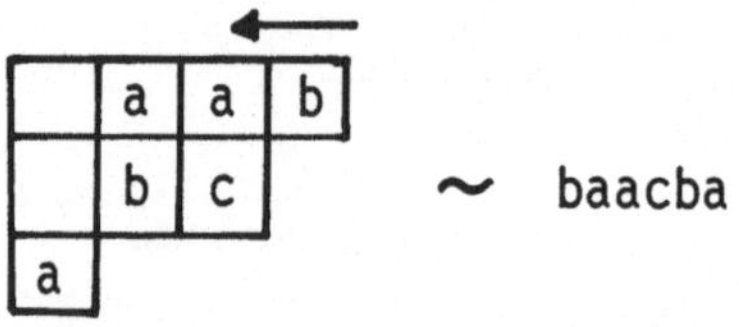

Fig. A.4   Letter sequence associated with the
Young Tableau.

Assign two sets of numbers to each box,

(i) Starting from the top left box write N, and increase by 1 for each box in the first row. In the second box of the first column write N - 1 and then increase by 1 along the second row etc.

(ii) Starting from the far right, move along a particular row to a chosen box, and then move down to the end of the column containing this box. Label the chosen box by the number of boxes through which you have passed on your journey. Repeat for all boxes.

The dimension is now given by the product of the first set of numbers divided by the product of the second set of numbers (see example fig. A.2). This technique, going under the name of "the Hook Rule" compares favourably with having to memorise the usual formula [24].

The product of two Young diagrams follows the rules:

(i) In one of the diagrams put an a in each box of the first row, b in each box in the second row etc. (fig. A.3).

(ii) Enlarge (i) by adding $\boxed{a}$'s from (2) in all possible ways subject to:

(a) the diagram remaining regular (i.e. $n_1 \geq n_2 \geq n_3$ ..)

(b) no two $\boxed{a}$'s in same column

(c) maximum number of rows is N .

(iii) Repeat (ii) with $\boxed{b}$'s and $\boxed{c}$'s and in the final diagram scanning from right to left starting on the top row and continuing down, arrange the letters in one sequence (fig. A.4). Then the only allowed diagrams are those which form a lattice permutation i.e. up to any point in the sequence, the number of a's $\geq$ numbers of b's $\geq$ number of c' $\geq$ ... (the example fig. A.4 is not a lattice permutation whereas abacab is). One final point to note is that a column of N boxes can be dropped from the final diagram since such a column is just a singlet. As can be seen from the example in fig. A.2 for the case of SU(3).

APPENDIX 2

The Lie algebra of SU(N) is specified by taking the set of $(N^2 - 1)$ tracelesse hermitean N × N matrixes g_i with product law and completeness expressed by

$$g_i g_j = (d_{ijk} + i f_{ijk}) g_k + \frac{2}{N} \delta_{ij} \quad \text{sum over repeated indices}$$

$$(g_i)^L_K (g_i)^P_R = 2(\delta^P_K \, \delta^L_R - \frac{1}{N} \delta^P_R \, \delta^L_N) \, .$$

SU(N) has (N - 1) Casimirs, we are however interested only in the quadratic Casimir for the purposes of the lectures. The latter equation gives us directly the quadratic Casimir in the basic representation

$$\delta^P_K C^N(N) = \frac{1}{4}(g_i g_i)^P_K = \frac{(N^2 - 1)}{2N}\delta^P_K$$

where the label N in parenthesis indicates the dimension of the representation. Generally in the representation $D_N(n_1, n_2 \ldots n_N)$ where n_i labels the number of boxes in the i-th row of a Young tableaux [25],

$$C^N(D_N) = \frac{1}{2}\left[\sum_{i=1}^{N} n_i(n_i + N + 1 - 2i) - \frac{1}{N}\left(\sum_{i=1}^{N} n_i\right)^2\right]$$

Of particular importances is the relationship between outer products of different generators (as we meet in the one gluon exchange process, or basic projection operator techniques [26]) and the Casimir operator. For two representations $F^{(1)}_i$ and $F^{(2)}_i$ we can use the total Casimir to write

$$C^N(\text{Total}) = \sum_i (F^{(1)}_i + F^{(2)}_i)^2 = C^N(\text{Rep.}1) + C^N(\text{Rep.}2) + 2F^{(1)}_i F^{(2)}_i$$

e.g. N = 3, (1) and (2) are both in the basic representation then

$$3 \times 3 = 3^* + 6$$

$$C^3(3) = C^3(3^*) = \frac{4}{3}, \quad C^3(6) = \frac{10}{3}$$

$$F^{(1)}_i \cdot F^{(2)}_i = \frac{1}{4}\lambda^i_1 \lambda^i_2 = -\frac{2}{3} \text{ in the } 3^* \text{ representation}$$

$$\frac{1}{3} \text{ in the 6 representation.}$$

These results are trivially generalisable to the higher groups and for any number of representations.

REFERENCES

1. R.H. Dalitz, Proceedings of the XIII International Conference on High Energy Physics, University of California Press (1967). Reprinted in J.J. Kokkedee, The Quark Model, W.A. Benjamin (1969).

2. H. Fritzsch, Proceedings of the XVIII Internationale Universitätswochen für Kernphysik, Schladming (1978)

3. O. Nachtmann, These proceedings

4. D. Robson, Nucl.Phys. B130, 328 (1977)

5. E. Ma, Phys.Lett. 58B, 442 (1975); Phys.Rev.Lett. 36, 1573 (1976)
F. Wilczek and A. Zee, Phys.Rev. D16, 860 (1977)
Y.J. Ng and S.H.H. Tye, Phys.Rev.Lett. 41, 6 (1978)

6. History
D. Lichtenberg, Phys.Rev. 178, 2197 (1969)
P.H. Dondi, Karlsruhe Report 1975: Multi-Quark-Antiquark Bound States and the New Resonances
J. Iizuka, Prog.Theor.Phys. 54, 1178 (1975)
M.I, Pavković, Phys.Rev. D14, 3186 (1976)

$qq\bar{q}\bar{q}$

6.1 R.L. Jaffe, Phys.Rev. D17, 1444 (1977)

6.2 Chan Hong-Mo and H. Høgaasen, Nucl.Phys. B136, 401 (1978); Phys.Lett. 72B, 121 and 400 (1977)

6.3 Chan Hong-Mo, Proceedings of the IV European Antiproton Symposium 1978, TH.2540 - CERN
S.T. Tsou, Nucl.Phys. 141, 397 (1978)
A.W. Hendry and I. Hinchcliffe, Preprint LBL 7597
S. Ono, Preprint University of California at Davis

Other multiquark states

R.L. Jaffe, Phys.Ref.Lett. 38, 195 (1977)
P.J.G. Mulders, A.T.M. Aerts and J.J. de Swart, Phys.Rev.Lett. 40, 1543 (1978); Phys.Rev. D17, 260 (1978)
M. Fukugita, K. Konishi and T.H. Hanson, Phys.Lett. 74B, 261 (1978)
B.G. Wybourne, Austr.J.Phys. 31, 117 (1978)
Chan Hong-Mo et al., Phys.Lett. 76B, 634 (1978)
H. Høgaasen and P. Sorba, Preprint TH-2500 - CERN
M. De Crombrugghe, H. Høgaasen and P. Sorba, Preprint TH-2537 CERN

7. D. Horn and J. Mandula, Phys.Rev. D17, 898 (1978)

8. Proc.Roy.Soc., 318, 243; and J.L. Rosner, Phys.Rep. 11C, 189 (1974)

9. General Review. L. Montanet, V International Conference on Experimental Meson Spectroscopy 1977, CERN/EP/PHYS-77-22

10. M. Coupland et al., Phys.Lett. 71B, 460 (1977). See especially fig. 2.

11. A.A. Carter, Phys.Lett. 76B, 634 (1978)

12. C. Evangelista et al., CERN Preprint (1978)

13. G. Flügge, XIX International Conference on High Energy Physics Physics, Tokyo (1978)

14. P. Pavlopoulos, Phys.Lett. B72, 415 (1978)

15. L. Gray et al., Phys.Rev.Lett. 26, 1491 (1971)

16. T.A. Armstrong et al., Phys.Lett. D77, 447 (1978)

17. T. De Grand, R.L. Jaffe, K. Johnson and J. Kiskis, Phys.Rev. D12, 2060 (1975)

18. K. Johnson and C.B. Thorn, Phys.Rev. D13, 1934 (1976)

19. R.L. Jaffe, Phys.Rev. D15, 267 and 281 (1977)

20. A. De Rujula, H. Georgi and S.L. Glashow, Phys.Rev. D12, 147 (1975)

21. See J.L. Rosner, ref. 8

22. A.A. Carter et al., Phys.Lett. 67B, 117 (1977)

23. M. Hamermesh, Group Theory, Addision Wesley (1962)

24. A. Pais, Rev.Mod.Phys. 38, 215 (1966)

25. B.G. Wybourne, Classical Groups for Physicists, John Wiley & Sons (1974)

26. S. Gasiorowicz, Elementary Particle Physics, John Wiley & Sons (1967)

WEAK NEUTRAL CURRENTS UNVEILED

Helmut Faissner

III. Physik. Inst. Tech. Hochschule Aachen

Aachen, Germany

À GARGAMELLE, l'heroine des courants neutres.

Abstract

The history of Weak Neutral Currents is briefly reviewed. Neutrino-electron scattering is treated in some detail: the data about three channels are now mutually consistent, and fix the weak neutral electron current to either pure axial vector A, or to vector current V. Parity violation in high energy ed scattering singles out the A-solution of the Salam-Weinberg type. A corresponding analysis of the space-time structure of the quark weak neutral currents reveals almost pure V - A. Assessing the isospin structure wants data from exclusive or semi-inclusive reactions, like elastic neutrino-proton scattering or single-pion production. The result is essentially unique, and leaves the standard Salam-Weinberg $SU_2 \times U_1$ model, with a mixing parameter $\sin^2\theta_W = 0.25\pm0.03$ as the sole possibility to describe weak neutral currents.

1. The Rise of Neutral Currents

The history of Neutral Currents is short. Since 1971 the big heavy liquid bubble chamber GARGAMELLE had been exposed to the neutrino and antineutrino beams from the CERN proton-synchrotron (PS). All during 1972 the GARGAMELLE Collaboration of

Aachen, Bruxelles, CERN, Ecole Polytechnique (Paris),
Milano, Orsay, and University College London

were struggling to see if partons had the properties of quarks [1]. Neutral currents belonged to the unpleasant duties, one was supposed to take care of in one's spare time. The rumors that some theoreticians would like to see them existing, added only an air of bitter scepticism.

But around Christmas 1972 the Aachen Group were lucky enough to find in their batch of antineutrino film an isolated electron of moderate energy, which had been emitted within the accuracy of the measurement, exactly in beam direction (Fig. 1). This is of course precisely what one expects for the recoil from a high-energy neutrino-electron scattering. And if the neutrino has been of the muon-neutrino variety, indeed, the interaction between $\overset{(-)}{\nu}_\mu$ and e had to go through an intermediate neutral current. Therefore, a single unambiguous $\overset{(-)}{\nu}_\mu e$ event suffices to establish the existence of this new force of Nature with certainty*).

---

*) The situation is not too different from the discovery of the Ω^-, although the single e-signature is simpler. Probably for that reason the notorious Hamburg magazine "Der Spiegel" called the Aachen electron the "event of the century". (Characteristically, the "Spiegel" did not mention the discoverers, confused the relevant theoreticians, and printed, in place of Fig. 1, a completely irrelevant, messy ν_e background photo ...)

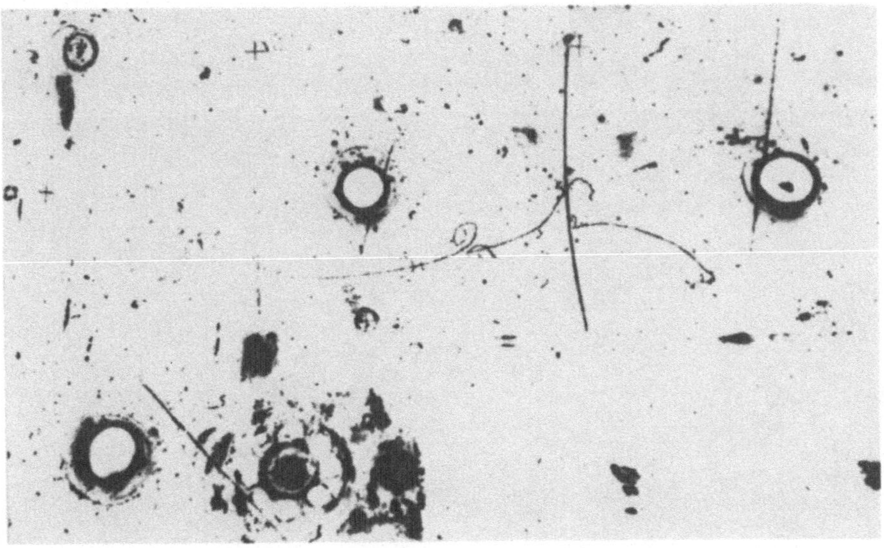

Fig. 1: The Aachen $\bar{\nu}_\mu e$ scattering event - the first evidence for neutral currents. (Neutrinos from the left)

Of course it must be proven that the event in question cannot be explained otherwise, for instance by the < 1% ν_e ($\bar{\nu}_e$) contamination in the beam, or by asymmetric e^-e^+ pairs from converted photons. A few months of discussions and calibration measurements convinced us that these backgrounds have, at most, a chance of 3% to simulate the event observed. Consequently it was presented to the Bonn Electron-Photon Symposium, in August 1973 [2,3], and shortly afterwards published [4].

In the meantime, the Orsay-Group under the inspiring leadership of the late André Lagarrigue, got evidence that some of the inelastic muon-less hadronic events were in fact neutrino-induced. This was an enormous breakthrough, considering that such messy events, like the one in Fig. 2, had already been noticed in the 1963/64 CERN neutrino bubble chamber [5] and sparkchamber [6]

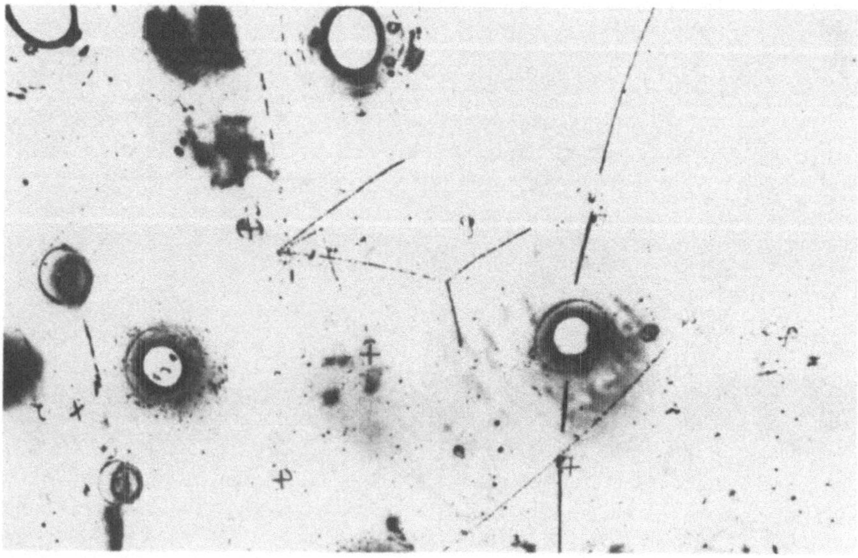

Fig. 2: Muon-less hadron star, as observed in GARGAMELLE.

experiments. (Having seen these photos myself, and too lightly dismissed them as neutron stars, I cannot blame this slip on unfit instrumentation: Our old sparkchambers were well suited for the interaction analysis [7]. That we did not believe what we saw, was an unfortunate conspiration of mental blocking, by theoretical prejudice, and experimental mischief.) The separation of neutron and neutrino induced muonless events took the GARGAMELLE (GGM) Collaboration many months of hard work. It implied the analysis of every track in every event, the meticulous analysis of the spatial and energy distributions [8], experimental studies of neutron-induced events, lengthy Monte Carlo calculations, and hadronic cascade studies initiated in GGM by protons of known energy [9]. Only then the GGM-Group published their evidence for hadronic weak neutral currents [3,8] alongside with their electron.

2. The Theoretical Significance of Neutral Currents

When this happened at the Bonn 1973 Conference [3], the Physics Community was actually eagerly waiting for the weak neutral currents to appear. Almost unnoticed for many years, an old, attractive idea had been gaining ground: that the weak and the electromagnetic interactions are one and the same, and that the weak interaction appears only weak at low energies, because its field quanta are so massive. There are many roots to such a

Unified Gauge Theory:

the idea of unity of forces [10-13], the important question of renormalizability [12-14], and at last the hope of generalizing such a theory eventually to all interactions ...

From all these fascinating aspects we need only consider the analogy between weak and electromagnetic interactions: The latter is known to act between neutral vector currents V_α, by exchange of a mass-less field-quantum, the photon γ. A one-dimensional gauge group insures the universality and conservation of the coupling constant e, the charge. The pre-1973 weak interaction, instead, was short-ranged and acted between charged, left-handed currents $(V - A)_\alpha$, and this called for massive, charged intermediate vector bosons $W^\pm$. Enlarging the e.m. gauge group in the simplest possible way leads to a group

$$G = SU(2) \times U(1) \quad , \qquad (1)$$

where the (mathematical) fields of force are the generators of the group, namely

(W^+, W^o, W^-) for SU(2) - "weak isovector"

and

B^o for U(1) - "weak isoscalar" .

The weak boson triplet $\underline{W}$ plays the same rôle for weak isospin as e.g. the pion does for strong isospin. The appearance of a neutral intermediate vector boson amongst the weak force fields is a mathematical necessity for the group (1). It is not so obvious, though, which neutral particle is going to couple to which current. If G is broken, the physically relevant combinations are, as first realized by Glashow [10] and taken up by Salam [11,12] and Weinberg [13]

$$\begin{aligned} Z^o &= \cos\theta_w \, W^o + \sin\theta_w \, B^o \quad , \\ \gamma &= -\sin\theta_w \, W^o + \cos\theta_w \, B^o \quad . \end{aligned} \tag{2}$$

Z^o is the massive (= weak) neutral current quantum, γ the familiar, massless photon.

This mixing of fields implies a correspondingly mixed weak neutral current [15]:

$$J_\alpha^{NC} = I_3 \, (V - A)_\alpha - 2Q \sin^2\theta_w \, V_\alpha^{em} \quad , \tag{3}$$

where I_3 and Q are, respectively, the 3rd component of weak isospin and the charge of the particle in question. θ_w is the weak mixing angle (as before) introduced by Glashow [10], and now generally named after Salam [11,12] and Weinberg [13]. As one sees, the mixing parameter $x_w = \sin^2\theta_w$ measures the admixture of ambidextrous e.m. current to left-handed weak current, and determines thus also the ratio of left-handed (V - A) to right-handed (V + A) current pieces. Note that the result of this mixing does depend on the type of particle (via the quantum numbers I_3 and Q).

According to the spirit of unification the effective Fermi coupling constant G is essentially e^2/M_w^2, in the simplest models:

$$G/\sqrt{2} = e^2 / 8 M_w^2 \sin^2\theta_w \quad .\qquad (4)$$

Similarly, the Glashow-Salam-Weinberg $SU_2 \times U_1$ model gives for the coupling strength of the _neutral_ weak _current_ in general:

$$G^{NC}/\sqrt{2} = \kappa e^2 / 8 M_Z^2 \sin^2\theta_w \cos^2\theta_w \quad , \qquad (5)$$

whereby in the _minimal model_ of Salam and Weinberg [12,13]:

S-W: $M_w = \cos\theta_w M_Z$ and $\kappa = 1$ i.e. $G^{NC} = G.$ (6)

Glashow [10] left the overall normalization κ of neutral and charged coupling constants open. It is clear that one can build a large variety of models, by employing larger groups G, thereby introducing more than one massive intermediate neutral boson Z, several mixing angles θ_ν and so on. A recent count revealed more than 500 proposed models.

Besides, there is the question how one should group the known elementary particles, i.e. quarks and leptons, into multiplets of G. We shall follow the simple prescription of Glashow, Iliopoulos and Maiani [16], and place all left-handers into doublets, all right-handers into singlets.

It is a remarkable achievement of experimental physics, that the structure of weak neutral currents has been widely cleared up within five years. It is just the minimal model of Salam and Weinberg [12,13]. In the following I shall briefly review the experimental data which lead to this conclusion, starting (sec. 3) with neutrino-electron scattering, and adding the beautiful γ-Z^o-interference experiment at SLAC. Then I shall pass over to inclusive (sec. 4) and semi-inclusive (sec. 5) hadronic neutrino interactions, and end with some exclusive ν-reactions (sec. 5 and 6) with nucleons. At the end of each section I shall summarize, what kind of knowledge it has given. There is not much time to dwell with experimental details. But occasionally I shall hint at difficulties which might grow into worries ...

3. Neutrino-Electron Scattering and the Weak Neutral Current of the Electron

As remarked already, a muon-(anti-)neutrino can scatter off an electron only by virtue of a neutral current - to first order, that is, and to the extent that muon-number is conserved. And unlike the many other processes of the neutral current type, neutrino-electron scattering has the advantage of being predictable without any theoretical ambiguity or uncertainty, since the two particles involved are point-like and have no strong interactions. This makes the study of the two reactions

$$\nu_\mu e \to \nu_\mu e \quad (7) \qquad \text{and} \qquad \bar{\nu}_\mu e \to \bar{\nu}_\mu e \quad (\bar{7})$$

an experimentum crucis of weak interaction physics.

Its cross section is easily derived: since a neutrino is left-handed, its interaction (A) with a left-handed electron current proceeds through a spin singlet state. Clearly, the distribution of electron center-of-mass (c.m.) recoil angles θ_e^* is isotropic, yielding a flat electron energy distribution in the lab. (B) Similarly, a neutrino encounter with a right-handed electron leads to a spin triplet state, an anisotropic angular distribution in the c.m., and a lab electron energy distribution which peaks at low energies. In terms of energy transfers

$$y = T_e/E_\nu \approx E_e/E_\nu \quad (8)$$

the distributions become:

(A) Singlet[+] $S = 0 : W(\theta^*) \sim 1 \rightarrow W(y) \sim 1$ (9_1)

(B) Triplet[+] $S = 1 : W(\theta^*) \sim (1 - \cos\theta^*)^2 \rightarrow W(y) \sim (1 - y)^2$ (9_3)

In deriving these formulae we have excluded spin-flip, i.e. the "bad" Dirac variants S, P, T.

If we denote by A and B, respectively, the (relative) probability of the weak neutral electron current to be left-handed (V - A) or right-handed (V + A), the differential cross section becomes (for $E_e >> m_e$):

$$\frac{d\sigma}{dy} = \kappa^2 G^2 \frac{2}{\pi} m_e E_\nu \left\{A + B(1 - y)^2\right\} . \tag{10}$$

The appearance of the (general) weak neutral current coupling constant is clear; the rest is phase-space.

Antineutrinos differ from neutrinos <u>only</u> by being right-handed. (This is called a Majorana neutrino theory [17]). Thus the same expression holds also for them, but with A and B interchanged:

$$\frac{d\bar{\sigma}}{dy} = \kappa^2 G^2 \frac{2}{\pi} m_e E_{\bar{\nu}} \left\{B + A(1 - y)^2\right\} . \tag{$\overline{10}$}$$

The total cross sections are obtained by integration over y:

$$\sigma = \sigma_e E_\nu (A + B/3) , \tag{11}$$

$$\bar{\sigma} = \sigma_e E_{\bar{\nu}} (A/3 + B) , \tag{$\overline{11}$}$$

+) Strictly speaking we mean spin projected on momentum direction.

with $\sigma_e = \kappa^2 \times 17.2 \times 10^{-42}\ cm^2/GeV$, a rather frightening small number. Soon after 't Hooft [18] had first derived these relations, they attracted the interest of many authors, including Rosen, Sehgal and others [19]. They observed that the two real numbers A and B, which contain all the physics there is, could be extracted in many different ways: from σ and $\bar{\sigma}$, from σ and $\langle y \rangle$, from $d\sigma/dy$ and so on. I have suggested some formulae [20], which use as an input the ratio of measurable quantities, like $\bar{\sigma}/\sigma$ or $\langle \bar{y} \rangle / \langle y \rangle$, which should be less susceptible to experimental backgrounds and biases than the quantities themselves.

Even so, it is not an easy matter to separate neutrino-electron scattering from background, considering that its cross section is tenthousand times smaller than the total neutrino nucleon cross section. Fortunately the majority of these $\overset{(-)}{\nu}_\mu$-reactions has a muon in the final state - a long, non-interacting track, which will never be confused with the short electro-magnetic cascade, into which an electron will develop in the dense ($\rho \approx 1.5\ g/cm^3$) media used (see fig. 1). Two potential backgrounds remain:

a) NC induced muon-less hadronic reactions, with only neutral hadrons, and a π^0, of which one decay gamma had been lost.

b) The ν_e contamination in the $\overset{(-)}{\nu}$-beam ($\approx 0.5\%$) can produce genuine electrons, and fairly often the accompanying hadron(s) is (are) not visible.

Both types of background can be rather well rejected, by taking advantage of the peculiar kinematics of genuine neutrino-electron scattering. It is characterized by the smallness of the target mass, which makes it hard to get energy onto the electron - hence the factor m_e in the cross section $\overset{(\longleftrightarrow)}{(10)}$! By the same token one cannot get much transverse momentum p_T imparted - hence a lab electron angle of order

$$\Theta_e \approx p_T/E_\nu \lessapprox \sqrt{s}/E_\nu \approx \sqrt{2m_e/E_\nu} \approx 30 \text{ mrad} \quad \text{for} \quad E_\nu = 1 \text{ GeV} .$$

The exact formula is [21]:

$$E_e \Theta_e^2 = 2 m_e (1 - y) \quad . \tag{12}$$

The measurable left-hand side of this equation is essentially the mass of the struck particle. This test quantity was first exploited by the Aachen-Padova Group [21,22], and taken over by most other νe scattering groups. Under favourable background conditions it may suffice to require an angle smaller than a few degrees (say 3° to 5°). Also, since $\langle y \rangle$ lies between 1/4 and 1/2 for a V,A theory, one expects $\langle E_e \rangle \approx 0.37 \langle E_\nu \rangle$, and can profitably reject electron energies higher than $\langle E_\nu \rangle$. This means an energy cut at $\approx$ 2 GeV for the CERN PS, and reduces the background from inverse β-decay virtually to zero [23-26].

Various groups have succeeded in measuring $\nu_\mu e$ or $\bar{\nu}_\mu e$ scattering. The cross sections, or upper limits, obtained are given in Table I, along with some information on event numbers and signal-to-noise. Some comments on methods and difficulties may be in order:

a) The GARGAMELLE Collaboration, who had discovered the first $\bar{\nu}_\mu e$ event [2-4], continued to collect events at a steady pace of 1 per year, (one at Orsay, the last one at Bruxelles). Then the allotted antineutrino beam time was exhausted, and the $\bar{\nu}_\mu e$ experiment was declared finished [23]. The background was quite small ≈ 0.5 events. It consisted mainly of asymmetric $\bar{e}$ e pairs from NC produced π^0 decay gammas.

Table I: Measured cross sections of elastic $\bar{\nu}_\mu e$ and $\nu_\mu e$ scattering

Group	[Ref]	Beam	$\langle E_\nu \rangle$ GeV	No. of candidates total	good	Cross section 10^{-42} cm^2 GeV^{-1}
GGM	[23]		2.0	3	2.5	$1.0^{+2.1}_{-0.9}$
AC-PD	[28,29]		2.0	8[a)]	6.3	2.2±1.0
BEBC	[35]	$\bar{\nu}$	20	1	0.5	< 3.4
FNAL 15'	[35]		30	0	0	< 2.1
GGM	[36]		20	0	0	< 2.7
World average $\bar{\nu}_\mu e$			2 - 30	11	8.8	1.8±0.9
GGM	[24]		2.2	1	0.7	< 3.0
AC-PD	[28,29]		2.2	11[a)]	7.1	1.1±0.6
BNL-COL	[34]	ν	30	8	7.5	1.8±0.8
GGM	[33]		25	9	8.6	$2.4^{+1.2}_{-0.9}$[b)]
World average $\nu_\mu e$			2.2 - 30	28	23.2	1.6±0.5

a) With tight selection criteria, b) Originally [32]: $7.3^{+4.6}_{-3.8}$.

Hence the existence of $\bar{\nu}_\mu e$ scattering is established with high statistical confidence. Considering the small event numbers, this may sound surprising. But the question is not, whether the signal (here 2.5) could statistically have been zero. Rather one has to ask if the well measured background gives the signal observed by a statistical fluctuation. The confidence of the effect being established hinges, therefore, on the predicted variance of the background. The error of the cross section, instead, comes mainly from the statistics of the observed event numbers, and may be large.

Also $\nu_\mu e$ scattering has been searched for in GGM, and one event has been found (again at Aachen). Background is more severe in neutrinos than in antineutrinos (see below): the observed event should be confronted with 0.5 background events expected [24]. Clearly this does not establish $\nu_\mu e$ scattering, and gives for the cross section only a limit.

b) The Aachen-Padova (AC-PD) Collaboration decided to build a large ($\approx$ 20 ton) multiplate sparkchamber, in which $\bar{\nu}_e e$ and $\nu_\mu e$ scattering could be detected, and also measured to some accuracy [20-22]. The plate thickness was chosen to 1 cm aluminum. This corresponds to a radiation length X_0 of 20 cm, and gives an electron around 1 GeV typically a shower pattern like that of Fig. 3. The electron energy can be determined from the total number of sparks, at 1 GeV with an accuracy of 22%. The initial electron direction can be inferred, from the beginning of the track, to about $\pm 1^o$. This suffices for a quantitative analysis, although many of the details, as seen in GARGAMELLE (see Fig. 1), are lost. The real short-coming is the absence of a magnetic field: hence the generated secondary electrons are being spread out by multiple scattering only, and as a result a distinction between e^-, e^+ and an e^-e^+-pair (= "gamma") is not possible.

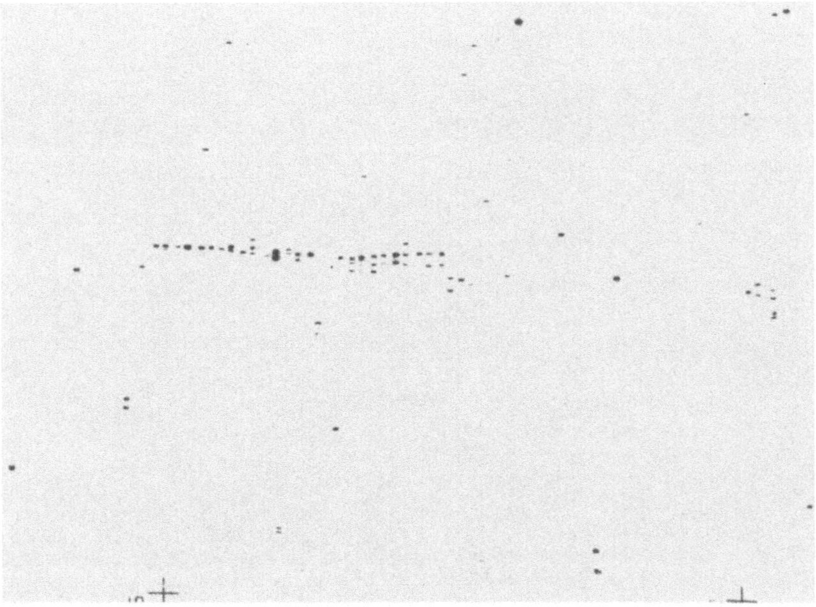

Fig. 3: $\nu_\mu e$ scattering candidate, registered in the Aachen-Padova multi-plate aluminum sparkchamber (ν's from the left).

This makes a sparkchamber more vulnerable by gammas than bubble chambers. Most of this 1γ-background stems from neutral current produced single π^0's, with one decay gamma lost:

$$\overset{(-)}{\nu}_\mu + N \rightarrow \overset{(-)}{\nu}_\mu + N + \pi^0 \quad , \qquad \pi^0 \rightarrow \gamma + (\gamma) \tag{13N}$$

$$\overset{(-)}{\nu}_\mu + P \rightarrow \overset{(-)}{\nu}_\mu + P + \pi^0 \quad . \qquad \pi^0 \rightarrow \gamma + (\gamma) \tag{13P}$$

An example of a "naked" π^0, with the two decay gammas visible, is given in Fig. 4. Most probably it is an example of NC π^0-production off a neutron (13N), but occasionally also the proton reaction (13P) can appear this way, if the recoil proton is too short, charge-exchanges, or gets absorbed. (We note in passing that these NC reactions had not yet been discovered, when the AC-PD experiment was proposed. 1γ-Background estimates were based on other peoples' upper limits - unfortunately we measured π^0 rates a factor of two higher! [25]).

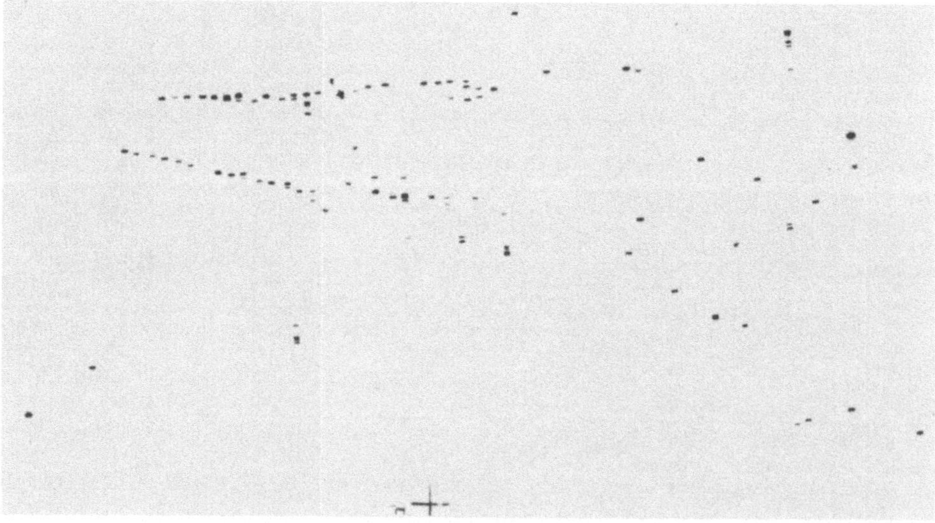

Fig. 4: "Naked" $\pi^0 \rightarrow 2\gamma$, initiated by hadronic neutral current in the AC-PD sparkchamber.

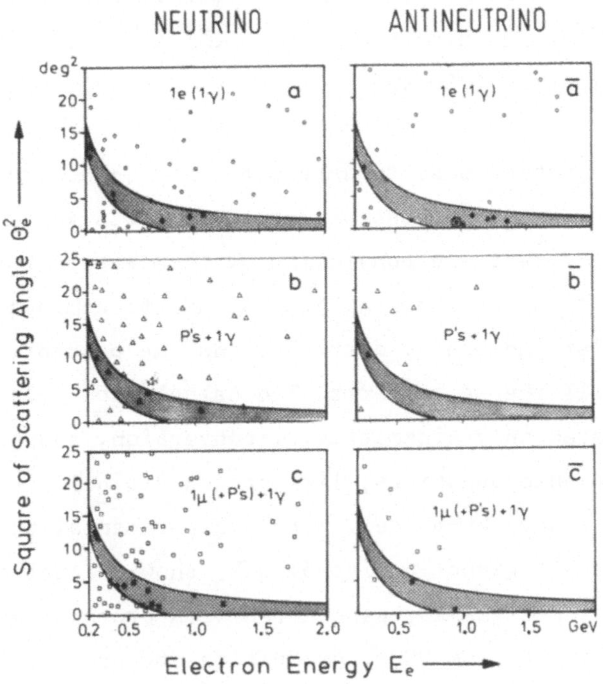

Fig. 5: Θ_e^2 E_e-scatter plot for

$\overset{(-)}{a}$) $\overset{(-)}{\nu}_\mu e$ candidates,

$\overset{(-)}{b}$) P associated single γ's (from $P\pi^0$),

$\overset{(-)}{c}$) μ associated single γ's (from CC $\mu\pi^0$).

How serious the 1γ-background is, may be seen from the main diagnostic tool in νe analysis, the Θ_e^2-E_e scatter plot of Fig. 5: a ($\bar{a}$) shows the single, isolated showers, i.e. the ν_μ ($\bar{\nu}_\mu$) e scattering candidates, which do comprise, however, a certain amount of gammas. The shaded domain is where νe scatterings should lie, with y > 0 and $E_{\overset{(-)}{\nu}}$ > 0.8 GeV imposed. But there are single showers scattered all over the place. Only for antineutrinos there is a distinct concentration of events inside the allowed band. This different behaviour of neutrino and antineutrino background is easily understood: the 0.8 M pictures, taken in either case, correspond to twice as many ν than $\bar{\nu}$ crossing the chamber, and the single π^0 NC cross section ratio $r = (\bar{\nu}\mathcal{N} \to \bar{\nu}\mathcal{N}\pi^0)/(\nu\mathcal{N} \to \nu\mathcal{N}\pi^0)$ was found to be close to 0.50 [26].

The main problem was, to determine the 1γ-background in either case, utilizing suitable control reactions. This was done in several ways. The most direct and convincing of them was to take several hundred naked π^0's, as shown in Fig. 4, and to compute from their observed angular and energy distributions the probability for one gamma to fulfill the νe criteria. The calculation [27], though lengthy, invested only kinematics, γ-conversion, and geometry - all subtle hadronic questions, like proton visibilities and cross section ratios, were by-passed. Purely experimental background determinations are given in Fig. 5: b($\bar{b}$) show scatter plots of $\nu(\bar{\nu})$ induced single gammas, associated with protons, and thus presumably stemming from NC π^0 production off protons (13P). This reaction must have a cross section quite similar to that with neutrons (13N), and thus, apart from known factors of order one, the 1γ-background to graphs a and $\bar{a}$ can be read off the bands in b and $\bar{b}$: perhaps 4 events for ν and only 1 for $\bar{\nu}$! A similar answer comes from CC single π^0 production, given in Figs. 5c and $\bar{c}$. Statistically more relevant numbers come from the events outside the allowed bands, in case of single showers extended to angles as large as 30^0, and extrapolated into the νe region with the help of the Monte Carlo simulation [27]. Within errors, the different methods agreed.

The final results were presented to the Hamburg Lepton Conference 1977 by Hans Reithler [28]. The two measured cross sections in the σ-$\bar{\sigma}$ plane evince the space-time structure immediately (Fig. 6): the AC-PD error ellipse, (and also the GGM rectangle), appear to favour a right-handed electron (V + A), but are compatible with an ambidextrous one (A or V). A pure left-hander

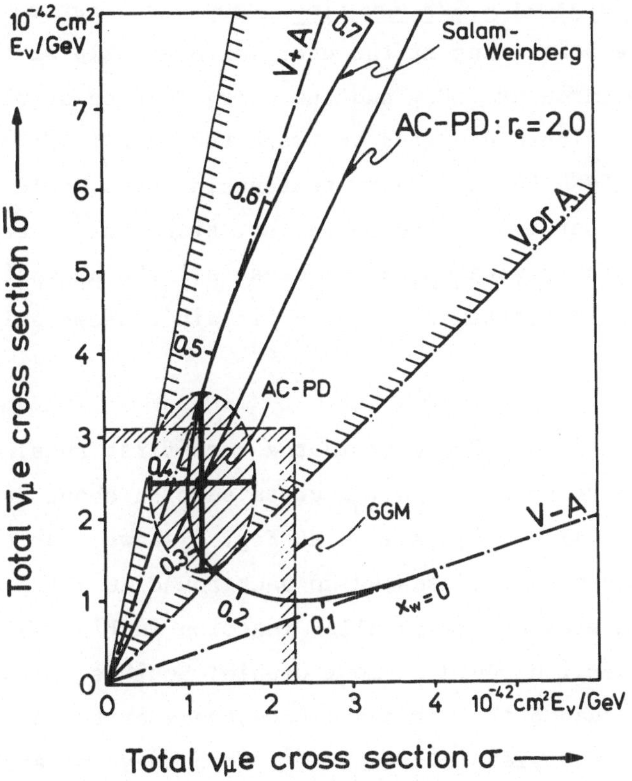

Fig. 6: Total $\overset{(-)}{\nu}_\mu e$ cross sections, measured around 2 GeV by Aachen-Padova and GARGAMELLE. Straight lines delineate simple V,A interactions; the curve is the Salam-Weinberg prediction.

(V - A) is excluded. Agreement with Salam-Weinberg is amazing, $\sin^2\Theta_W = 0.35 \pm 0.08$. Reithler also gave an answer to the non-trivial question, how one can compare the two energy-distributions, although both of them are based on only 7 good events. The answer is: one can - V + A being favoured over A or V, albeit only on a 1.5 to 2 standard deviation (st.d.) level [27,28]. The subsequent publication [29], and later Conference Reports [30] contain, in addition, the results of a maximum likelihood method by Milla Baldo-Ceolin and associates of Padova.

c) The Gargamelle SPS Neutrino Group was first to try measuring $\nu_\mu e$ scattering at the energies available at the Super-Proton-Synchrotron at CERN, and the corresponding accelerator at Fermilab (FL). These energies are more than ten times as high as those at the CERN PS, and every reasonable guess would have been that the νe experiment would be easier, since the νe cross section increases linearly with energy, whereas all the hadronic background would eventually flatten off. But - "it ain't necessarily so ..." [31].

The group found 10 events in the very first runs, and this would have corresponded to a $\nu_\mu e$ cross section about 7 times the Aachen-Padova value [32]. A look at Fig. 5 shows that this is impossible: even if one assumed no background at all - quite improbable in view of the parallel reaction in Fig. 5b! - the AC-PD cross section could at most double! Besides, the new GGM group had abandoned the stringent acceptance criteria, consistently applied at PS energies [20-24,27-30]. Then they collected a large fraction of high energy electrons. Not all of them could come from νe scattering, lest $\langle y \rangle$ should approach unity! - This internal inconsistency convinced me more than anything else that the majority of these electrons had nothing to do with $\nu_\mu e$ scattering. In principle they could have come from a new process - unlikely, as it is, at a c.m. energy of typically 300 MeV.

As it turned out, the high new GGM result seems to have been a combination of too liberal cuts and of a statistical fluke. The cross section went monotonously down in time, and the last value reported [33] is perfectly compatible with the Aachen-Padova measurement, and with the old GARGAMELLE limit [s. Table I]!

d) Even before the rise and fall of the high energy $\nu_\mu e$ cross section came to a good end, a Columbia-FL-Group, under Baltay, looked into $\nu_\mu e$ scattering at Fermilab. They used the 15' bubble chamber, filled with 64 atomic % Ne-H_2, applied the kinematical procedures described, and wound up with perfectly normal distributions in energy and 'target mass' ($\approx \Theta_e^2 E_e$). From 10 events found - at a very low background level - they deduced a cross section of 1.8 ± 0.8 in our unit of 10^{-42} cm^2 GeV^{-1} [34]. (This is, incidentally, the minimal νe cross section in the Salam-Weinberg model.) This cross section is in flat contradiction with the original high GGM SPS value, but it agrees nicely with the AC-PD result, the difference being a 0.7 standard deviation.

e) $\bar{\nu}_\mu e$ has been tried by a Gargamelle Group at the CERN SPS - without success thus far. Not a single candidate showed up. This sets a rather low limit on the $\bar{\nu}_\mu e$ cross section [36]. Similar results were obtained with BEBC and at FNAL [35].

f) Low energy $\bar{\nu}_e e$ scattering has been measured by Reines et al. at a nuclear reactor [37]. In contrast to muon-neutrinos, the scattering of electron-neutrinos can proceed through both, neutral and charged currents. Reines found a cross section equal [38] to that of the classical V - A theory [39]. Thus, taken by itself the $\bar{\nu}_e e$ measurement does not reveal any neutral currents. But in conjunction with the $\overset{(-)}{\nu}_\mu e$ data it will be most significant.

Summary of νe:

Table I summarizes what we know about νe cross sections. The world averages over the three ν_μ, and the two $\bar{\nu}_\mu$ measurements, respectively, fall pretty close together: the $\bar{\nu}_\mu/\nu_\mu$ asymmetry, suggested by the early data, has vanished. This may well be due to slight statistical fluctuations: all PS values are within half a st.d. of the present averages! But it ought to be checked by energy measurements and y-distributions.

Taken at their face values the ν_μ and $\bar{\nu}_\mu$ cross sections imply an almost pure A or V neutral current for the electron. In the framework of the minimal model [12,13], it must be A, and it would correspond to a mixing angle

$$\nu_\mu e \text{ and } \bar{\nu}_\mu e: \quad \sin^2\Theta_W = 0.24 \pm 0.06 \quad . \tag{14}$$

This value suits also the $\bar{\nu}_e e$ data quite well [36].

A model-free analysis was attempted by Reithler |28| and myself [20], and carried further by Milla Baldo-Ceolin [30]. The idea was to free oneself from the limitations and pre-set tracks of a particular model, and to use the data in their own right [19]. Apart from the SU(2)-U(1) mixing, described by $\sin^2\Theta_W$, we want to gain insight into the Higgs sector [12,13] characterized by the isospin of the Higgs particle I_ϕ, and finally into the multiplet structure, mentioned before, which concentrates on the isospin of the right-handed electron I_R.

Ross and Veltman have suggested a practical way how to do that [40]: They observe that $x_W = \sin^2\Theta_W$ can be extracted from the ratio r_e of $\bar{\nu}_\mu e$ and $\nu_\mu e$ cross sections $\bar{\sigma}$ and σ. Taking the only measured r_e (by AC-PD), this gives $x_W = 0.35 \pm 0.10$, in reasonable agreement with the values extracted from the cross sections themselves. We used this mixing parameter, (and assumed $I_R = 0$), to predict the sum of the ν_μ and $\bar{\nu}_\mu$ cross sections, and compared it to our experimental result. This gives directly the NC normalization constant κ of eqs. (5) and (6):

$$\kappa = 1.04 \pm 0.15 \quad , \tag{15}$$

and with the Ross-Veltman identification

$$\kappa = 1/2\ I_\phi \tag{16}$$

the Higgs isospin to 1/2. To the best of my knowledge this is the first experimental statement about the elusive Higgs particle at all.

Taken this simplest Higgs structure for granted, the final analysis is best performed in the plane of the relative NC axial-vector and vector coupling constants, C_A and C_V, (Fig. 7), which are related to our chiral coefficients by:

$$A = \frac{1}{4}(C_V + C_A)^2 \quad , \quad B = \frac{1}{4}(C_V - C_A)^2 \quad .$$

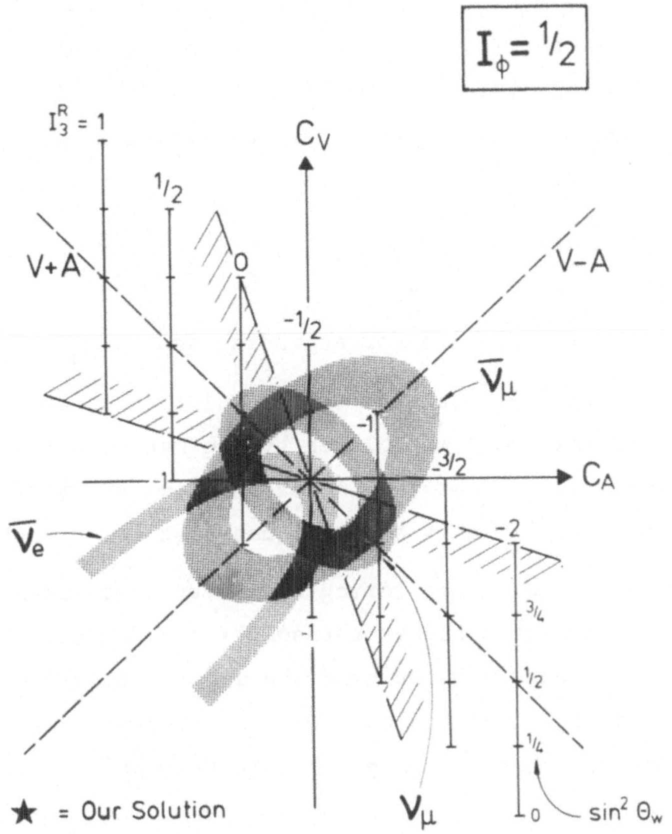

Fig. 7: Constraints imposed by the measured $\nu_\mu e$, $\bar{\nu}_\mu e$ and $\bar{\nu}_e e$ total cross sections on the vector and axialvector coupling constants C_V and C_A [20]. Vertical lines allowed by different gauge models.

Hence, and from eqs. (11) and $(\overline{11})$ one sees at once that a fixed total cross section gives in the C_V-C_A diagram an ellipse; the errors make it an elliptical ring. It is clear that the ellipses have their main axes respectively parallel and orthogonal to the V - A and V + A directions. Thus, any attempt to determine C_A and C_V from σ and $\bar{\sigma}$ faces, in general, a fourfold ambiguity. From the AC-PD data alone the $\sigma - \bar{\sigma}$ overlap (dark-dashed) is fairly large. But, as Reithler was first to show [28], when combined with the $\bar{\nu}_e$e data [37], they reduce the allowed regions for C_A and C_V to the tiny black triangels in Fig. 7. They exclude all gauge models, which are characterized by the different vertical lines - except two: a Salam-Weinberg [12,13] solution close to pure A, and a vectorlike ("hybrid") solution close to pure V. Milla Baldo-Ceolin [30] arrived at this conclusion also by a simultaneous fit on I_ϕ and I_R^3. Numerically from Fig. 7:

$$\begin{aligned} &\text{A-like: } C_A = -0.57 \pm 0.08, \quad C_V = +0.09 \pm 0.08 \; ; \\ &\text{V-like: } C_A = +0.09 \pm 0.08, \quad C_V = -0.57 \pm 0.08 \; , \end{aligned} \tag{17}$$

with the A-like S-W-solution corresponding to $\sin^2\Theta_W = 0.29 \pm 0.04$.

It was argued [20] that the vector solution is improbable, since it is outside the limit placed by the measured $\langle\bar{y}\rangle/\langle y\rangle$ (dashed straight lines). In view of the increasing similarity of ν_μ and $\bar{\nu}_\mu$ cross sections (Tab. I) this argument has lost in strength.

But there is the eD scattering asymmetry measurement at SLAC [41] now, which excludes the vector solution altogether (Fig. 8). Note, by the way, what a broad band the many scattered electrons still admit, and how incisive the 60 νe scatterings are. By contrast, not much light was shed on the problem by the optical experiments [42].

Nevertheless, after 7 years of experimentation, the structure of the weak neutral electron current gradually looms up: it is Salam-Weinberg with $\sin^2\Theta_W$ close to 1/4.

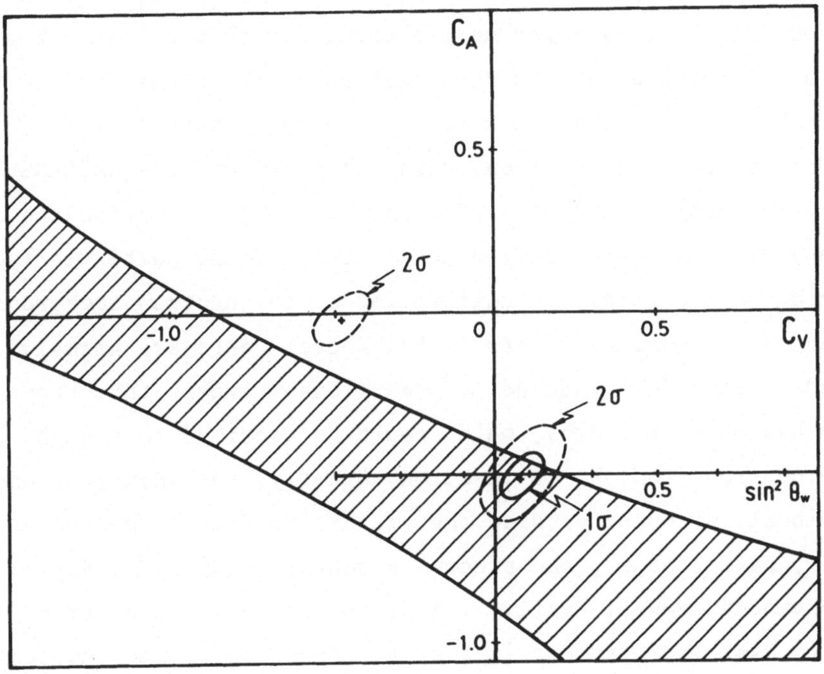

Fig. 8: Values of C_V and C_A allowed by the SLAC polarized eD scattering experiment (dashed band). Contour lines follow from the νe scattering data of Fig. 7.

4. Inclusive Muon-less ν_μ $(\bar{\nu}_\mu)$-Reactions (Truly Elastic ν_μ $(\bar{\nu}_\mu)$-Quark Scattering)

It sounds like a historical accident that the neutral weak current (NC) of hadrons was discovered in inclusive neutrino interactions, where the final hadron state (X) could be anything, from a single pion (π) to a rather complicated meson-nucleon cascade:

$$\overset{(-)}{\nu}_\mu + \mathcal{N} \to \overset{(-)}{\nu}_\mu + X \quad . \tag{18}$$

Indeed, it is relatively easy to prove that there was no muon produced (see Fig. 2) - and no electron, for that matter - but it is not trivial at all to show that these reactions were induced by neutrinos, and not by high-energy neutrons. As mentioned in the introduction, this analysis took the GARGAMELLE (GGM) Group many months of hard work. Let me just mention one of the many arguments proposed, which I always found particularly simple and convincing: the calibration of the neutron background by means of "associate events" (AS). These events are normal, charged current (CC) induced $\overset{(-)}{\nu}_\mu$-reactions - except that they show a hadron star nearby, which could be ascribed to a high-energy neutron from the CC-reaction. Clearly, such neutrons can be copiously produced <u>outside</u> the visible volume of the bubble chamber, may enter it, and produce a neutron star (N^*). But as one convinces oneself easily, the frequency ratio of neutron stars to associated events is a simple function of the neutron interaction length and of the chamber geometry; in fact for the actual GARGAMELLE disposition [43] $N^*/AS \approx 1$. Thus, having observed $AS/NC \approx 10\%$, the GARGAMELLE group could safely trust their effect [3,8], and check the elaborate N^* subtraction [8] by a direct physical measurement.

None such detailed background study was possible in the next generation of experiments, which used large counter-sparkchamber arrangements, [44,45] exposed to (anti)neutrino beams from the 400 GeV proton-synchrotron at Fermilab (FNAL). The decision, whether or not a particular track was due to a muon, was made on the basis of range only, and it took the Harvard-Pennsylvania-Wisconsin (HPW) team some time until all systematic effects were under control: Penetrating pions faked muons, and large angle (or short) muon tracks tended to get lost in the hadron splash. Besides, the wide-band neutrino beams employed by HPW were quite often rather ill defined, which made the interpretation of the data fairly difficult. Yet in 1974 the HPW Group managed, after some fluctuation, to confirm GARGAMELLE'S discovery [3,8] of inclusive muon-less neutrino events, and obtained - despite the higher energies - practically the same NC/CC cross section ratios [44].

A Caltec Group at Fermilab (CITF), using a similar apparatus but a better defined narrow-band beam, obtained almost the same results some time later [45]. By then weak neutral hadronic currents had been seen by three different groups, working at different labs, with different techniques, in different beams, and at very different energies. No doubt this new force of Nature was established.

The next step would be to clarify, which structure these neutral currents have - or, to put the question in a more fundamental way: are the observed muon-less events really of the V,A-current type, inherent of unified gauge theories? - There are many aspects to that: a very basic one is, if the neutral particle in the final state is really identical to the primary neutrino. This is very badly known, but according to Sehgal it can be tested by NC-CC interference in ν_e e-scattering [46]. And even taking this for granted - do ν_μ and ν_e behave the same also in neutral current interactions? This is not known at all, but Sehgal and Sakakibara [47] expect only small differences. Lacking anything more definite, we accept these prejudices.

Even so, the new interaction could have any amount of S,P,T mixed into our preferred V,A-current structure. Under favourable conditions this might be seen from the $\bar{\nu}/\nu$ cross section ratio r, which could exceed then the V,A limits [48]:

$$\text{V,A:} \quad \frac{1}{3} \leqq r \leqq 3 \quad ; \quad \text{but} \quad \text{S,P,T:} \quad \frac{1}{7} \lessapprox r \lessapprox 7 \, . \qquad (19)$$

Nothing can be concluded instead, if the V,A-bounds happen to be respected: According to a well known "Confusion Theorem" [49] any V,A cross section can be reproduced by a suitable combination of S,P,T. In principle one can distinguish them, say by polarization experiments - but this is presently out of reach. Thus, all the groups worrying about the Lorentz structure of the neutral weak current assumed V,A, and checked how well this would fit the data. With some inner reluctance I shall follow this path.

The theoretical treatment of deep inelastic neutral current neutrino nucleon encounters is then quite easy, provided one accepts the quark-parton picture of the nucleon [50]. This was conceived on the basis of deep inelastic electron nucleon scattering experiments, at SLAC [51], which showed essentially constant q^2-distributions, i.e. a behaviour as if the hit objects were points! That these point-partons had all the properties of quarks, was shown subsequently by the GARGAMELLE collaboration. While we were still striving to get the first results [52], it occured to me that they could possibly be explained by the action of the 3 valence quarks of the nucleon alone [53]*. This valence quark dominance (VQD) was found to hold good to ≈ 90% at energies of a few GeV [52], and still to be valid to ≈ 80% at some 100 GeV [54]. The remaining percent are supplied by antiquarks, from the meson cloud (or "sea"). An overall normalization factor of 1/2 is ascribed to "gluons", and will not bully us here.

---

* Although this little paper contained, for the first time, the "naive" quark-parton $\nu\mathcal{N}$ and $\bar{\nu}\mathcal{N}$ cross sections, as used by all groups ever since (and a first estimate of the effective quark-parton mass too), it was rejected by "Physics Letters". When Rein re-submitted its substance, embellishing it with lots of formulae and integrals, it was (of course) accepted.

A point-like quark q will have essentially the same cross section for elastic ν $(\overline{\nu})$ scattering as an electron, namely eqs. (10) and $(\overline{10})$, respectively. But quarks are no real particles, and thus the target rest mass m has to be replaced by x M, where M is the proton mass. The new scaling variable x is defined between 0 and 1, and can, in a way, be connected with an effective quark-parton mass [55]. More commonly, however, x is identified with the fraction of the nucleon's <u>momentum</u>, P this particular quark carries [50]. Then it is easy to show that x is given essentially by the ratio of four-momentum transfer squared, q^2, to energy transferred ν:

$$x = -q^2/2M\nu \quad . \tag{20}$$

The distribution of x is some quark density $\rho_q(x)$ in momentum space, normalized to 1. It has repeatedly been measured [51,52,54].

In analogy to νe scattering (10) the νq cross section is:

$$\frac{d^2\sigma_q}{dxdy} = \kappa^2 G^2 \frac{2}{\pi} M E_\nu \; x \rho_q(x) \quad \left\{ A_q + B_q (1-y)^2 \right\} , \tag{21}$$

$$\equiv \sigma_q \quad x \rho_q(x) \quad \left\{ A_q + B_q (1-y)^2 \right\} , \tag{21'}$$

where q can be either:

a (proton-like) up-quark (u), or

a (neutron-like) down-quark (d).

The cross section $\overline{\sigma}_q$ for (right-handed) antineutrinos $\overline{\nu}$ is obtained by interchanging A and B. Since all what matters is the <u>relative</u> spin orientation between projectile and target, this is also the cross section of (left-handed) neutrinos ν against (right-handed) antiquarks $\overline{q}$:

$$\sigma_{\overline{q}} = \overline{\sigma}_q \ , \quad \text{and (of course)} \quad \sigma_q = \overline{\sigma}_{\overline{q}} \quad . \tag{22}$$

With these relations the $\overset{(-)}{\nu}P$ and $\overset{(-)}{\nu}N$ cross sections can easily be written down, in full generality. We just note the practically important case of an isoscalar target, i.e. of a nucleus with A = 2Z, and take for example the deuteron. With complete valence quark dominance:

$$\frac{d^2\sigma_D}{dxdy} = \frac{1}{2}\sigma_q\, x \left\{2\rho_u(x) + \rho_d(x)\right\} \left\{(A_u+A_d) + (B_u+B_d)\,(1-y)^2\right\} \qquad (23)$$

Again it would make no trouble to include antiquarks as well.

Lalit Sehgal, at Aachen, who introduced this sort of analysis [56], is a strong believer in VQD, and therefore neglected antiquarks for a start. He also used chiral coupling constants in place of their squares:

$$\begin{aligned} A_u &= u_L^2 \;, \quad B_u = u_R^2 \;; \\ A_d &= d_L^2 \;, \quad B_d = d_R^2 \;; \end{aligned} \qquad (24)$$

and connected them with the axialvector and vector coupling constants, as usual:

$$\begin{aligned} u_L &= \frac{1}{2}(C_V^u + C_A^u) \;, \quad u_R = \frac{1}{2}(C_V^u - C_A^u) \;; \\ d_L &= \frac{1}{2}(C_V^d + C_A^d) \;, \quad d_R = \frac{1}{2}(C_V^d - C_A^d) \;. \end{aligned} \qquad (25)$$

In the standard model for the neutral current [12,13]:

$$C_A^i = I_3^i \;, \qquad C_V^i = I_3^i - 2Q^i \sin^2\Theta_W \;, \qquad i = u,d \;, \qquad (26)$$

where I_3 is the 3rd component of (weak) isospin I, Q the quark's charge. Now one computes easily the single ν $(\bar{\nu})$ q total cross sections [in units of $\frac{\sigma_q}{4}$, defined by eq. (21')] :

$$\sigma_{\nu u} = 1 - \frac{8}{3} \sin^2\Theta_w + \frac{64}{27} \sin^4\Theta_w \quad , \tag{27u}$$

$$\sigma_{\nu d} = 1 - \frac{4}{3} \sin^2\Theta_w + \frac{16}{27} \sin^4\Theta_w \quad , \tag{27d}$$

$$\sigma_{\overline{\nu}u} = \frac{1}{3} - \frac{8}{9} \sin^2\Theta_w + \frac{64}{27} \sin^4\Theta_w \quad , \tag{$\overline{27u}$}$$

$$\sigma_{\overline{\nu}d} = \frac{1}{3} - \frac{4}{9} \sin^2\Theta_w + \frac{16}{27} \sin^4\Theta_w \quad . \tag{$\overline{27d}$}$$

Hence any NC cross section can be calculated, provided one knows the quark (and antiquark) densities ρ (and $\overline{\rho}$). In the NC/CC ratios R_i they may cancel in part, and we give some popular R's for $\rho_u = \rho_d$, and complete valence quark dominance:

$$R_P = \frac{3}{4} - \frac{5}{3} \sin^2\Theta_w + \frac{4}{3} \sin^4\Theta_w \quad , \tag{28P}$$

$$R_N = \frac{3}{8} - \frac{2}{3} \sin^2\Theta_w + \frac{4}{9} \sin^4\Theta_w \quad , \tag{28N}$$

$$\overline{R}_P = \frac{3}{8} - \frac{5}{6} \sin^2\Theta_w + 2 \sin^4\Theta_w \quad , \tag{$\overline{28P}$}$$

$$\overline{R}_N = \frac{3}{4} - \frac{4}{3} \sin^2\Theta_w + \frac{8}{3} \sin^4\Theta_w \quad . \tag{$\overline{28N}$}$$

By averaging one obtains R ($\overline{R}$), valid for an isoscalar target. It is clear that $\sin^2\Theta_w$ can be obtained from either of them. From the early GGM data [43] Sehgal obtained the first value for this mixing angle, namely

$$x_w \equiv \sin^2\Theta_w = 0.39 \pm 0.09 \quad . \tag{29}$$

Considerable progress has been made since then. Not only did the GGM, HPW, and CITF groups continue their efforts: we had also BEBC [54], and the CERN-Dortmund-Heidelberg-Saclay-Collaboration [57] entering the scene. The CDHS-collaboration had an unusually large apparatus constructed, consisting of 5 cm thick magnetized iron plates, interleaved with wire chambers for the measurement of muon angles and momentum, and with scintillation counters to register the deposited hadron energy. The set-up was operated in the CERN SPS narrow-band neutrino beam, with tunable energies between 50 and 200 GeV. The arrangement is ideally suited for studying charged

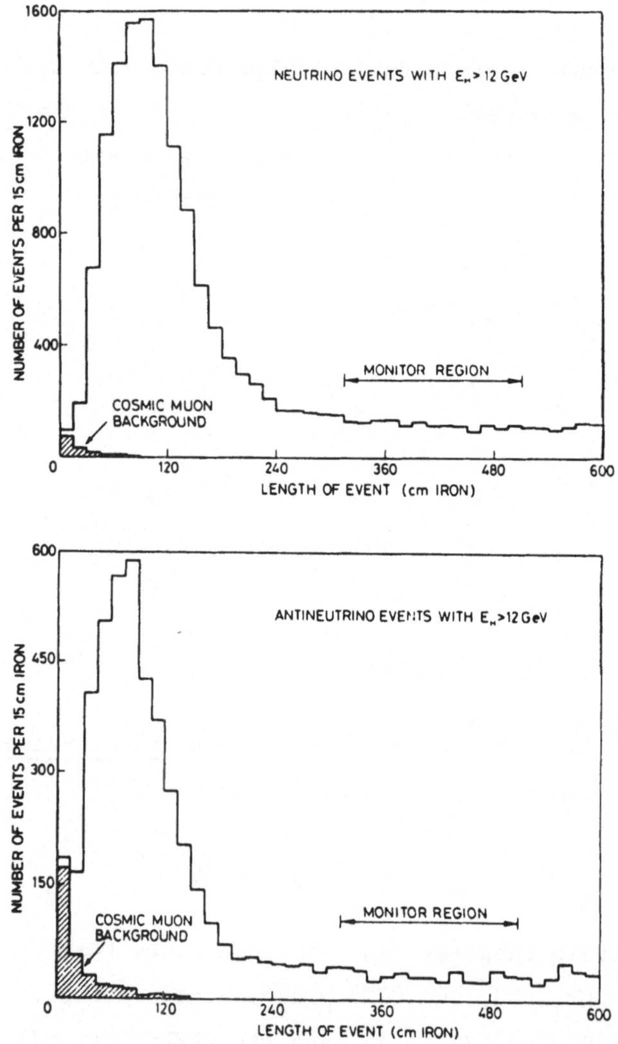

Fig. 9: Distributions of projected range ("event length") of inclusive events: top for neutrinos, bottom for antineutrinos.

current inclusive processes. In order to single out neutral current induced inclusive reactions as well, the CDHS group resorted to a technique, originally devised by CITF [45]: It is based on the projected range, or "length" ℓ, of each event, and assumes all very long events to contain muons. Then one extrapolates the number of charged current events, found in this "monitor region", back to short events, which are mainly due to neutral currents (Fig. 9). Compared with the meticulous track-by-track analysis of GARGAMELLE [8], the CDHS method is simple, fast and elegant. Neutron stars could be a problem, but the enormous iron thicknesses adequate to these high energies make them much less important than at PS energies. Of course, the whole CC-subtraction hinges on the

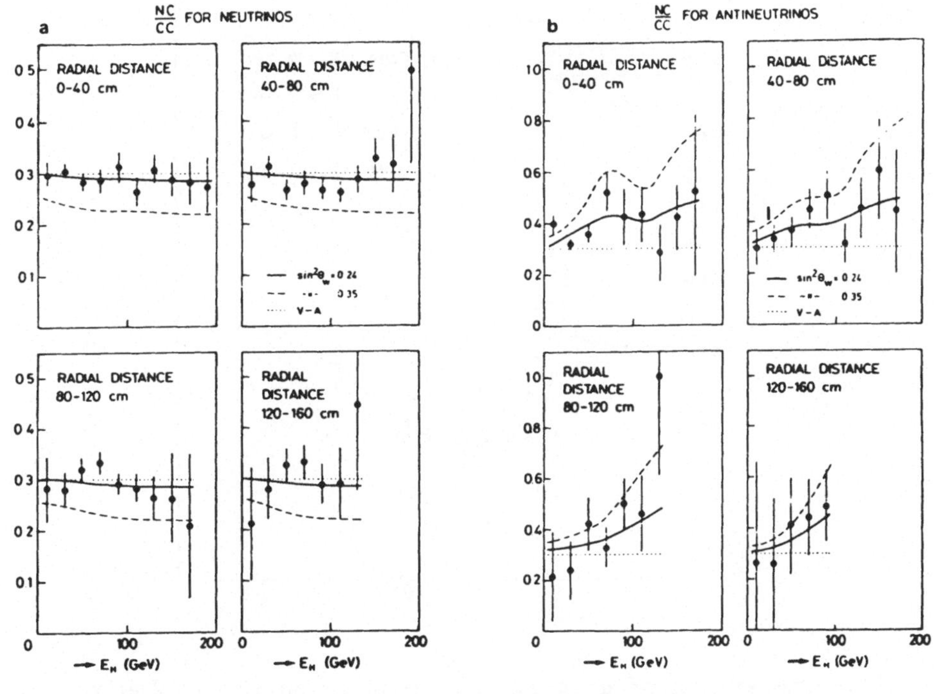

Fig. 10: Ratio of neutral to charged current event numbers, as a function of hadron energy for different radial bins, for ν and $\bar{\nu}$. - Curves are minimal model fits with x_W = 0.24 (——), 0,35 (- - -), and 0,0 = V-A (.....).

quark-parton model assumed, but the authors are confident that they do it right.

This systematic reliability, combined with enormous statistics, enabled CDHS to solve a number of problems, which had not been completely settled before. The most salient of them is the proof that the hadronic neutral current is definitely not pure V - A, i.e. not completely left-handed as the charged current is. This is equivalent to saying that the mixing constant $\sin^2\Theta_W$ does not vanish. This had certainly been surmised right from the start [3,8], but now it is proved with > 3 standard deviation (st.d.). In addition, the CDHS data suggest that the previous

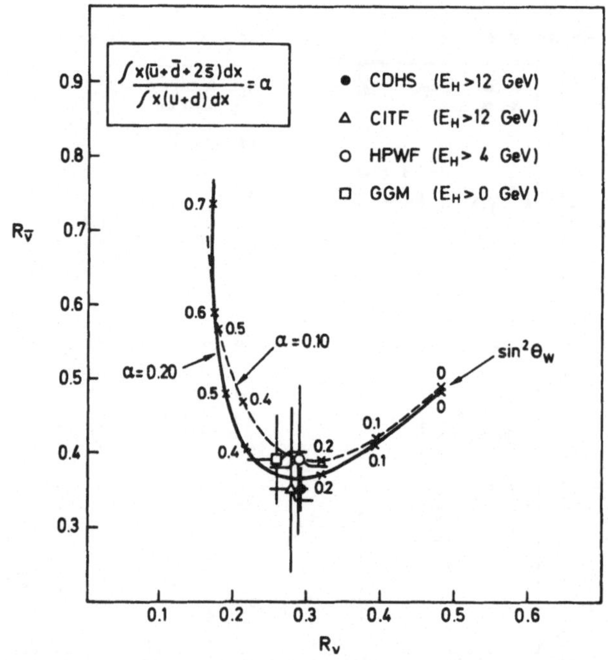

Fig. 11: Comparison of experimental results on the inclusive NC/CC ratios, R and $\overline{R}$, with the minimal model prediction after Sehgal [56] with $\overline{q}/q$ = 10% (dashed line) and 20% (full line).

values of $\sin^2\Theta_W$ were too high (by ≈ 1 st.d.). All that is demonstrated in Fig. 10, where the experimental NC/CC ratios R ($\bar{R}$), for the nucleon $\mathcal{N}$, are plotted against hadron energy, and in bins of radial distance in the apparatus (which determines the ν ($\bar{\nu}$) energy).

A summary of the R and $\bar{R}$ values obtained so far by the different groups [43-45, 57] is given in Fig. 11. They are plotted in the R-$\bar{R}$-plane, together with the prediction of the minimal model [56] ("Weinberg's nose"). The mutual agreement of the experiments is quite good. Their consent with Salam-Weinberg could not be better. The mixing constant $\sin^2\Theta_W$, which was around 1/3 before CDHS, is now brought down close to 1/4:

$$\text{Inclusive world average:} \quad \sin^2\Theta_W = 0.26 \pm 0.02 . \tag{29}$$

Within the frame-work of a more general phenomenological analysis, of course, the inclusive measurements of R and $\bar{R}$ impose just two constraints on the squares of the coupling constants: $u_L^2 + d_L^2$, and $u_R^2 + d_R^2$, must each be a constant. In the $u_L - d_L - (u_R - d_R)$-planes this defines a circle, i.e. a ring, if errors are included. This is depicted in Fig. 12, with the cross-hatched areas fore-shadowing already future developments. In order to resolve this circular ambiguity we have to resort to less inclusive reactions.

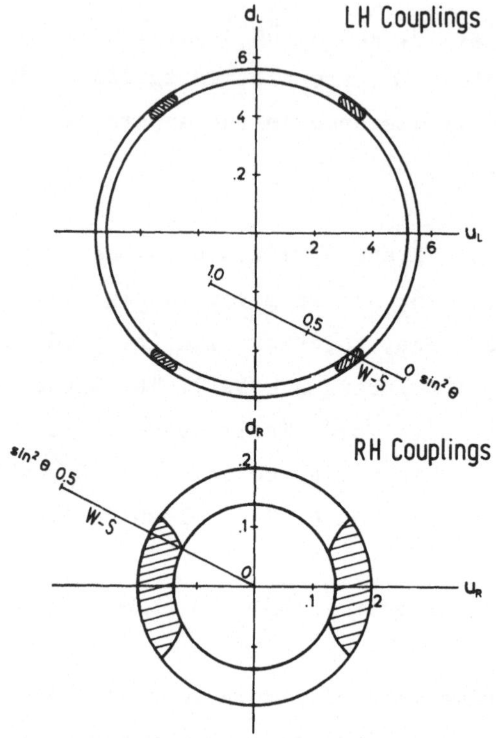

Fig. 12: Left- and right-handed u- and d-quark neutral current coupling constants. The rings are allowed by the inclusive measurements of CDHS et al., the dashed regions are singled out by pion production data (see Sec. 5)

5. Muon-less Pion Production and the Isospin Structure of the Weak Neutral Current

Indeed, the isospin structure of the weak neutral current could be extracted from inclusive measurements, if one compared the cross section off a proton with that of the neutron, (or an isoscalar nucleon for that matter). Because of the different number of u- and d-quarks involved, and because of the incoherent action of the single quarks [50-54], two such measurements on different targets would give the u- and d-quark cross sections separately. If they came out different, the neutral current could not be an isoscalar. Measurements with liquid hydrogen have been started [58], but are not yet finished.

At this stage Sehgal decided, to derive the isospin-properties from comparing the frequency of the different charge (i.e. I_3) states of semi-inclusive pions [59]. These pions are assumed to stem from the quarks hit in the first place [50,54]. A detailed analysis of these reactions implies thus some knowledge about the quark's transformation (or "fragmentation") into ordinary hadrons. This can be gathered from several weak and electromagnetic reactions.

The experimental study of muon-less semi-inclusive pion production by neutrinos and antineutrinos:

$$\overset{(-)}{\nu}_{\mu} + \mathcal{N} \rightarrow \overset{(-)}{\nu}_{\mu} + \pi^{\pm,0} + X \qquad \overline{(30)}$$

was done, at Aachen and Bruxelles, by a small sub-group of GARGAMELLE [60], using film from the exposure with freon (CF_3 Br). The type of target is important in these more detailed studies, since re-interactions of the produced pions in the parent nucleus may change their observable numbers (and charges) in a charge-dependent way! The treatment of these nuclear effects is intricate. It requires extended Monte Carlo calculations, and numerous experimental checks. But it has been done by two groups [61,62]. The results indicate effects of typically 20% for pion absorption (and/or charge-exchange), such that the corrected experimental charge ratios should be systematically good to better than 10%.

These charge ratios were deduced by Kluttig, Morfin and Van Doninck [60] from a raw sample of about 500 pions $\pi^{\pm,0}$, in each of the six channels. They are given in Table II, both, uncorrected and corrected after Pohl [62]. Also listed are a few trivial theoretical predictions: If the weak neutral current was an isoscalar ($I_{NC} = 0$), then all the three charge states of the pion must occur with the same weight. In the case of a pure isovector ($I_{NC} = 1$) this equality still holds for the two charged pions. Of course, all these equalities should be true for ν and $\bar{\nu}$.

A glimpse at Tab. II reveals that the $\pi^{\pm}/\pi^{0}$ ratios are completely off from unity: a pure isoscalar neutral current is thus absolutely excluded. The case against pure isovector is not quite as strong, but if one takes the totality of data, still suggestive enough. Thus we may conclude that the weak neutral current is a mixture of isovector and isoscalar.

In order to proceed further, Sehgal restricted the analysis to the "current-fragmentation region", where disintegrating quark and ensuing hadrons are most directly connected. In the data this

Table II: Ratios of numbers of pion charge states: a) observed, b) corrected, c) predicted by isoscalar, and d) isovector neutral current.

Frequency ratio:		π^{+}/π^{0}	π^{-}/π^{0}	π^{+}/π^{-}	$(\pi^{+}+\pi^{-})/\pi^{0}$
a) Observed for	ν	0.74±0.13	0.72±0.13	1.03±0.23	1.46±0.20
	$\bar{\nu}$	0.74±0.13	0.58±0.13	1.28±0.34	1.32±0.20
b) Corrected for	ν	0.77±0.16	0.60±0.15	1.29±0.38	1.37±0.24
	$\bar{\nu}$	0.76±0.16	0.43±0.15	1.77±0.67	1.18±0.23
c) Theory IS		1.0	1.0	1.0	2.0
d) Theory IV		-	-	1.0	-

selection could not be made exactly, since muon-less events are not sufficiently constrained. Instead, the experimenters [60] asked simply for an energy transfer $\nu > 1$ GeV, and for a value of the fragmentation variable $z_\pi = E_\pi/\nu$ between 0.3 and 0.7. By comparing with charged current data the authors concluded that the NC sample, collected this way, contains 80% current-fragmentation events, indeed.

The charge-ratios derived from this restricted sample (Table III) do differ from those of the unselected events: Again, the sheer fact that these numbers differ from each other (and from unity) demonstrates, without any model assumption at all, that the hadronic neutral current is a mixture of isovector and isoscalar. Besides, the $(\pi^+ + \pi^-)/\pi^0$ ratio of ≈ 2 confirms the idea that the pions observed stem from the "decay" of quarks with isospin $I_q = 1/2$. And as Sehgal remarked [59], "the direction of

Table III: Frequency ratios (%) of semi-inclusive pion charge states, with $\nu > 1$ GeV, and $0.3 \leq z_\pi \leq 0.7$, experimental and minimal model (with $x_W = 0.40$)

Frequency ratios in %	ν exp.	ν theory	$\bar{\nu}$ exp.	$\bar{\nu}$ theory
a) π^+/π^-	77 ± 14	71 ± 5	164 ± 36	143 ± 11
b) π^+/π^0	82 ± 17	83 ± 4	126 ± 21	118 ± 3
c) π^-/π^0	107 ± 22	116 ± 4	77 ± 21	82 ± 3
d) $(\pi^++\pi^-)/\pi^0$	190 ± 35	200 !	203 ± 45	200 !

the π^+/π^- asymmetry, in the ν and $\bar{\nu}$ data, holds an important message concerning the structure of the current", (namely the sign of the isoscalar-isovector interference term).

The analysis proper is quite straight-forward: Sehgal wrote the charge ratios in terms of the NC coupling constants, and of the quark fragmentation functions D_q^π for ν and $\bar{\nu}$ [59]:

$$\left(\frac{\pi^+}{\pi^-}\right)_\nu = \frac{(u_L^2 + \frac{1}{3} u_R^2) D_u^{\pi^+} + (d_L^2 + \frac{1}{3} d_R^2)\, D_d^{\pi^+}}{(u_L^2 + \frac{1}{3} u_R^2) D_u^{\pi^-} + (d_L^2 + \frac{1}{3} d_R^2)\, D_d^{\pi^-}} , \qquad (31)$$

$$\left(\frac{\pi^+}{\pi^-}\right)_{\bar{\nu}} = \frac{(u_R^2 + \frac{1}{3} u_L^2) D_u^{\pi^+} + (d_R^2 + \frac{1}{3} d_L^2)\, D_d^{\pi^+}}{(u_R^2 + \frac{1}{3} u_L^2) D_u^{\pi^-} + (d_R^2 + \frac{1}{3} d_L^2)\, D_d^{\pi^-}} . \qquad (\overline{31})$$

From charge symmetry:

$$D_u^{\pi^+} = D_d^{\pi^-} , \quad \text{and} \quad D_u^{\pi^-} = D_d^{\pi^+} . \qquad (32)$$

Thus the π^+/π^- ratios can be written just in terms of the two $D_u^{\pi^\pm}$ - fragmentation functions - in fact, they depend on their ratio only. This ratio is known from charged current and electroproduction data; the experimental cuts have been taken into account.

Inserting the experimental π^+/π^--ratios of Table III into eqs. (31) and ($\overline{31}$) gives two equations for the unknown coupling constants (squared). Two additional equations are known from the inclusive data already, namely (from GGM):

$$u_L^2 + d_L^2 = 0.24 \pm 0.04 \; ; \quad u_R^2 + d_R^2 = 0.055 \pm 0.030 . \qquad (33)$$

Solution of this system of four equations is easy, and gives:

$$u_L^2 = 0.082 \pm 0.035 \quad , \quad d_L^2 = 0.158 \pm 0.035 \quad ; \qquad (34L)$$

$$u_R^2 = 0.055 \pm 0.025 \quad , \quad d_R^2 = 0.001 \pm 0.025 \quad , \qquad (34R)$$

where the (strongly correlated) errors come mainly from $D_u^{\pi^+}/D_u^{\pi^-}$. Despite these limitations something remarkable has been achieved: for the first time the strength of the four fundamental N.C. quark couplings has been disentangled! From the inclusive data we know already left-quarks to dominate. Now we see that it is mainly the d_L-quark which couples. Right-handed couplings, instead, are very much smaller, and, if any, it is the u-quark which does the job ...

This considerable increase of knowledge is illustrated in Fig. 13: the bands in the $u_L^2 - d_L^2 - (u_R^2 - d_R^2)$ - planes come from the inclusive measurements (33), the ellipses from the semi-inclusive results (34). The unavoidable comparison with the minimal model [12,13] reveals again stupefying agreement, and suggests, for ν $(\bar{\nu})$ a most probable mixing constant $x_W = \sin^2\Theta_W$ of 0.31 ± 0.06 (0.34 ± 0.05). Accordingly, the Salam-Weinberg predictions for the charge-ratios in Table III were computed with $x_W = 0.30$. The spectacular agreement with experiment means, amongst other things, that the sign for the isovector-isoscalar interference term is right.

Sehgal's analysis has been resumed, modified, and extended by Sakurai [63], Ecker [64], Abott & Barnett [65] and others [66]. There is no time to describe all the tricks invented, nor all the data used, in order to make the analysis of neutral currents complete. Let us just register the main conclusions: all these authors agree essentially on the set of coupling constants

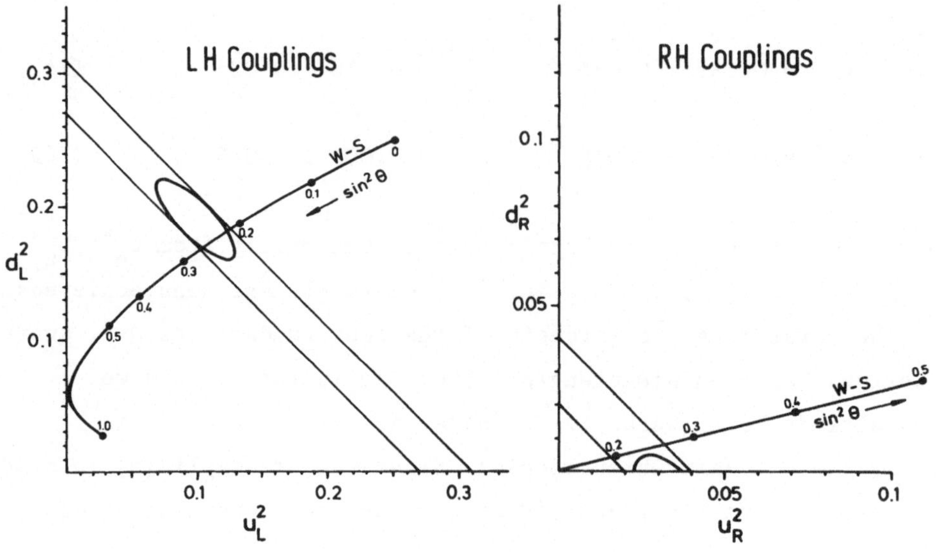

<u>Fig. 13</u>:   Restrictions imposed by semi-inclusive NC pion
            production (ellipses) on left-handed and right-handed
            u and d couplings from inclusive data (bands). Curve
            minimal model.

(squared), as given in eq. (34). As we have seen, they follow
from the minimal model with $x_w \approx 0.25$ to 0.3.

   We close this chapter with a remark on NC induced <u>exclusive</u>
pion production:

$$\overset{(-)}{\nu} + \mathcal{N} \;\rightarrow\; \overset{(-)}{\nu} + \mathcal{N} + \pi \;\; , \tag{35}$$

where $\mathcal{N}$ can be either a proton P or a neutron N, and π a pion
of any charge permitted by charge conservation. We have mentioned
the π^0-channels already in eqs. (13), and we stressed their
importance as νe background. As demonstrated by the sparkchamber
event of Fig. 4, exclusive single π^0 production, giving one or
two visible decay gammas (and perhaps a proton at the origin) is

a very characteristic process. It was amongst the first NC reactions to be discovered: by GARGAMELLE at first [67] and then in quick succession by Aachen-Padova [25] and Columbia Illinois-Rockefeller [68]. Charged pion production was studied by an ANL group [69], under grave difficulties with statistics and neutron background, and again by GARGAMELLE [70]. Finally, a breakthrough was achieved, when the GARGAMELLE chamber was filled with propane offering free protons as targets, and when the reactions off a proton were seperated from those off a neutron, so that all NC 1π channels could be studied simultaneously [71].

It was quite natural that Abott and Barnett [65] used exclusive one-pion data, and in particular the π^0 results, in their analysis. This has been criticized, in particular by Evelyn Monsay [72], because the data show systematic discrepancies, the worst one being in the measured π^0 NC/CC ratio for neutrinos R_0', which was found to be 0.40 ± 0.06 by AC-PD [25], and barely half as high in the early GGM freon data [67], and by CIR [68]. The reason for that lies probably in the very different experimental biases and cuts, which in turn imply substantial differences in the nuclear corrections. For example: AC-PD asked for two visible π^0 decay-gammas, whereas all other groups were content with one of them. In this way AC-PD put a very high energy threshold (≈ 300 MeV) into their acceptance. This experimental hardening of the π^0 spectrum, decreases the nuclear effects, so that the measured NC/CC ratio R_0' should be not far off the true value R_0.

But this remains to be proven by detailed Monte Carlo calculations. Until that is accomplished we shall not use the old ratios. Instead, we illustrate the power of exclusive data by some high-lights from the recent GARGAMELLE propane experiment [71]: Fig. 14 shows $P\pi$-invariant mass distributions, from top to bottom: in the NC $P\pi^0$ and $P\pi^-$ channels, and below in the CC $P\pi^0$ channel for comparison. In all three channels the Δ resonance

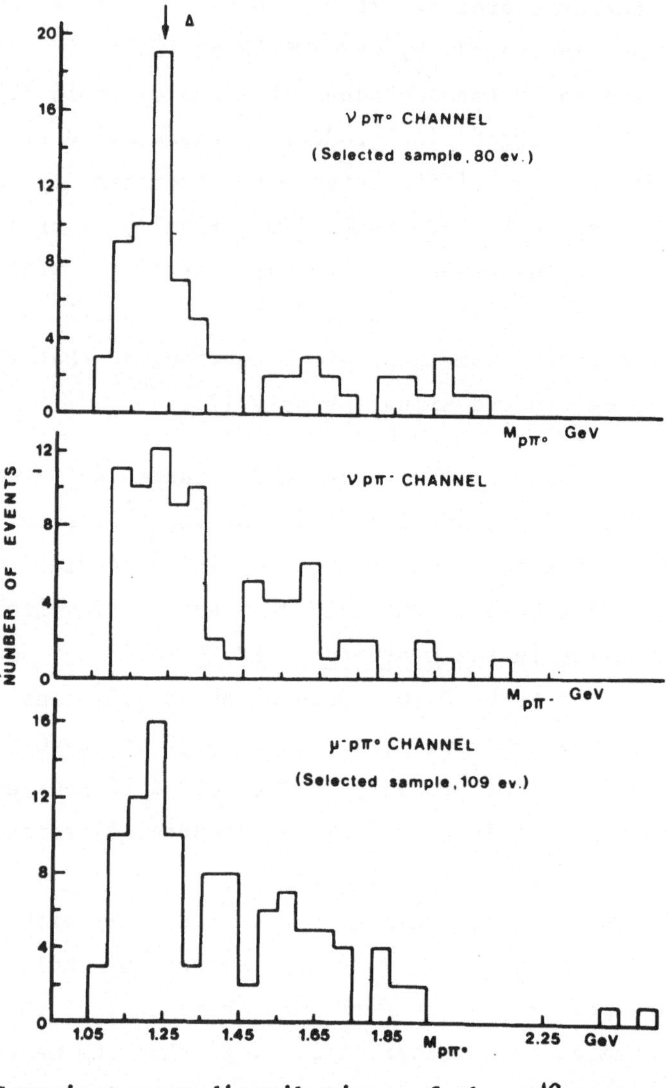

Fig. 14: Invariant mass distributions of the $\pi\mathcal{N}$ system in the final states: $\nu_\mu P\pi^0$, $\nu_\mu P\pi^-$; and (for comparison) $\mu^- P\pi^0$.

is clearly visible, (actually in the π^0 channels more cleanly than with π^-). This is the direct, and long awaited proof [73] that the neutral current does contain an isovector piece. And, comparing NC with CC, one has the impression that this isovector component dominates in both!

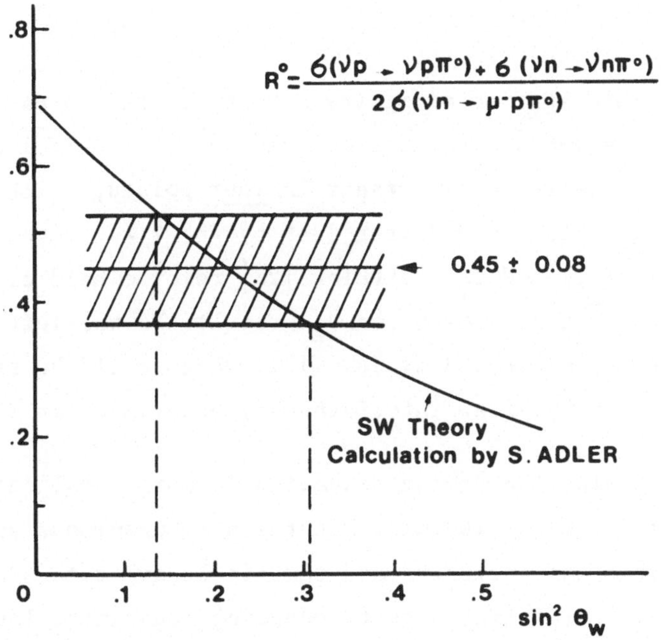

Fig. 15: Exclusive single π^o ν-induced NC/CC ratio R^o vs. $\sin^2\Theta_W$. Calculation by Adler, experimental band (dashed) from the GARGAMELLE propan experiment.

Finally, let us have a glimpse at the NC/CC ratio from the GARGAMELLE propane experiment [71], and its comparison to the Salam-Weinberg model [12,13], as worked out by Adler [61] (Fig. 15). The value of R_o = 0.45 ± 0.08 may signal hope to AC-PD. Besides, the mixing angle comes out comfortingly low.

6. Elastic Neutrino-Nucleon Scattering and the Signs of the Neutral Current Coupling Constants

The pion data yielded the magnitude of the NC coupling constants u_i and d_j, but they did not fix their sign. These ambiguities imply considerable disparities in the physical meaning of the different sign combinations ("solutions"). In order to remove them, one has to perform some sort of interference experiment.

We had this sort of ambiguity already in the νe data of Fig. 7: the $\nu_\mu e$ and $\bar{\nu}_\mu e$ total cross sections, even if measured with infinite precision, intersect in four points, which are completely equivalent and differ just by the 2 × 2 sign combinations of C_A and C_V. Reithler [28] removed half of them by combining the $\overset{(-)}{\nu}_\mu e$ data with the $\bar{\nu}_e e$ result. He was left with the A vs. V dilemma, expressed by eq. (17), which could be resolved by the eD asymmetry (i.e. interference!) experiment at SLAC [41].

In principle, the hadron situation is more complicated, since hadrons carry (strong) isospin. But there is unanimous agreement [59,63-66] that the inclusive and semi-inclusive data admit just four physically different sets of coupling constants, labeled "Solution A to D". Their leading couplings are, respectively:

$$A = (A, iV), \quad B = (V, iV) \quad ; \quad C = (A, iS), \quad D = (V, iS) \qquad (36)$$

Solution A is practically identical with the Salam-Weinberg model, solution B is its vector-like alternative, and the two remaining, isoscalar dominated, solutions are rendered already quite improbable by the semi-inclusive and exclusive pion production results.

There are many ways to decide amongst these four possibilities. Quite a few of them have been tried in the analyses mentioned. It is comforting that the conclusions are nearly the same ... We shall limit ourselves to a very simple and obvious reaction, namely to elastic ν_μ ($\bar{\nu}_\mu$) scattering off protons (and neutrons!):

$$\nu_\mu + P \rightarrow \nu_\mu + P \quad (37P) , \qquad \bar{\nu}_\mu + P \rightarrow \bar{\nu}_\mu + P \; ; \quad (\overline{37P})$$

$$\nu_\mu + N \rightarrow \nu_\mu + N \quad (37N) , \qquad \bar{\nu}_\mu + N \rightarrow \bar{\nu}_\mu + N \; . \quad (\overline{37N})$$

It is intuitively clear that these truly elastic $\nu \mathcal{N}$ scattering processes entail interference between different amplitudes, as the quarks act coherently. And the detailed analysis [14] of these processes reveals that their discrimination power between different models and solutions is remarkable.

Experimentally, $\nu_\mu P$ scattering has been looked for long before the structure of neutral currents was under dispute - actually long before neutral currents were discovered [75]. In the very first CERN neutrino beam I had a scintillation counter sandwich running [76], which was rather sensitive to low energy proton recoils from (37P) - unfortunately also to a very heavy background from neutrons:

$$N + P \rightarrow P + N \; . \quad (38)$$

Consequently only a very liberal limit could be placed on the NC/CC cross section ratio R_P between (37P) and the quasi-elastic ν_μ-reaction:

$$\nu_\mu + N \rightarrow \mu^- + P \quad (39) , \qquad \bar{\nu}_\mu + P \rightarrow \mu^+ + N \; . \quad (\overline{39})$$

(To the $\bar{\nu}_\mu$ reaction we shall be getting later). A terribly stringent limit on R_P of $\lesssim 3\%$ was reported by the CERN heavy liquid bubble chamber [77]. It discouraged many people (including myself), but turned out to be wrong! A more solid limit from the same data was derived much later by Cundy <u>et al.</u> [78]: $R_P \lesssim 10\%$, a number which coincides actually with today's value.

Truly elastic $\nu_\mu P$ scattering was discovered by the Columbia-Illinois-Rockefeller (CIR) Collaboration [79], and independently by the Harvard-Pennsylvania-Wisconsin (HPW) Group [80], both working at the Brookhaven National Lab (N.Y.). CIR used an aluminum spark-chamber, similar to that of Aachen-Padova. HPW employed a scintillator counter sandwich interleaved with drift-chambers, which was specially taylored to detect and to measure ν_μ- and $\bar{\nu}_\mu$-P scattering, and made ample use of counter pulse-height and time-of-flight information. They have recently obtained very impressive data on $\nu_\mu P$, and were the only ones to report on $\bar{\nu}_\mu P$ [81]. Comparison with the GARGAMELLE measurement in propane [82] shows that here (for a change) counter techniques are superior.

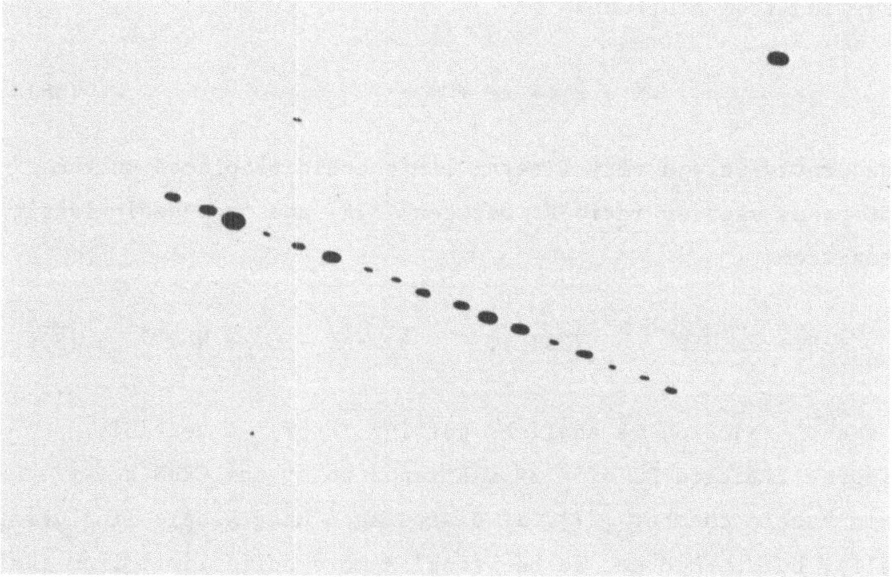

Fig. 16: Recoil proton in the Aachen-Padova sparkchamber.

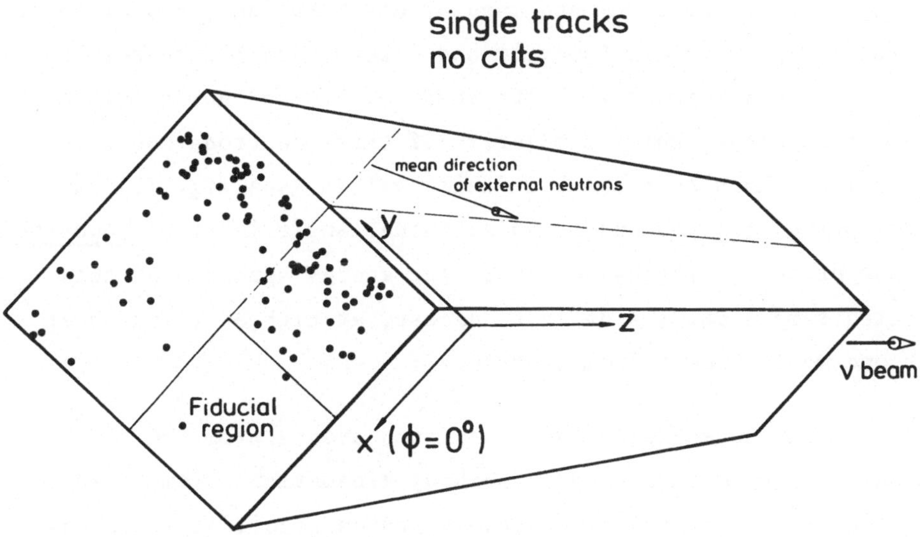

Fig. 17: Spatial distribution of proton starting points (projected on the front face) in the AC-PD sparkchamber.

Soon after the discovery of $\nu_\mu P$, the Aachen-Padova Group saw a recoil proton signal appearing in their spark chamber [83]. I shall spend the rest of my review describing this experiment, partly, because it is not yet as well published as the others are, but mainly because it gave me great pleasure, to watch our students, Uli Samm and Elisabeth Frenzel, overcoming slowly but steadily all difficulties. Indeed, to observe a single straight proton track is easy (Fig. 16). But the distribution of their starting points, projected onto the front side of the chamber (Fig. 17), is highly non-uniform. Together with a very strong azimuthal angular asymmetry it proves that most of these protons stem from neutrons entering the chamber from above, and with an average angle as indicated in Fig. 17.

A detailed study of these neutrons was the first task, notably of their spectrum and angular distributions, and of their interaction properties. Spectrum and interaction length were found in good agreement with the study of Fry and Haidt [9]. It became also clear, how the majority of these neutrons could be removed: by choosing a small fiducial volume (see Fig. 17), and by accepting only protons in an azimuthal angle interval <u>opposite</u> to that of the impinging neutrons. The energy spectrum of this selected sample shows a definite excess, at high energies, over what one would expect from neutrons.

In order to separate these neutrino events from the remaining neutron background, two-body kinematics is employed, specifically the correlation between proton recoil angle Θ_P and kinetic energy T_P. For a proton initially at rest:

$$E_P \; (1 - v_o \; v_P \; \cos\Theta_P) \; = \; M \; (1 - y) \quad , \qquad (40)$$

where the initial projectile's velocity $v_o = 1$ for ν, and $= P_N/E_N$ for N. Fermi motion and/or nuclear interactions disturb this relation, but on average it will still hold [84]. A scatter plot of range ($\sim T_P$) against $\cos\Theta_P$ is thus a reasonable diagnostic means to tell the rams from the sheep, as it were. Fig. 18a shows the $\nu_\mu P$ candidates selected as described, Fig. 18b the protons from the quasi-elastic CC reaction (39) where there is no neutron background at all, and Fig. 18c gives the full single proton sample, consisting almost entirely of neutron recoils. Hence one sees confirmed, what one could have anticipated from the start: neutron recoils cluster at small angles <u>and</u> small energies! No such cluster is seen in the genuine neutrino events (Fig. 18b), but the $\nu_\mu P$ candidate sample (Fig. 18a) shows both, large angle-high energy neutrino-like events, and the bad neutron blob near the origin. I cut it out by demanding a (reconstructed) neutrino

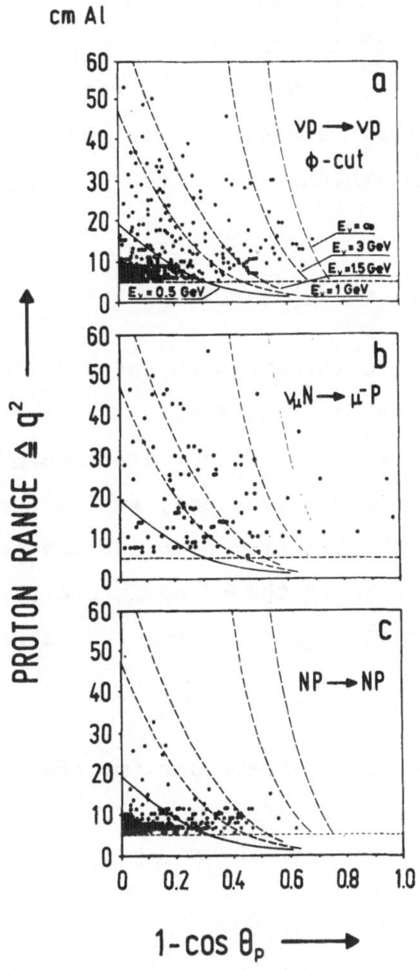

Fig. 18: Proton range and emission angle Θ_P scatter plots for a) $\nu_\mu P$ candidates, b) elastic CC ν-induced P's, c) protons from neutron background (AD-PD).

energy > 0.5 GeV. The remaining neutron background can safely be subtracted, since the spectrum is known.

In order to derive a meaningful NC/CC ratio R_P, the protons from the quasi-elastic sample (Fig. 18b) have been treated in exactly the same way as the $\nu_\mu P$ scattering candidates. Thus, nuclear corrections tend to cancel. Even so, it took several months of hard work, and all the machinery provided by Adler [61], Pohl [62] and others, to convince ourselves that all the pion contaminations in either sample, and all the proton and neutron absorption and charge exchange reactions were under control. The result is a raw ratio

$$R_P' = (15 \pm 4)\% \quad .$$

This result is in good agreement with the older values published [79,80,82]. However, as Samm was first to notice, it is really due to a mixture of $\nu_\mu P$- and $\nu_\mu N$-scatters! Depending on the ratio of νN to νP scattering cross sections, and on the experimental conditions, the contribution from neutrons may be quite large. Conventionally one subtracts the νN admixture by assuming a certain $\nu N/\nu P$ cross section ratio r_N, and tries to reach some conclusion by iteration.

Instead, we decided to accept our imperfections, and to take advantage of them.* Thus, we admit measuring a mixture of $\nu_\mu P$ and $\nu_\mu N$, and hence an experimental NC/CC ratio

$$R_P' = R_P + f_{PN} R_N \quad , \tag{41}$$

where f_{PN} is the (relative) probability that a neutrino produced neutron makes an acceptable proton. This transformation may happen:

a) in the parent nucleus,

b) somewhere in the chamber.

Case a) can be treated by looking at protons in the quasi-elastic $\bar{\nu}_\mu$ reaction $(\overline{37})$, case b) by investing the interaction

---

* This attitude was called "Faissner's Principle" by Mitter.

properties of neutrons, and the geometry of the set-up, in lengthy Monte-Carlos. Samm got the combined result [84]:

$$f_{PN} = (33 \pm 4)\% \; .$$

With this number plugged into (41), our experimental NC/CC ratio defines a band in the R_P-R_N plane (Fig. 19). The four solutions are also shown, and there is no doubt that we favour A, and exclude C. Since D is out, for the pions' sake, we are again left with the A-V ambiguity first encountered in νe. Fortunately, the new Harvard et al. data on $\nu_\mu P$, and even more incisively those on $\bar{\nu}_\mu P$, eliminate the vector-like solution B at a three st.dev. level [81]. Hence we can take solution A - or right away Salam-Weinberg - and determine the relevant parameters from the upper intersection point of our experimental line with "Weinberg's nose", (the second solution is unphysical):

$$R_P = 0.10 \pm 0.03, \quad \sin^2\Theta_W = 0.32 \,{}^{+0.18}_{-0.09} \; ; \quad R_N = 0.15 \pm 0.04 \qquad [85]$$

The first two numbers are in perfect agreement with those of HPW [81]. The neutron ratio R_N is new, and I emphasize that it was as directly (or indirectly) derived as R_P was.

Little remains to be said: from its origin with the fairy GARGAMELLE the weak neutral current has grown into a veritable stream, fecundating such different regions as electron scattering and annihilation, atomic and nuclear physics, astronomy and cosmology. Its structure has been unveiled, and it is improbable that the still controversial experiments, like those in optics, will force us to resort to more complicated, parity conserving gauge models, like that of Fritzsch and Minkowski [86].

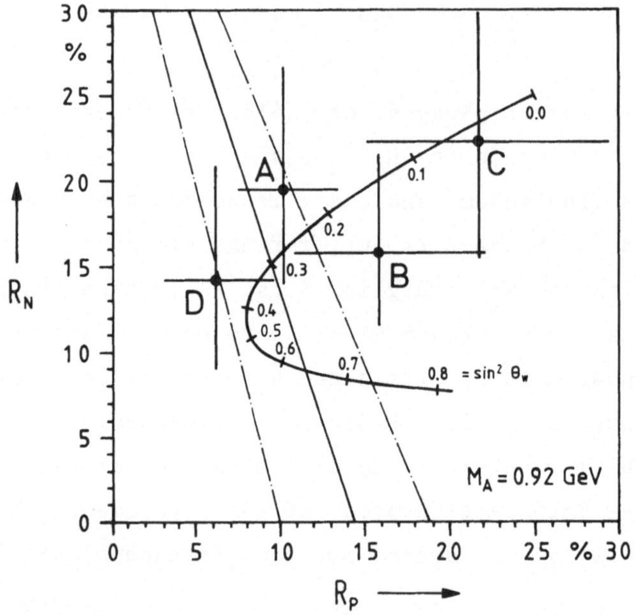

Fig. 19: Model-free comparison of the AC-PD $\nu_\mu P + \nu_\mu N$ scattering result (band) with the four coupling constant solutions A-D, and Weinberg-Salam.

This is not to say that we should be idle. Many experiments have to be improved, some must be undertaken anew. Let me close with my personal list of "things to be done":

- all exclusive reactions have to be measured an order of magnitude better than now (i.e. with a factor of 1/3 in error!)
- e.m.-weak interference ought to be done at varying q^2,
- the atomic and nuclear mess has to be cleaned up,
- I want to see if ν_e's have the same neutral current as ν_μ,

- to study $\nu_e e$ (with interference term!) would be particularly nice,
- nuclear spectroscopy by neutrino induced N.C. looks extremely attractive [87]
- the neutral current of the muon must be explored by $\bar{e}e \rightarrow \bar{\mu}\mu$ (say at PETRA).

Last not least:

- coupling constants, like $\sin^2\Theta_w$, could be a function of q^2, and, therefore, of energy. This should be explored by all means, and might explain some of the discrepancies which occured in comparing experiments done at different energies [88].

This ends my list of urgent experiments, I do not feel competent to adjoin an analogous list for theoreticians, I add just one wish: that they continue to look for symmetries beyond our minimal $SU_2 \times U_1$ group, striving for the ultimate unification of all forces of Nature.

Acknowledgements

This review would not have been written without the advice, the suggestions, and the corrections, from the part of the many colleagues and friends at Aachen Tech, who worry about the different aspects of weak neutral currents. I am particularly indebted to Lalit Sehgal, who guided me through all the relevant aspects of theory, and in addition was kind enough to do the proof-reading. For help in collecting and understanding the data, I wish to thank my friends Deden, Hasert, Krenz, Morfin, Pohl, Reithler and Samm. Irene Gojdie typed patiently and accurately the manuscript, Hubert Schulz did a fine job with the figures, and with the reproduction, and Michael Grimm worked miracles to get everything together. Last not least, I wish to thank Prof. Fries, who was so kind to invite me, and who pushed me to get the talk written up.

References

[1] P. Musset and J.-P. Vialle, Neutrino Physics with Gargamelle, Physics Reports 39C, No. 1 (1978) p. 1

[2] H. Faissner et al., Contribution No. 273 to: Proc. 6th Internat. Symposium on Electron and Photon Interactions at High Energies, Bonn 1973, Ed. H. Rollnik and W. Pfeil (North-Holland 1974) p. 550

[3] G. Myatt, ibid. p. 389

[4] H.J. Hasert et al., Phys. Letters 46B, 121 (1973)

[5] E.C. Young, Oxford Thesis (1966), and Yellow Report CERN 67-12
M. Paty, Paris Thesis (1965), and CERN 65-12

[6] M. Holder, Aachen Thesis (1967) and Aachen Report PITHA-19 (1967)

[7] H. Faissner et al., Nucl. Instrum. and Meth. 20, 213 (1963)
H. Faissner, Acta Physica Austriaca, Suppl. I (1964) p. 189

[8] H.J. Hasert et al., Phys. Letters 46B, 138 (1973)

[9] W.F. Fry and D. Haidt, CERN-Report 75-1 (1975)
D. Haidt, Phys. Blätter 31, 654 (1975)

[10] S. Glashow, Nucl. Phys. 22, 579 (1961)

[11] A. Salam and J.C. Ward, Phys. Letters 13, 168 (1964)

[12] A. Salam, Elementary Particle Theory, ed. N. Svartholm (Almquist and Wiksells, Stockholm 1968) p. 367

[13] S. Weinberg, Phys. Rev. Letters 19, 1264 (1967); 27, 1688 (1971)
see also Phys. Rev. D5, 1412 (1972)

[14] G. 't Hooft, Nucl. Phys. B33, 173 (1971) and B35, 167 (1971)
G. 't Hooft and M. Veltman, Nucl. Phys. B44, 189 (1972)

[15] see E.S. Abers and B.W. Lee, Phys. Reports 9C, No. 1 (1973)
S. Weinberg, Aix-en-Provence Conf. (1973), J. Physique, Suppl. 10, 34, C1 (1973)
L.M. Sehgal, Aachen Report PITHA-68 (1973)

[16] S.L. Glashow, J. Iliopoulos, and L. Maiani, Phys. Rev. D12, 1285 (1970)

[17] K.M. Case, Phys. Rev. 107, 307 (1957)

[18] G. 't Hooft, Phys. Letters 37B, 195 (1971)

[19] P. Rosen, Proc. Neutrino Conf. Downington 1974 (AIP, New York 1974) p. 5
L.M. Sehgal, Nucl. Phys. B70, 61
M. Gourdin, Proc. Neutrino Conf. Aachen 1976, ed. H. Faissner, H. Reithler, and P. Zerwas (Vieweg, Braunschweig 1977) p. 234

[20] H. Faissner, Proc. B.W. Lee Memorial Conf. 1977, ed. D. Cline (FNAL, Batavia, 1978) p. 85

[21] see H. Faissner, Proc. Varenna 1977 School, Nuovo Cim. Suppl., in press

[22] H. Faissner _et al._ pres. by F. Bobisut, Proc. Neutrino Aachen 1976, loc.cit., p. 223
H. Faissner _et al._, Proc. Neutrino Conf. Baksan 1977, ed. M.A. Markov _et al._ (Nauka, Moscow, 1978) Vol. 2, p. 142

[23] J. Blietschau _et al._, Nucl. Phys. B114, 189 (1976)
see also F.J. Hasert, _La physique du neutrino à haute energie_ (Paris, 1975) p. 257

[24] J. Blietschau _et al._, Phys. Letters 73B, 232 (1978)

[25] H. Faissner _et al._ pres. by T. Hansl, Proc. Neutrino 1976, loc.cit., p. 278

[26] H. Faissner _et al._, Phys. Letters 68B, 377 (1977)

[27] H. de Witt, Diplomarbeit T.H. Aachen (1978)

[28] H. Reithler, Proc. Int. Conf. Lepton Physics High Energies, Hamburg 1977, Ed. F. Gutbrod (DESY 1977) p. 343

[29] H. Faissner _et al._, Phys. Rev. Letters 41, 213 (1978)

[30] H. Faissner _et al._, presented by M. Baldo-Ceolin, Proc. Neutrino Conf. Purdue, 1978, ed. E.C. Fowler (Purdue, West Lafayette, Ind. 1978) p. 387
E. Radermacher, Proc. Tokyo High Energy Conf. 1978, ed. S. Homma, M. Kawaguchi, and H. Miyazawa (Phys. Soc. Japan 1979) p. 321
M. Baldo-Ceolin _et al._, Proc. Tokyo, loc.cit., contr. paper No. 1087
G. Puglierin, Proc. Conf. Neutrino Phys. Acc. Oxford 1978, ed. A.G. Michette and P.B. Renton (Rutherford, 1978) p. 279

[31] G. Gershwin, Porgy and Bess, Aria of "Sportin' Life"

[32] P. Alibran et al., Phys. Letters 74B, 422 (1978)

[33] N. Armenise et al., Phys. Letters 81B, 385 (1979)

[34] A.M. Cnops et al., Phys. Rev. Letters 41, 357 (1978)
C. Baltay et al., Proc. Neutrino Conf. Purdue, loc.cit. p. 413
C. Baltay, Proc. Tokyo Conf. loc.cit. (1979) p. 882

[35] N. Armenise et al., Phys. Letters 81B, 385 (1979) - BEBC
J.B. Berge et al., Phys. Rev. Letters 41, 357 (1978) - 15'

[36] D. Bertrand et al., Phys. Letters (1979) in press, Aachen Report PITHA-79/09

[37] F. Reines et al., Phys. Rev. Letters 37, 315 (1976) and Proc. Aachen Conf. 1976, loc.cit., p. 217

[38] Fred Reines, in his publications [37], gave the cross section only as a function of energy. Upon my insistence he also averaged it, and telephoned the result plotted in Fig. 7 from California.

[39] R.P. Feynman and M. Gell-Mann, Phys. Rev. 109, 193 (1958)
E.C.G. Sudarshan and R.E. Marshak, Phys. Rev. 109, 1860 (1958)

[40] D.A. Ross and M. Veltman, Nucl. Phys. 95B, 135 (1975)

[41] C.Y. Prescott et al., Phys. Letters 77B, 347 (1978)

[42] see N. Fortson, Proc. Neutrino Conf. Purdue, 1978, loc.cit. p. 417
L.M. Barkov and M.S. Zolotorev, ibid., p. 423
L. Wilets, ibid., p. 437

[43] F.J. Hasert et al., Nucl. Phys. B73, 1 (1974) - Final discovery
J. Blietschau et al., Nucl. Phys. B118, 218 (1977) - Last data

[44] A. Benvenuti et al., Phys. Rev. Letters 32, 800 (1974)
B. Aubert et al., ibid. 32, 1454 (1974)
T.Y. Ling, Proc. Neutrino Conf. Aachen 1976, loc.cit., p. 296
P. Wanderer et al., Phys. Rev. D17, 1679 (1978) - data

[45] B.C. Barish et al., Phys. Rev. Letters 34, 538 (1975)
L. Stutte, Weak Interaction Workshop, Ohio State Univ. (1975), unpublished
F.S. Merritt et al., CALT 68-601 (1977) - data
B.C. Barish et al., CALT 68-605 (1977) - data

[46] L.M. Sehgal, Phys. Letters 55B, 205 (1975)

[47] S. Sakakibara and L.M. Sehgal, Phys. Letters 83B, 77 (1979)

[48] S.P. Rosen, Proc. Neutrino Conf. Downingtown 1974, ed. C. Baltay (AIP, New York, 1974) p. 5
S.P. Rosen et al., Proc. Neutrino Conf. Balaton 1975, ed. A. Frenkel and G. Marx (Budapest 1975), Vol. I, p. 186
M. Gronau, Haifa Tech preprint (1975)
S. Pakvasa and G. Rajasekaran, Phys. Rev. D12, 113 (1975)
J.J. Sakurai, Neutrinos-1974, loc.cit. p. 57

[49] B. Kayser et al., Phys. Letters 52B, 385 (1974)
see also R.L. Kingsley, Phys. Rev. D10, 2216 (1974)
T.C. Yang, Phys. Rev. D10, 3744 (1974)
C. Jarlskog, Nuovo Cim. Letters 4, 377 (1970)

[50] R.P. Feynman, Photon-Hadron Interactions (Reading, Mass., 1972); Proc. Neutrino Conf. Balaton 1972, ed. A. Frenkel and G. Marx (Budapest 1972) Vol. 2, p. 75
J.D. Bjorken and E.A. Paschos, Phys. Rev. 185, 1975 (1969)

[51] E.D. Bloom et al., Phys. Rev. Letters 23, 930 (1969)
M. Breidenbach et al., Phys. Rev. Letters 23, 935 (1969)
see also W.K.H. Panofsky, Proc. Int. Conf. High-Energy Phys. Vienna 1968, ed. J. Prentki and J. Steinberger (Vienna 1969) p. 23

[52] T. Eichten, H. Faissner et al., Phys. Lett. 46B, 274 (1973)
H. Deden et al., Nucl. Phys. B85, 269 (1975)
see also W. Krenz, Neutrino-1976 Aachen, loc.cit., p. 385 and Habilitationsschrift Aachen (1978)

[53] H. Faissner, Aachen Report PITHA-60 (1972)
D. Rein, Phys. Letters 43B, 209 (1973)
The concept of valence quarks was introduced by J. Kuti and V.F. Weisskopf, Phys. Rev. D4, 3418 (1971); see also S.D. Drell and K. Johnson, Phys. Rev. D6, 3248 (1972)

[54] P.C. Bosetti et al., Phys. Letters 70B, 273 (1977) - CC
P.C. Bosetti et al., Phys. Letters 76B, 505 (1978) - NC
H. Deden et al., Nucl. Phys. B149, 1 (1979) - NC, and Proc. Oxford Conf. (1978), loc.cit. p. 292

[55] H. Faissner et al., Proc. Neutrino-1976, Aachen, loc.cit., p. 393
H. Faissner and B. Kim, to be published

[56] L.M. Sehgal, Nucl. Phys. B65, 141 (1973); see also R.B. Palmer, Phys. Letters 46B, 240 (1973)
A. de Rújula et al., Rev. Mod. Phys. 46, 391 (1974)

[57] M. Holder et al., Phys. Letters 71B, 222 and 72B, 254 (1977) Also F. Dydak, Facts and Prospects of Gauge Theories, Schladming Lectures (1978), ed. P. Urban (Wien & New York 1978) p. 436 and Habilitationsschrift Heidelberg (1978) J. Steinberger, Hadron Structure and Lepton-Hadron Interactions, Cargèse 1977, ed. M. Lévy et al. (New York & London 1979) p. 201

[58] Traudl Hansl (CDHS), private communication

[59] L.M. Sehgal, Phys. Letters 71B, 99 (1977); and Proc. Neutrino Purdue (1978) loc.cit. p. 253

[60] H. Kluttig, J.G. Morfin, and W. Van Doninck, Phys. Letters 71B, 446 (1977)

[61] S.L. Adler, S. Nussinov, and E.A. Paschos, Phys. Rev. D9, 2125 (1974)

[62] M. Pohl, Aachen Thesis (1979)
C. Longemare, Orsay Thesis (1978)

[63] P.Q. Hung and J.J. Sakurai, Phys. Letters 63B, 295 (1976); 69B, 323 (1977); and 72B, 208 (1977)
see also J.J. Sakurai, Proc. Neutrino Conf. Aachen 1976, loc.cit. p. 515; Proc. Neutrino Conf. Baksan 1977, loc.cit. Vol. 2, p. 242; Proc. Conf. Neutrino Acc. Oxford 1978, loc.cit. p. 338; Current Trends in the Theory of Fields, (Dirac-Symp.) ed. J.E. Lannutti and P.K. Williams (AIP, New York 1978) p. 38

[64] G. Ecker, Phys. Lett. 72B, 450 (1977)
G. Ecker and H. Pietschmann, Acta Phys. Austr. 45, 313 (1976)

[65] L.F. Abbott and R.M. Barnett, Phys. Rev. D18, 3214 (1978)

[66] P. Langacker and D.P. Sidhu, Phys. Letters 74B, 233 (1978)

[67] F.J. Hasert et al., Phys. Letters 59B, 485 (1975)

[68] W. Lee et al., Proc. Neutrino Conf. Aachen 1976, loc.cit., p. 319

[69] S.J. Barish et al., Phys. Rev. Letters 33, 448 (1974) and 36, 179 (1976)
see also P. Schreiner, Proc. Neutrino Conf. Aachen 1976, loc.cit., p. 333

[70] G.H. Bertrand-Coremans, P. Musset et al., Phys. Letters 61B, 207 (1976)

[71] W. Krenz et al., Nucl. Phys. B135, 45 (1978), and M. Pohl, Ref. [62]

[72] E.H. Monsay, Phys. Rev. Letters 41, 728 (1978)

[73] see B.W. Lee, Proc. Neutrino Conf. Aachen 1976, loc.cit., p. 704

[74] C.H. Albright et al., Phys. Rev. D14, 1780 (1976)
E. Fischbach et al., Phys. Rev. D15, 97 (1977)

[75] see H. Faissner, Proc. Neutrino Conf. Balaton 1975, loc.cit., Vol. I, p. 85

[76] H. Faissner et al., Nuovo Cim. 32, 782 (1964)

[77] M.M. Block et al., Phys. Letters 12, 281 (1964) - On p. 285 the statement is even: G^{NC}/G^{CC} ($\Delta S = 0$) < 3%!
see also H.H. Bingham et al., Proc. Siena Conf. 1963, ed. G. Bernardini and G.P. Puppi (Bologna 1963), Vol. I, p. 555

[78] D.C. Cundy et al., Phys. Letters 31B, 478 (1970)

[79] W. Lee et al., Phys. Rev. Letters 37, 186 (1976), and Ref. [68]

[80] D. Cline et al., Phys. Rev. Letters 37, 252 (1976) and 37, 648 (1976);
brilliant: L.R. Sulak et al., Proc. Neutrino Conf. Aachen 1976, loc.cit., p. 302

[81] L.R. Sulak et al., Proc. Neutrino Conf. Purdue 1978, loc.cit., p. 355
H.H. Williams, Proc. Conf. Neutrino Acc. Oxford 1978, loc.cit., p. 315
A. Entenberg et al., Phys. Rev. Letters 42, 1198 (1979)

[82] M. Pohl et al., Phys. Letters 72B, 489 (1978)

[83] H. Faissner et al., Proc. Neutrino Conf. Baksan 1977, loc.cit., Vol. 2, p. 164

[84] U. Samm, Diplomarbeit Aachen (1979)

[85] H. Faissner et al., Proc. Neutrino Conf. Purdue 1978, loc.cit., p. C89 and Phys. Rev., to be published

[86] H. Fritzsch and P. Minkowski, Nucl. Phys. B103, 61 (1976)

[87] T.W. Donnolly and R.D. Peccei, Phys. Reports 50, 1 (1979)

[88] Traudl Hansl, in the discussion following her Aachen talk [25], had envisaged such an energy dependence. (She was well ahead of her time ...)

CHARM SPECTROSCOPY

G. Buschhorn

Max-Planck-Institut für Physik und Astrophysik
München

ABSTRACT

In three lectures the experimental situation on the resonances of the J/ψ-family and charmed mesons obtained from e^+e^--reactions in the region of 3-5 GeV total energy is reviewed and discussed in light of the quark model. Following a survey of the relevant detectors at the SPEAR- and DORIS-storage rings the following topics and results are discussed in detail:

1) Total hadronic annihilation cross section σ_{had}^{tot}. A threshold in σ_{had}^{tot} is observed at about 3.7 GeV with narrow resonances J/ψ(3095), ψ'(3685) below it and a broad resonance ψ''(3772) and several resonance-like structures at 4.03, 4.16 and 4.41 GeV above it. The behaviour of σ_{had}^{tot} is consistent with the threshold for the production of a new heavy quark (charm quark c) with a mass of about 1.5 GeV and a charge of (2/3)e.

2) J/ψ-family ($c\bar{c}$-states). On the basis of their measured quantum numbers and properties the narrow resonances J/ψ(3095) and ψ'(3685) are assigned to the 1^3S_1-resp. 2^3S_1-state of a bound $c\bar{c}$-system (charmonium). Radiative decays of ψ' into intermediate states χ(3415), P_c/χ(3510) and χ(3550) have been measured and are in good agreement with their assignment to the $^3P_0$-, $^3P_1$-, $^3P_2$-charmonium states. Measurements of radiative decays into further states, X(2820) and χ(3455), are in serious disagreement with their assignment to 1^1S_0-resp. 2^1S_0-charmonium states. The ψ''(3772)-state, decaying predominantly into $D\bar{D}$, is assigned to the 1^3D_1-charmonium state; possible assignments for the structures in the 4.0-4.4 GeV region are discussed.

3) Charmed mesons. The evidence for the D-mesons as $c\bar{q}$-resp. $\bar{c}q$-bound states (q = u, d-quark) is discussed on the basis of the Glashow-Iliopoulos-Maiani model. The experimental determination of the masses, quantum numbers and decay branching ratios of D- and D*-mesons is discussed in detail. Furthermore, the evidence for parity violation and semileptonic decay modes of the D-mesons is discussed. Evidence for the F-mesons as $c\bar{s}$-resp. $\bar{c}s$-state is derived from the threshold behaviour of inclusive η-production, fits to the ηπ-invariant mass, and from its observed semileptonic decay. The observed threshold behaviour of $(p+\bar{p})$- and $(\Lambda+\bar{\Lambda})$-production in the region 4.5-5.5 GeV may be associated with charmed baryon production.

INDEX

Anomalous dimension 288
Asymptotic freedom 85,162
Bag 354
Baryon Antibaryon 346
Bjorken scaling 267
Bjorken variables 216
Cabibbo-angle(s) 8,37
Cabibbo-universality 36
Callan-Gross relation 89
Charmonium 164,181
CP-violation 33
Colour 95
Cross section of elastic ν-scattering 282
Cross section of weak leptonic neutral current processes 379
Deep inelastic scattering 84
Dimensions 59,79
Diquarks 346
Drell-Yan-processes 55,85,89, 246,256
Drell-Yan-model 59
Duality 346
Electroweak unification 12
Exclusive π-production by neutral currents 411
Exotic states 340
Flavour 338
Gauge group, minimal 303
Gauge invariance 4,5,303
Gauge theories 3
g_e-2,g_μ-2 Measurements 121
Glashow-Iliopoulos-Maiani (GIM) mechanism 7,34
Gluons 41,188,339
Glueball 340
Goldstone-Boson 20
Grand unification 39,305
Harmonic oscillator 319
Higgs mechanism 13
Higgs particles 21,46,310,390
Hyperfine splitting 330
Inclusive π production by neutral currents 405
Inclusive muonless ν-reactions 391
'Internal' conservation laws 4
Jets 197,242,245
Kobayashi Maskava matrix 36
Lamda parameter (QCD) 288
Leptons 99,115
Lepton masses 31,110,127,133
Leptoproduction 216
Moments of Quark densities 222,286
Nachtmann moments 286
Nachtmann variables 282

Neutral currents 15,1oo
Neutral heavy leptons 15o
Ortholeptons 129,146
Paraleptons 129
Parton densities 22o
Parton model 83
Potential models for Quarkonia 164
Proton decay 53
P-wave states in Charmonium 185
Quantum chromodynamics 217
Quantum flavour dynamics (QFD) 7
Quarks 28
Quark masses 31,11o
Quark structure functions 71,92
Quarkonium 161
R-ratio 73,87
Running coupling constant 22o,288
R = σ_L/σ_T 269,271
Scaling 62,66,89,271
Scaling violation 226,234,278
Second class currents 1o3
σ_L/σ_T QCD predictions 241
σ (order α_s) 218
Spectral condition 317
Spin orbital angular momentum configuration 32o
Spontaneous symmetry breaking 17
- in gauge theories 23
- in standard model 25
Standard SU(2) U(1) model 8
Structure function 266
SU(5) grand unification 39
SU(6) 338
τ-lepton 1o1,132
Two photon processes 144
Vector coupling constant of neutral currents 391
Weak charged gauge bosons 12
Weak isoscalar 375
Weak neutral gauge bosons 13
Weak neutral currents 15,1oo, 376
Weak mixing angle 14,44
Weinberg Salam angle 296
Yang Mills theory 11
Young diagram technique 365